Alternative Energy Technologies

Nano and Energy

Series Editor: Sohail Anwar

PUBLISHED TITLES

Alternative Energy Technologies: An Introduction with Computer Simulations
Gavin Buxton

Advanced Nanoelectronics
Razali Ismail, Mohammad Taghi Ahmadi, and Sohail Anwar

Computational Nanotechnology: Modeling and Applications with MATLAB®
Sarhan M. Musa

Nanotechnology: Business Applications and Commercialization
Sherron Sparks

Nanotechnology: Ethical and Social Implications
Ahmed S. Khan

Alternative Energy Technologies

An Introduction with Computer Simulations

Gavin Buxton

ROBERT MORRIS UNIVERSITY
MOON TOWNSHIP, PA, USA

CRC Press
Taylor & Francis Group
Boca Raton London New York

CRC Press is an imprint of the
Taylor & Francis Group, an **informa** business

CRC Press
Taylor & Francis Group
6000 Broken Sound Parkway NW, Suite 300
Boca Raton, FL 33487-2742

Printed on acid-free paper
Version Date: 20141028

International Standard Book Number-13: 978-1-4822-1703-2 (Hardback)

Visit the Taylor & Francis Web site at
http://www.taylorandfrancis.com

and the CRC Press Web site at
http://www.crcpress.com

Contents

Preface

Alternative energy sources are becoming increasingly important in a world striving for energy independence, clean air, and a reprieve from global warming. Solar cells, wind power, and biofuels are some of the competing alternative energy sources hoping to gain a foothold in our future energy mix, and the economic advantages of these technologies are continually increasing as costs are reduced and efficiencies increased. This book introduces the basic science behind a wide range of alternative energy technologies, albeit in a unique way. Alternative sources of energy are introduced through some simple computer models which can be used to capture their behavior. Computers have permeated throughout our entire way of life, and revolutionized scientific research, opening up a whole new way in which we do science and engineering. This book not only introduces the science behind alternative energy sources (as well as some of the environmental needs for alternative energy sources), but also the role of computer simulations in elucidating the science of these systems. In particular, simple models implemented within spreadsheet environments are discussed which makes the simulations straightforward and accessible to students with no prior programming experience. The computer models introduced in this book could be taught as part of a regular class, used as a spring board for student research, or used by researchers not familiar with computer modeling to explore the power and versatility of simple computer simulations.

List of Figures

1

Introduction to Alternative Energy Sources

CONTENTS

We owe much of our modern standard of living, our prosperity and comfort, to the industrial revolution and the subsequent accelerated growth of industry and technology. This is predicated, or built, upon the unsustainable extraction and consumption of fossil fuels to supply our society with the necessary energy to power such industrial growth. However, time is running out. Dwindling supplies and environmental concerns mean it is imperative that we find future energy sources which are both clean and sustainable. In other words, alternative energy sources.

How much time we have before we need to switch to alternative energy sources is open to debate, with opinion often swayed by financial interests and political pretense and misinformation. Why, if the situation is so dire, would people try to hinder our transition to a more sustainable and cleaner future? Partly because the severity of the situation is still debatable, and also because the industrial status quo still have profits to make (and politicians to bribe via lobbying and campaign funds). But the tide is changing and alternative energy sources are becoming increasingly more economically viable and more widely adopted.

First, let's look at why alternative energy technologies are so important to our future (and that of the planet), before turning our attention to the different forms of alternative energy sources which are coming together to hopefully meet our energy needs.

1.1 Global Warming

In the US there is a curious divide between how conservatives and liberals perceive global warming. It might seem strange that science is politicized (being based on objective and observable truths), but the American media is often partisan and antagonistic which leads to polarized and biased perceptions. In particular, conservative media refer to an "academic liberal conspiracy" where they believe scientists are lying about global warming, and seek to undermine climate scientists and their research. In effect, the conservative media (and the industrialists and businessmen who fund such media from profits derived from fossil fuels) have made it difficult for conservative politicians to advocate adopting alternative energy sources (at least if they hope to be re-elected). However, global warming is obviously real and here we will briefly consider the science behind global warming and the potential environmental impacts that global warming might have upon our world.

The earth's climate is variable, not just on a local scale, but globally across the entire planet. At the center of this variability is an energy balance between the energy incident on the earth from the sun, and the energy radiated back out to space. In other words, there is a radiation budget between the incident radiation (which will depend on factors such as how reflective our planet is and how bright our sun is) and radiation being emitted back out to space (depending on factors such as the temperature of the planet and how much outgoing thermal radiation is trapped by the greenhouse effect). The components that influence this energy balance can have severe and devastating impacts on our planet's climate and ecology, and is at the heart of our existence on this planet.

The energy coming in to our planet comes from the sun. This is primarily in the form of light in the visible spectrum and ultraviolet. The high temperature of the sun means that it emits these higher frequency (shorter wavelength) and more energetic electromagnetic radiation. In particular, the temperature of the surface of the sun is roughly 5500 ^oC and the sun gives off a tremendous amount of energy (of which only a very small fraction is directed towards the earth). Some of this light is reflected back out into space, from clouds in the sky or from ice on the ground, for example, but some of this energy is absorbed by the earth, heating up the surface of our planet. The fraction of light which is reflected back out to space is called the albedo. Different regions of our planet's surface will have different albedos and, furthermore, the albedo can change over time. For example oceans can absorb on the order of 90% of the incoming solar energy, and reflect very little, resulting in albedos of the order of 0.1. However, clouds (depending on thickness and altitude), snow and ice covered regions of the planet, and deserts can reflect much more of the solar energy back into space. On average, over the entire planet, the earth has an albedo of approximately 0.3, meaning that 30% of the solar energy incident on the earth is simply reflected back out in to space. The 70% of energy absorbed by our planet results in the planet warming up, but the earth also gives off thermal radiation in the form of infrared light. This is because the surface of the earth is much colder than the sun, and the radiation emitted is, therefore, lower in frequency and longer in wavelength. Some of this emitted radiation goes back out into space, but some of it becomes trapped within our atmosphere. In particular, the outgoing long wavelength infrared radiation can be absorbed by atoms and molecules in our atmosphere, causing these atoms and molecules to become excited. As the atoms and molecules return to their ground state, the absorbed long wavelength energy is re-emitted. However, rather than this infrared radiation being emitted out into space, as the radiation emitted from the surface of the earth might be, some of this radiation is now emitted back towards the earth. Therefore, not all of the energy emitted back out to space from the earth is able to leave the earth and this causes the earth to become warmer. This effect is called the greenhouse effect, which just to confuse everyone has very little to do with how an actual greenhouse functions (which is through the suppression of convection between the warm air inside the greenhouse and the colder air outside). The gases which absorb this outgoing long wavelength radiation are called greenhouse gases. The greenhouse effect is most definitely a good thing and without the greenhouse effect our planet would be a barren frozen world devoid of life. Having just the right amount of greenhouse effect makes our planet warm and our climate hospitable. As a species, however, we are drastically influencing this greenhouse effect and causing global warming.

Within the scientific community, the climate scientists who study the effects of global warming overwhelmingly endorse the notion that anthropogenic emissions (mainly CO_2) are causing our planet to heat up. Recent surveys suggest that 97% of climate scientists are of the consensus that humans are causing global warming, which is in stark contrast to the general public where 57% of the US disagree with the basic science of global warming or are completely unaware of the scientific consensus that the earth is warming due to human activity. It is worth noting that this is not new or controversial science that we are considering. The idea that our atmosphere could limit heat losses from our planet had been suggested around 1800 by the French physicist Fourier. The basic science of global warming has been known, and rigorously established, since 1896 when Svante Arrhenius (a Swedish physicist and chemist, who received the Nobel Prize for Chemistry in 1903) predicted that CO_2 emissions from burning fossil fuels could cause the planet's temperature to rise. More recently, the Intergovernmental Panel on Climate Change (IPCC) expressed with 95% confidence (which for a scientific body is really very high, as scientists are inherently skeptical) that humans are the main cause of the current global warming. Natural variation on their own are simply insufficient to describe the current warming trends, and it is only when the effects of greenhouse gas emissions are taken into consideration that we can account for global warming. While there is substantial complexity inherent in the earth's climate, which makes long term predictions difficult, the basic science of global warming is irrefutable and the evidence of our planet's warming insurmountable. However, this does not mean that the debate is over, as there will always be debate and skepticism within the scientific community about any subject, but the debate has shifted away from whether or not human emissions of greenhouse gases cause global warming, to how bad the warming might get and what the consequences of this warming might be.

If we were to double the amount of CO_2 in the atmosphere then we might expect to see the earth warm up by around 1 ^{o}C. However, as the planet warms up other factors which influence the energy balance, how the incoming solar radiation balances with outgoing thermal radiation, might change as well. These changes cause feedback in our climate and amplify the heating of our planet. For example, as we emit more CO_2 and cause the planet to warm we might expect to see more water vapor in the atmosphere (as more water will evaporate in a warmer climate). Water vapor is another greenhouse gas, and this can further add to the warming of the planet. (The reason this is referred to as a feedback mechanism is because the warmer it gets, the more water evaporates, which makes it warmer, which makes more water evaporate, and so on, with the amount of water vapor in the atmosphere and the temperature of the planet both increasing as a consequence of each other.) Another example of a feedback mechanism is the melting of ice. As the planet warms up, the ice will melt and the albedo of the planet will reduce as high albedo snow and ice is replaced by low albedo water and land. A reduction in the planet's albedo means that more of the solar energy from the sun will be absorbed, and this will further warm our planet. In fact, in the earth's past the reverse has occurred, with a cooling planet becoming increasingly covered with ice and the subsequent reflection of solar energy off the ice, and back into space, causing the planet to cool further. This resulted in the 'snowball earth', times in our planet's history when the planet was entirely frozen (or at least slushy). The build-up of carbon dioxide from volcanoes eventually warmed our planet enough to return it from its brief frozen state.

The last remaining argument of the few scientists skeptical about global warming is in terms of increased cloud cover. In particular, the "adaptive iris" theory hopes that an increase in cloud cover (in response to global warming and increases in water vapor) would increase the earth's albedo, reflect more sunlight back out to space, and cool the earth. This is the last significant feedback mechanism which scientists predict might curtail rapid increases in the earth's temperature in response to greenhouse emissions. However, this theory is still rigorously debated with some proposing that the effects of increase cloud cover might might further increase the temperature of our planet. Before considering this theory in detail, let's review some earlier skeptical scientific arguments surrounding global warming.

One of the now largely discredited theories surrounding global warming was that solar output (which can vary over time) was driving global warming and not anthropogenic (human-based) emissions of greenhouse gases. The theories themselves were scientifically valid and the scientists offered plausible and well thought out explanations for global warming. It is simply that over time, these theories failed to match the observations (principally that the solar output has reduced over recent years, while the planet continued to warm). In fact recent studies seem to indicate that the worst case scenario, for solar output alone driving global warming, has increasing solar output accounting for around 0.15^{o}C of the observed degree or so warming we have experienced over the past 300 years. When considering solar output, however, we should first turn our attention a little earlier to the 'faint young sun problem'.

In paleoclimatology, the faint young sun problem refers to a time in our earth's history when the sun was less luminous than it is today. Looking at the typical evolution of stars, it is predicted that around 4 billion years ago our sun was around 70% as luminous as it is today. This significant decrease in solar output would result in a frozen world over the first couple of billion years of our planet's history. However, geological records suggests that there was liquid water over the surface of our planet during our planet's early history, with water-related sediments having been discovered dating back 3.8 billion years and life starting to form as far back as 3.5 billion years ago. The leading hypothesis, for this warmer than expected earth, is the greenhouse hypothesis which predicts increased CO_2 concentrations caused early global warming which counteracted the reduced heating from the faint sun.

More recently, the 'little ice age' which occurred during the 17th century has been attributed to reduced solar activity. There was a period of roughly 70 years from around 1645 to 1715, where observations at the time describe a reduction in the number of sunspots, which in turn indicates that the solar output from our sun was slightly reduced. This is also known as the Maunder Minimum, and during this time extremely cold winters were reported across Europe and the US. The fact that solar output can be directly related to the temperature of our planet is well established (and kind of obvious). Hypothesizing that the warming observed in the 80's was due to increased solar output, rather than CO_2 concentrations, is therefore perfectly plausible.

In 1991 Eigil Friis-Christensen and Knud Lassen reported a close match between data concerning the length of solar cycles (and, hence, solar output) and the land air temperature in the northern hemisphere. However, the most recent data point in their paper was not averaged correctly, and subsequent data seems to indicate that solar cycle length has had very little correlation with the global warming observed since around 1975. Natural variations in solar output appear to no longer dominate the long-term variation in our planet's temperature. To put it bluntly, there is no increased level of solar activity that can explain recent global warming.

Henrik Svensmark reported in 1998 a theory that cosmic rays from space are capable of increasing cloud cover, and suggested that there was a direct correlation between cloud cover and cosmic ray concentrations. The amount of cosmic rays interacting with the earth is dependent on the solar output, with increases in solar output causing larger magnetic fields around the sun which in turn deflect cosmic rays. While the correlation between cosmic rays and cloud cover is certainly not universally accepted, the possibility that small changes in solar output could cause more dramatic increases in the planet's temperature (than expected from simple solar forcing) is certainly interesting. However, the solar output has dropped since the 80's and this theory has failed to account for continuing warming while the solar output has been dropping. To put it bluntly, again, there is no increased level of solar activity that can explain recent global warming.

The majority of the debate surrounding global warming shifted quite quickly from solar activity to the impact of increased cloud cover, as a consequence of global warming, on the energy balance between solar radiation being absorbed by our planet and thermal radiation leaving our planet. In 2001 Richard S. Lindzen presented a theory called the adaptive infrared iris. As the planet heats up (due to an accepted scientific consensus that global warming is occurring) an increase in cloud cover is likely to occur. The clouds could reflect incoming solar radiation back into space, and reduce the amount of incoming solar radiation absorbed by the earth. Less incoming solar radiation absorbed by the earth will indubitably result in less warming. This could lead to a negative feedback in the global climate and help stabilize the earth's temperature and reduce future warming. At least that was the theory. However, the feedback from clouds is not so simple (our climate never is).

In particular, there is a large question remaining as to how various cloud types might influence the energy balancc of the earth. Low stratocumulus clouds are relatively thick and, as such, are not transparent to visible light. As a consequence they do not let a significant amount of solar energy through to the earth, and instead reflect some of this energy back into space (increasing our planet's albedo). While these clouds might also absorb thermal radiation emitted from our planet, and stop it from going out into space, as they exist lower in the atmosphere the clouds themselves are relatively warm and the emitted thermal radiation out into space is similar to what might be emitted from the earth in the clouds absence. As a consequence, low stratocumulus clouds tend to cool the surface of the earth. Clouds higher in our atmosphere, in contrast, are thin and primarily transmit solar radiation. Furthermore, these clouds absorb outgoing thermal radiation from the earth and result in less energy escaping from our planet. The higher clouds are colder (because they're higher) and as a consequence the thermal radiation emitted from the clouds themselves is less than the thermal radiation absorbed by the clouds from the warmer earth below. In essence, less energy is emitted out to space and as a consequence the surface of the earth is warmed by these higher clouds. Very thick cumulonimbus clouds (stretching from near the surface of the earth to an altitude of around 10 km) are thick enough to reflect a significant amount of solar radiation, while at the same time are of higher altitude (at least at the top) that the cold cloud tops radiate less thermal radiation into space than is emitted from the warmer earth below. Cumulonimbus clouds, therefore, both contribute to warming and cooling (and are widely considered to cause neither a net warming nor cooling of our planet). While an increase in cloud cover might be expected upon increasing the planet's temperature, as a consequence of global warming, it would appear that some contributions will exacerbate the problem while others might alleviate our planet's warming. Unfortunately, recent studies indicate a positive cloud feedback, thereby amplifying global warming due to anthropogenic influences, but this remains an active and rigorously debated aspect of climate science.

Further skepticism about global warming has arisen recently as a consequence of a 'hiatus' in warming trends. Global temperatures rose alarmingly in the 80's and 90's, but appear to have plateaued since around 2000 (at the time of writing this). This leveling off of the global temperature is in contrast to predictions based on the unabated increase in greenhouse gas emissions. A couple of theories have arisen to account for this hiatus. Ironically, given that the role of solar activity in causing global warming is no longer considered important, it has been proposed that a reduction in solar activity could be causing the hiatus. This reduction in solar energy output from our sun could temporarily mask the global warming trend (and with a constant solar output we might have expected to see the temperatures rise rather than the leveling off actually observed). As solar activity increases again then the temperature increases would presumably rise much more rapidly

A more recent explanation suggests that temperatures have continued to rise (just not on the land). The heat contained within our oceans and the sea levels have been observed to continue rising over the hiatus. This suggests that the additional heat in our climate, as a consequence of global warming being continually driven by greenhouse emissions, is being absorbed by our oceans. The oceans, therefore, have seen an increase in their temperatures (and the sea-levels are continuing to rise because of thermal expansion) while the surface temperature of our planet has experienced a hiatus in temperature rise. In effect, vertical ocean circulations are taking warmer waters deeper and bringing colder water to the surface. In other words, the same amount of heat is being absorbed by our planet (and greenhouse gases are still trapping more heat), but this is not reflected (at present) by surface temperatures. This is part of a natural decadal cooling phenomena like La Nina (when cold ocean surface temperatures are observed, as opposed to El Nino where high ocean surface temperatures are observed). The long-term trend, however, is still expected to be a warming of the surface of our planet as we continue to emit greenhouse gases into our atmosphere.

The question of global warming isn't whether or not it is occurring, it's how bad will things get. The important aspect of our climate to bear in mind is that it's a dynamic system with a lot of complex (and predominantly positive) feedback mechanisms. In other words, even if we were to replace all polluting fossil fuel sources of energy with alternative energy sources, we are already committed to driving the climate towards temperatures greater than they are at present. The thermal inertia already present in our climate system will result in continued temperature increases and sea-level rise, regardless of what we do today. Increases in temperature on the order of 4°C by the end of this century are routinely predicted, naturally with huge levels of uncertainty, along with increasingly dire predictions if we do nothing. Let's consider some of the possible consequences of business as usual.

Hundreds of millions of people already live in coastal areas around the world which flood in the event of extreme weather. Global warming is expected to only further exacerbate this flooding. First, the increase in ocean temperatures will result in an increase in the frequency and severity of hurricanes and cyclones. Second, the sea-levels are rising with global warming (primarily due to thermal expansion at the moment) and this will obviously make flooding more likely. The coupled consequence of these two effects will make coastal communities around the world increasingly more vulnerable to extreme weather phenomena. In addition, warmer climates can hold more precipitation resulting in potentially more rainfall or snow accumulation, which can result in more rivers flooding. As global warming increases, therefore, we can expect more flooding to occur around the world.

Another extreme weather event, likely to become more severe as a consequence of global warming, is droughts. Droughts are characterized by prolonged dry periods, with little precipitation, and can have a detrimental impact on agricultural crops or result in reduced reservoir and ground-water levels. Global warming may exacerbate water scarcity, in areas where droughts already naturally occur, even as precipitation increases in other parts of the world. It is expected that global warming will result in droughts setting in quicker than they might without anthropogenic warming effects, and that the severity of the drought will be more intense.

Coral reefs are biologically productive areas, often in otherwise unproductive waters, which make them crucial to ocean ecology. As the temperatures of the sea water increases, coral bleaching can occur. The photosynthetic algae that lives in the tissue of corals (zooxanthellae) can become increasingly vulnerable to damage by light as the temperature of the water increases. As these algae die or leave the coral host (the process known as coral bleaching) the corals tend to die in great numbers. Increasing temperatures in the ocean are expected to have a detrimental impact on these beautiful and diverse ecosystems. It is worth noting that thermally resistant strains of zooxanthellae can sometime thrive after bleaching events and even if we curb global warming trends, anthropogenic stresses such as overfishing and pollution will continue to devastate coral reefs.

The ecological effects of global warming can already be seen on land. As the planet warms up species tend to migrate further towards the poles, and spring events occur earlier. A recent study predicted that by 2050, an expected 15 to 37% of species could be 'committed to extinction'. Predicting our planet's future climate, let alone the ecological response to global warming, is dubious to say the least. However, it would appear that global warming will indubitably result in the mass extinction of a large number of species and irrevocably change our planet's ecosystem. Of course some life on our planet may thrive as a consequence of global warming.

It is worth noting, from a humanitarian perspective, that food and water shortages throughout the world, along with the mass displacement of coastal populations, will be particularly devastating to poorer people. The effects of global warming, while financially expensive and disruptive in developed countries, will contribute to a large number of deaths in developing countries. It is difficult to point at a particular flooding in China or a given drought in Africa and blame the resulting deaths on global warming, but it is expected that the frequency and severity of these extreme weather patterns will only increase with global warming. In addition, it might be difficult to point to wars and conflicts around the planet, and specifically blame global warming for the scarcity of critical resources that are often fought over. However, it already known that wars are fought over resources such as water and rain-fed agriculture, and that these resources will be in increasingly short supply with increasing global warming. The CNA Military Advisory Board has described global warming as a "severe risk to national security" and that global warming will act as a catalyst for global political conflict. As we discuss the costs associated with alternative energy technologies, therefore, it is worth remembering the financial and human costs of global warming and the continual burning of fossil fuels.

1.2 Pollution

The growth of our industrialized society, and our associated energy use, has been crucial to prolonging life expectancies and providing many health benefits. However, the different sources of energy that we use will always entail some health risks. Coal is a major energy source, accounting for around 40% of the electricity generated in the world, and is a major source of anthropogenic CO_2 emissions. Besides the health risks with mining coal, which can be a dangerous endeavor, the US relies heavily on mountaintop removal and strip mining to acquire its coal. The ecological damage from removing the tops of mountains can be quite devastating. The contamination of rivers and streams (the origin of drinking water resources for millions of people) has a large health toll on communities living near coal-mining operations, with elevated rates of cardiovascular, pulmonary, and kidney diseases reported. However, the real environmental damage from coal occurs after combustion. Particulate matter, carbon monoxide, nitrogen oxides, sulfur dioxide and mercury, can have a detrimental impact on the health of residents near coal-fired power plants and the planet in general. Petroleum is also a major source of pollution, accounting for more than 90% of transportation fuel, and releasing a cocktail of pollutants similar to the combustion of coal. In addition, secondary ozone formation can occur, from nitrogen oxides and volatile organic compounds from burning petroleum chemically combining with oxygen to form ozone during warm summer months. Natural gas is increasingly important to our energy mix, especially with the advent of hydraulic fracturing. Natural gas combustion generates half the CO_2 of coal combustion and fewer pollutants. Furthermore, the drilling for natural gas is environmentally safer than the mining and extracting of coal. (There has been controversy surrounding hydraulic fracturing, but any negative environmental impacts of natural gas extraction could be mitigated through responsible practices). Nuclear energy is also considered a conventional source of energy (although newer developments in nuclear energy technologies could be considered alternative). Nuclear energy only supplies around 11% of global electricity (although France and other countries rely much more heavily on nuclear energy). There has historically been concern expressed over a link between nuclear power plants and childhood cancers near these nuclear power plants, although recent studies suggest that no such link exists. Nuclear accidents, such as Chernobyl and Fukushima, have resulted in the resettlement of many people away from these power plants, with enormous social and economic tolls, and the release of radioactive pollution into the environment. The health risks from these events are much less than the health risks from coal (and it should be remembered that these extreme events which would be much less likely to occur in newly built nuclear reactors, while coal pollution is routine) and the number of people affected is relatively small. The main concern with nuclear energy is the stockpiling of radioactive waste, which is currently contained and managed in interim storage facilities, with no long term solution to this storage problem (although the volume of this waste is much less than other industrial processes). However, more modern nuclear power plants use a much greater proportion of the nuclear fuel (and potentially some of the previously generated waste) and could provide relatively pollution free (certainly carbon free) power production. Let's look in more detail at some specific problems associated with pollution.

The effects of pollution on our health is not a new concern. Historically, pollution in Belgiums Meuse River Valley (1930), in Donora, Pennsylvania (1948), and in the Great Smog of London (1952) have been particularly disastrous events. In particular, temperature inversions (where warm air blows in over the land) limited the convection of polluted air up and away from near the earth where the pollution was being emitted. As a consequence, the air became increasingly polluted within these localized regions and caused a significant loss of life (over 4000 people are believed to have died during the Great Smog of London). It is partly in response to such tragedies that governments around the world have regulatory agencies to monitor and legislate for better air quality. In effect, legislation has largely eliminated most of the visible air pollution of over 50 years ago, but other pollutants have gained prominence and still pose significant health risks.

Fine particulate matter is a pollution given off through the combustion of fossil fuels. Typically, fine particulate matter refers to particles measuring less than 2.5 μm in diameter (PM 2.5). The chemical composition of fine particulate matter is diverse, and consists of soot, acid droplets, and metals. Fine particulate matter is produced through combustion, from sources such as power plants and transportation vehicles, with older coal powered power plants being particularly polluting (the average age of coal-fired power plants is over 45 years in the US). Concentrations of particulate matter are especially high in the eastern United States, but huge swaths of the world experience persistently elevated levels of fine particle pollution. These pollutants are extremely disconcerting as their small size allows them to be inhaled through the lungs' alveoli and into the circulatory system. Once the fine particulate matter is absorbed into the bloodstream it can be transported throughout the body to vital organs. In other words, fine particulate matter is small enough that it can evade the lungs' natural defenses. As a consequence, fine particulate matter can lead to premature death (especially the elderly or people with heart or lung disease), heart attacks and irregular heartbeats, strokes, lung cancer, aggravated asthma, acute respiratory symptoms (including aggravated coughing and difficult or painful breathing), chronic bronchitis, and a general decrease in lung function (shortness of breath). Fine particulate matter is, therefore, not only having a detrimental impact on human health, but the economic impact from health care costs and a loss of worker productivity are significant factors when assessing the financial costs of alternative energy sources in relation to fossil fuels.

Ozone is photochemically produced by reactions of nitrogen oxides, carbon monoxide, volatile organic compounds and methane. Emissions of these elements, the precursors to the reaction, from coal combustion and internal combustion engines are a major contribution to ozone formation. Furthermore, as the pollution is not ozone directly, this secondary pollution can be formed away from the source of the pollution (furthermore, ozone can exist for around two to four weeks and be transported around the world). Higher concentrations of ozone at the earth's surface is particularly troubling as high ozone concentrations can cause difficulty breathing (especially in the young and the elderly), inflame and damage airways, aggravate lung diseases (such as asthma, emphysema, and chronic bronchitis), and make the lungs more susceptible to infection.

The combustion of coal also results in the emission of mercury into our environment. The mercury then enters the water cycle, raining down, flowing through rivers and streams to the ocean, where it accumulates in the food chain. Ultimately, it can affect human health as we eat fish contaminated with high levels of mercury (not to mention the poor whales and other aquatic mammals exposed to very high levels of this toxic metal). Mercury is a neurotoxin (hence the saying 'as mad as hatter', after hat makers who used mercury in the process of making hats) causing damage to the brain and nervous system. Pregnant women and young children are especially susceptible to the detrimental effects of mercury, which can cause developmental problems and learning disabilities. A possible cause for optimism is the new protections introduced in the US in 2011, which are targeted at reducing mercury by over 90%. However, gas emission control technology, which can reduce mercury emissions, is expensive, and uncommon in China, which emits over 40% of the worlds mercury. In fact, the US only emits around 8% of the world's mercury so while this is a welcome step it is unlikely to immediately help reduce atmospheric mercury emissions.

Sulfur dioxide (and to a lesser extent nitrogen oxides) can mix with water in the atmosphere to form sulfuric (or nitric) acid. The resulting precipitation is known as acid rain. Acid rain damages forests and aquatic ecosystems. The effects of acid rain are primarily observed in the reduced biodiversity of lakes and streams, the elimination of fish species, and forest degradation. Almost the entire anthropogenic emissions of sulfur dioxide comes from coal-fired power plants. Flue gas desulfurization 'scrubbers' are significantly reducing these emissions in the US as a consequence of federal and state enforcement of provisions of the Clean Air Act.

The health costs associated with emissions form coal-fired power plants have been estimated to range from \$62 billion to \$523 billion annually. This translates to costs of from 3.2 cents to 28.9 cents per kilowatt hour (the average cost of electricity to the consumer in the US is around 12 cents per kilowatt hour). Therefore, the hidden health costs are certainly significant and potentially greater than the original cost of the electricity. Furthermore, this is not taking into consideration the economic damages sustained during extreme weather conditions, or the financial burden of mass population migration and conflicts that might arise, or increased food prices as agriculture suffers, all as a consequence of global warming.

Perhaps it is worth rethinking the phrase 'as cheap as coal'.

1.3 Solar Cells

The sun emits a tremendous amount of energy, of which around 1.8×10^{17} W of power is incident on the earth (to put this into perspective, the planet generates around 2.3×10^{12} W of power currently using a wide range of, often polluting, sources). This makes solar energy an abundant and plentiful power source that can be harnessed using solar cells, or photovoltaic cells. Solar cells directly convert sunlight into electricity. Typically, solar power generation employs solar panels which are composed of a number of individual solar cells. The first modern solar cell was developed in the 1950s at Bell Telephone Laboratories (originally intended for use as an energy supply for telephone repeater stations, but because of high costs found early applicability in satellites and space exploration). As mentioned, around 1.8×10^{17} W of power is incident on the earth and we currently use around 2.3×10^{12} W of power, which means (accounting for, say 15% efficiency in converting solar power to electricity) we need to use around one ten-thousandth of the earth's surface to generate electricity (or 43 000 km^2 - less than half of the great basin desert in Nevada, for example). The impact of land use, however, can be minimized by integrating photovoltaic systems into existing building structures and placing solar panels on top of existing roofing. Generating our electricity using solar cells would result in only a minimum impact on the environment (during manufacturing) as the technology is really quite elegant, from an environmental standpoint, and the conversion of sunlight to electricity occurs without any moving mechanical parts or environmental emissions.

The solar cell converts sunlight into electricity in a threefold process. First, a photon of light is absorbed by the semiconducting material allowing the electron to become separated from its hole. A hole is essentially the space where the electron resided before being excited by the photon, and interestingly by electrons hopping from one hole to the next, it can appear as if holes move through the semiconducting material in a manner similar to how electrons move (when we refer to charge carriers we refer to electrons and holes). In effect, the sunlight can be absorbed by the solar cell material and impart energy to electrons enabling them to escape from the localized atomic hole, and consequently create two carriers of electrical current - electrons and holes. This photovoltaic effect was first discovered in 1839 by Edmond Becquerel. Second, the energetic electron must preferentially move in a specific direction in the semiconductor. A typical solar cell consists of a doped inorganic semiconducting material (such as silicon). In particular, the semiconductor is doped (impurities are added) such that the semiconductor forms what is called a 'p-n junction'. The p side of the material contains an excess of holes, the positively charged charge carriers (hence the letter p for positive), while the n side of the material contains an excess of negatively charged electrons. There is a region near the interface between these two different p and n regions, called the depletion region, where an electric field is established. The different charge carriers, the electrons and holes, travel in different directions and are accelerated by this electric field towards the correct electrodes. The third and final step in the process of converting sunlight into electricity involves the moving charge (or electrical current) leaving the solar cells and traveling through an external circuit. Having appropriate electrodes is an important step in this process. Ultimately, energy can be lost throughout this process; sunlight can be reflected from the surface of the solar cell, or fail to be absorbed by the semiconductor material, the electrons and holes can become trapped or recombine with each other, and resistive losses can occur at the electrodes. The power conversion efficiency is simply defined as the ratio between how much electric power can be produced by the solar cell to how much solar power is incident on the cell. Different types of solar cells have different efficiencies, and we'll briefly consider some of the different types of solar cells here.

First generation solar cells remain the most popular type of solar cells and consist of single-crystal and polycrystalline silicon modules. Silicon is an obvious choice for solar cells as their absorption characteristics are a fairly good match to the solar spectrum, and the fabrication technology required to manufacture silicon solar cells is well established as a result of silicon's pervasiveness in the semiconductor electronics industry. The maximum theoretical power conversion efficiency (the Shockley - Queisser limit) of these first generation solar cells is around 33.7% (assuming a single silicon p-n junction), although in practice efficiencies of around 17 to 23% are typically observed. (It is possible to go beyond the theoretical Shockley - Queisser limit when considering tandem solar cells, for example, but we'll discuss this a little later.) It is estimated that around 32% of the cost of silicon solar cells stems from the cost of the silicon material, and a further 26% of the cost is associated with high-temperature fabrication processes. As a consequence, first generation solar cells can be considered too expensive to compete with fossil fuel power generation (obviously not accounting for the economic and humanitarian toll on society from burning fossil fuels).

A second generation of solar cell technology has emerged which are characterized by thin films of semiconductor material fabricated on non-silicon substrates using techniques such as sputtering and vapor deposition. The most common thin film solar cells are based on amorphous silicon (which has been used for many years in simple devices like calculators). Their efficiencies are relatively low, with power conversion efficiencies typically in the range of 4 to 6%. Other semiconductors have been proposed. For example, cadmium sulphide (CdS) was a popular candidate for thin film solar cells, but have been largely abandoned due to environmental concerns about the use of cadmium. Copper indium diselenide (CIS) is a semiconductor which has been used to create solar cells. The solar cells, however, have typically required a thin layer of cadmium sulphide between the solar cell and substrate which, again, raises environmental concerns about the use of cadmium. Furthermore, indium is in short supply and given the known reserves of indium the number of solar cells which could be manufactured would be pitifully small. Recently, polycrystalline silicon has emerged as a semiconductor for thin film solar cells. The polycrystalline layer is obtained through high temperature processing of an initially amorphous silicon layer. The fabrication costs of a polycrystalline thin film solar cell, however, might be a little more than other thin film solar cells which do not require such high-temperature processing. All of the second generation thin film solar cells use less material and can be deposited on flexible or curved substrates, making them cheaper (if less efficient) than first generation solar cells and more easily integrated into construction materials.

Third generation solar cells are attempting to create modest efficiencies while being very inexpensive. Typically, low cost materials (such as organic molecules or polymers) are used, and the fabrication process is low-temperature, high throughput and occurs in a regular atmosphere (rather than a clean room). To reiterate, polymer materials are inherently inexpensive. (Cheap plastic imitations of everything else are commonplace in our world, and we've replaced paper bags, glass bottles, metal piping, and virtually every toy with plastic alternates – why not solar cells?) Polymers have very high optical absorption coefficients meaning that polymer solar cells are typically only around 100 nm thick. The holy grail of polymer solar cells is to produce devices using roll-to-roll processing (like newspaper printing) or to create 'spray-on' polymer solar cells which can be deposited like paint from a spray can. Furthermore, the use of flexible substrates open up the possibilities of photovoltaic fabric with the color easily altered (for camouflaged tents or patterned clothing, for example). An early problem is that polymers can degrade under ultraviolet light or when exposed to the elements however these issues seem to have been addressed by the organic light emitting diode industry which regularly produce outdoor displays with similar materials and constraints as polymer solar cells. Another issue is the low efficiencies. Whereas light absorption in inorganic solar cells produce free electrons and holes immediately, in polymers the electrons and holes remain coulombically bound together in an excited state, called an exciton. This means polymer solar cells require an additional step to separate the charge carriers before they can be extracted from the solar cell. This dissociation typically occurs at the interface between two materials with different ionization potentials and electron affinities. The device morphology, therefore, requires additional thought.

An interesting new development is solar cells based on perovskites (a class of semiconductor material) which have rapidly been developed to exceed 15% power conversion efficiencies. The performance of these solar cells has increased so quickly that it is hard to ignore the potential of perovskite-containing solar cells. In addition, the voltage produced by perovskite solar cells (the open circuit voltage with nothing connected to the solar cell) is much greater than other types of solar cells. Furthermore, perovskites have a larger bandgap than silicon (hence the higher open circuit voltage) and, therefore, perovskites absorbs a region of the solar spectrum that silicon does not. This opens up the possibility of placing a silicon solar cell beneath a perovskite solar cell (to form a tandem cell, which we will discuss shortly) such that whatever light isn't of sufficient energy to excite the electrons in the perovskite solar cell can pass through and be absorbed by the silicon solar cell. There are, however, concerns about the long term stability of perovskites because they are water soluble (unlike silicon which is very stable) and the current perovskites contain lead which is toxic. Perovskites are a new development in solar cell technology, however, and perovskites containing non-toxic elements are likely to emerge in the future.

Given that solar cells can be quite expensive, one way to reduce costs is to have less solar cell material but funnel (or concentrate) the same amount of light onto a smaller area. This is known as solar concentration, and it essentially means we can capture the same amount of sunlight as we might with a large solar cell, but require much less of the expensive solar cell material. Obviously this is more important if you are using high-efficiency and high-cost silicon solar cells.

Conventional solar concentrators consist of some optical system which is used to concentrate, or focus, the sun's rays onto the solar cells. For example, a parabolic mirror can be used to capture a large area of the sun's rays and focus it on a smaller region. However, the absorption of concentrated sunlight often results in very high temperatures, which means the solar cell devices have to be cooled. This can be achieved in a hybrid photovoltaic/thermal solar system, which consists of a photovoltaic solar cell which generates electricity from the incident sunlight and a thermal system which extracts the heat from the device for practical use (additional electricity generation, water heating, or water purification, for example). Of course, thermal systems can be operated on their own in such a solar concentrator system, although the efficiency of the system is no higher than simply using solar cells and the costs can be higher. In a thermal system, typically a parabolic trough collector (a long reflective surface shaped into a parabola) is used to concentrate the sunlight onto a black tube(encased in glass and under vacuum to minimize heat losses). As the absorbing black tube heats up the fluid inside the tube heats up and, by pumping this fluid, the heat can be extracted from the system. Regardless of how the energy contained within the sunlight is to be extracted this type of solar concentrator requires the precise focusing of sunlight onto a small region of either solar cells or absorbing material. In other words, the parabolic reflector, or other optical system employed to concentrate the sun's rays, has to be oriented, or pointed, directly towards the sun. The sun, however, moves in the sky during the day. Such a system, therefore, must mechanically move also and track the position of the sun throughout the day. These required mechanical systems are expensive to maintain, and require extra space to maneuver and avoid shadowing neighboring solar concentrators. In other words, the inherent simplicity of a solar cell, which directly converts sunlight into electricity, is lost.

An alternative solar concentrator, termed a luminescent or organic solar concentrator, has recently emerged as an alternative to the optical systems which require complex solar tracking mechanics (technically they've been around since the 70's but only recently have practical devices been achieved). The luminescent solar concentrator consists of a waveguide (a plane or sheet of high refractive material) with dye molecules inserted inside (essentially a painted plastic sheet). The idea behind the luminescent solar concentrator is that sunlight will be absorbed by the dye and re-emitted into the waveguide, where it will bounce (through total internal reflection) all the way to the end, where high efficiency solar cells lie in wait. However, in the past the re-emitted light would be reabsorbed by other dye molecules before it reached the end of the waveguide and the energy would be lost (it really was just a colored plastic sheet). Recently, however, Förster energy transfer has been used to transfer the energy from the dye molecule to a luminescent molecule which emits light of slightly lower energy. The fact that the emitted light is of lower energy means that it cannot be reabsorbed by the dye molecules (not enough energy) and it can happily bounce all the way to the end of the waveguide. In other words, upon absorption of sunlight the dye acquires some of this energy and the luminescent molecule re-emits some of this energy isotropically; with a fraction of the light being emitted along the waveguide at a low enough angle for the light to undergo total internal reflection all the way to the end where it can be absorbed by a solar cell. The advantage of this system is that the waveguide (sheet of plastic) is large in area, but most of the solar energy absorbed by the sheet comes out through the very thin edges of the sheet. The area of solar cell material required to cover the edges of the luminescent solar concentrator is much less than would be required to capture the same amount of energy directly from sunlight, and this form of solar concentrator does not have to track the sun with cumbersome and expensive mechanical contraptions.

The cost of silicon solar cells is reasonably well known, as silicon solar cell technology is well-established and how the devices degrade over time is reasonably well understood. However, the costs associated with emerging solar cell technologies, which are less established, are difficult to estimate as the long-term stability of these devices has not been adequately determined. For example, polymer solar cells and luminescent solar concentrators are constructed from relatively new materials to the solar power industry. Organic light emitting diodes, on the other hand, are a more established industry, producing devices which can operate over large time scales and in direct contact with the outside environment and the dyes used in luminescent solar concentrators are the same molecules used in automotive paints (which we know to hold up sufficiently to the elements). So while emerging technologies are expected to have adequate lifetimes, we won't be entirely sure how long these devices will operate until we operate them over these large timescales.

The current cost of solar panels is around US \$0.70 per watt, and rapidly falling, which is competitive with how much electricity costs from the grid (grid parity). This is still more than it costs coal-fired power stations to generate the electricity, however, and the solar panels still have to be connected via inverters and other systems to the grid which significantly raises costs. Financial incentives have been introduced by governments around the world to encourage customers to purchase solar cells, to promote energy independence, reduce carbon emissions, and create high-tech jobs in what is expected to be a growing alternative energy jobs market. Additionally, these financial incentives cause more people to buy solar panels which further lowers the costs of solar panels through the economy of scale. Of course, governments often subsidize coal-fired power plants and encourage the burning of fossil fuels (as a way of stimulating job markets and appeasing political campaign donors). It could also be argued that the government (and tax-payers) inadvertently subsidize coal-fired power plants through health care costs as a result of pollution and clean-up costs for extreme weather events made increasingly more likely in our warming climate (Hurricane Katrina, for example, had estimated costs of \$148 billion which would buy a lot of solar panels). And so as solar power costs continue to compare more favorably to that of conventional power generation, government intervention can dramatically help solar power adoption, further lowering the costs of solar cells and help reducing the inadvertent environmental and financial costs of burning polluting fossil fuels.

1.4 Wind Power

It is perhaps strange to consider the wind as an alternative source of energy, especially as mankind has harvested this power for thousands of years to propel boats around the world using sails. However, in modern times we have relied almost exclusively on fossil fuels and nuclear power, wind energy only recently being used on a large scale to generate electricity.

Vertical-axis windmills were used in Afghanistan in the seventh century BC to grind grain. These early devices rotated about a vertical shaft (a similar wind turbine, a windmill connected to an electric generator, was built in 1887 in Scotland and is thought by some to be the first wind turbine). Horizontal-axis windmills were developed later and believed to be popular in Persia, Tibet, and China at about 1000 AD. As a consequence of the Crusades, these horizontal-axis windmills spread across Europe with the first horizontal-axis windmill appearing in England around 1150. These windmills are the classic large building with four large blades mounted onto the structure (think of the windmills of the Netherlands), and became lost to history (except in Holland) as the world moved towards an industrial revolution powered by the combustion of fossil fuels. As oil prices skyrocketed in the 1970's there was a resurgence of interest in wind power, but with the focus shifting towards generating electrical energy using wind turbines. Not only would wind turbines provide us with energy independence, but they are less polluting than fossil fuels. This is certainly more true today than it was in the 1970's, as global warming has become a more established threat to our way of life, and consequently the growth of power generation from wind turbines has increased dramatically.

A typical wind turbine (a horizontal-axis wind turbine) consists of a long slender tower rising into the air, supporting a rotor and blades and a nacelle. The nacelle houses the generator and gear box required to extract electrical energy from the wind. In particular, the gear box converts the low-speed high-torque mechanical power from the spinning turbines to the high rotating speeds required by the generator (the rotation of the turbines might by on the order of, say, tens of revolutions per minute while the generator shaft will rotate at thousands of revolutions per minute). The generator, in turn, converts this rotational mechanical energy into electrical energy. Controllers within the nacelle ensure the wind turbine is operating at its maximum efficiency and helps reduce structural loads in high winds to extend the lifetime of the wind turbine components and structure. For example, wind turbine blades can rotate along their longitudinal axes (essentially twist) to change the pitch of the blade and help control mechanical loads and aerodynamic torque. When the wind speed is high the turbine can be slowed or powered down to avoid exceeding safe operating loads. In addition, the electrical power being produced by the generator in the wind turbine must be converted to match the requirements of the electrical grid. Power electronic converters enable the electrical properties (such as frequency and voltage) of the energy being produced by the electrical generator to be matched to those of the electrical grid and enable seamless integration.

Most wind turbines which we see are horizontal-axis wind turbines (the rotor blades rotate about a shaft which is in the horizontal direction) and it is easy to forget that vertical-axis wind turbines exist at all (even though these were the first designs for both wind turbines and wind mills). Vertical-axis wind turbines have the rotor spin about a vertical-axis, and various designs for these wind turbines exist. Reported advantages of vertical-axis wind turbines include the fact that the vertical rotation of the blades enables the generator and gear box to be housed at the base of the structure, making the installation, operation, and maintenance of the wind turbine easier while reducing structural loads by removing the weight of the components from the top of the tower. Additionally, it's often claimed that vertical-axis wind turbines are quieter than horizontal-axis wind turbines, but with the size and scale of these systems still increasing it's difficult to define the noise characteristics of large-scale vertical-axis wind turbines yet. Lower elevations (which need not be the case in the future) result in lower wind speeds for the vertical-axis wind turbines, but this is believed to be off-set by using less expensive materials in its construction. Vertical-axis wind turbines can also more easily extract power from more turbulent winds where the direction of the wind changes. Whereas horizontal-axis wind turbines operate at much higher elevations to capture high speed laminar air flow, vertical-axis wind turbines could find a niche market in areas closer to the ground where the wind is more turbulent and lower speed. The more chaotic and turbulent environment in which vertical-axis wind turbines operate can result in more mechanical fatigue as the components are subject to variable forces and in various directions, and this makes these wind turbines less reliable. In fact, this turbulent environment is inherent of vertical-axis wind turbines regardless of the characteristics of the incoming wind. As the blades spin in a vertical-axis, the blade at the front of the wind turbine might be interacting with a clean laminar flow of air, but as soon as the air interacts with the blade it becomes turbulent and this turbulent air flow interacts with the blades towards the rear of the wind turbine. In other words, the vertical-axis wind turbine is inherently less efficient than horizontal-axis wind turbines at extracting energy from the wind, and vertical-axis wind turbines will always be limited to niche applications in regions of high turbulence.

Wind turbines can be erected on their own, in complete isolation, but the costs of transmitting electricity to the remote locations where wind turbines are located (often on the top of high mountains or increasingly at off-shore locations) it is more common to install wind turbines in large groups, or farms. These wind farms are carefully designed such that the turbulent air flow which can emerge from the wake of a wind turbine doesn't interact with another wind turbine, which would not only result in reduced power production (lower and more chaotic winds) but also could increase the mechanical loads on the structure and reduce the lifetime of the wind turbine. This is especially true for off-shore wind farms. Off-shore wind farms are increasingly popular because of the availability of large flat areas with higher wind speeds and less turbulence. However, there are are concerns relating to the engineering required to establish and maintain these structures in a marine environment, and the obvious increases in cost associated with the construction and the reduced access during routine maintenance.

The main criticism of wind turbines is the noise pollution, which is increasingly invading the previously tranquil rural areas that now house them. In particular, there are a large number of complaints about sleep disturbance. (To reduce transmission costs associated with transporting electricity from the wind turbines to the electrical grid, it is economically desirable to situate the wind turbines as close to homes and regions already connected to the electrical grid which only exacerbates potential noise pollution.) While the noise from wind turbines might be considered trivial, there has been a documented sleep disturbance of residents within close proximity of a wind turbine. As the wind turbines spin there is a gentle swishing noise as the blades move towards you, and occasionally thumps have been reported possibly as a consequence of stalling. However, noise (and aesthetic) concerns regarding wind turbines will become less important as wind turbines are moved off-shore (assuming off-shore wind turbines can operate more cost effectively than wind turbines on land, which currently they can't). Additionally, the continual development and refinement of the aerodynamics and mechanical components of wind turbines has seen a reduction in noise

Another potential concern of wind turbines has been the detrimental effects of wind turbines on wildlife, especially birds and bats. In particular, birds can be affected by wind turbines as they might seek to avoid them or inadvertently collide with the wind turbines. The effects of wind farms on the local bird population will depend, not only on the wind turbines, but also on the migratory volume and pattern of the birds and on the weather conditions (with birds more likely to crash into wind turbines during adverse weather conditions). In addition bats can also be negatively impacted by wind turbines. Lights at the top of wind turbines, to warn potential aircraft of their location at night, can attract insects which in turn attract the bats. While bats are thought to be less likely to crash into the wind turbines, the sudden drop in air pressure near the wind turbines can cause barotrauma and result in the expansion of the bats' lungs and internal hemorrhaging. However, the effects of wind turbines on bat mortality is negligible in comparison with the white nose syndrome that is killing bats in the US, and clear windows and cats kill many more birds than wind turbines. In fact, the pollution from coal-fired power plants kill more birds and bats than would be killed by the wind turbines that could replace them.

One potential obstacle to the adoption of wind energy is the intermittent nature of the wind. In other words, wind turbines only generate electricity when the wind blows. And it's not always windy. As changes in the electrical load are detected (for example, as large numbers of people turn on the air conditioning units on a hot day) the system operator in a power plant will ramp up electricity production to meet this demand. It is not possible to make the wind blow more at a given time, however, in the same way as perhaps it might be possible to start burning more coal. For this reason, wind power is always seen as a source of energy which has to be combined with other power sources (for example, nuclear power or hydroelectric power, if available).

To maximize the output from wind turbines these structures are being constructed larger and taller than ever before, while at the same time to minimize costs the wind turbine system has to be designed as lightweight and flexible as possible. The reliability of the wind turbine, and ultimately the lifetime that the wind turbine will operate, is crucial to extracting the most energy as possible from the wind during this lifetime and ensuring the maximum return on one's investment (furthermore, the remote location of wind turbines, on top of mountains and increasingly off-shore, makes maintenance and repair particularly daunting). However, wind turbines are soundly built and only extremely rarely do they collapse. This has led to the growth of wind turbines as a lucrative alternative energy source that can offer a quick turn on investment. In fact, at windy sites the cost of electricity from wind turbines is now competitive with the production price from coal-fired power plants. The cost of electricity from off-shore wind farms is still economically prohibitive, but it is likely that off-shore wind farms will become increasingly competitive in time.

1.5 Biofuels

Biomass is essentially material derived from recently living organisms such as plants, and could be an important renewable resource for the production of biofuels or biomaterials. Advances in biotechnology and genetic modification are allowing us to engineer plants with specific properties designed to enhance their conversion to biofuels and bioproducts. This is referred to as biorefinery, and the sustainable engineering of plant-based biomaterials is hailed by many as ushering in a new manufacturing paradigm. A future sustainable society could be built upon the complex molecules and materials that are biorefined from biomass, in a similar way to how modern society is so enslaved to the non-sustainable and non-renewable molecules processed from fossil fuels. For example, the US Department of Energy has set goals of replacing 30% of transportation fuel and 25% of industrial organics with biomass-derived substitutes by 2025. It would appear that biomass could become an increasingly valuable commodity in a new sustainable world.

Biomass energy is derived almost exclusively from photosynthesis, and so it is worth briefly reviewing the photosynthetic steps. Photosynthesis is a process by which energy from sunlight is converted into chemical energy. The process of photosynthesis is divided into two steps. The first step, referred to as the light dependent reaction, involves the absorption of light by photosynthetic pigments. This energy is subsequently stored chemically in a chemical called ATP. The second step, referred to as the light independent reaction, is where carbon dioxide is taken from the atmosphere, and the ATP is used to essentially make carbohydrates. However, photosynthesis is a very inefficient process. For example, 80% of the energy can be lost in high light intensity sunlight, which leads to damage of a protein subunit central to the photosynthetic apparatus. In general, 4.5% is considered the upper limit of photosynthetic efficiency with values of only around 1% being observed for rapidly growing trees. When compared, for example to solar cells, these efficiencies are incredibly low.

Biofuels are the main way in which energy is stored from biomass, and simply refers to a solid, liquid, or gaseous fuel that is derived from renewable biomass. While biofuels have gained a lot of attention recently as an alternative to petroleum fuels, it is worth noting that biofuels can be simply burned to produce heat, and that mankind has been burning biofuels (i.e., wood) since mankind discovered fire. In terms of modern transportation fuels, however, there are two popular liquid transportation fuels in ethanol and biodiesel. These both essentially consist of hydrocarbon molecules, with ethanol possessing shorter hydrocarbon chains and lower energy densities than biodiesel, that can both be used with little modification in existing internal combustion engines. Bioethanol is produced almost entirely from food crops, such as corn, and is commonly blended with petroleum-based gasoline and distributed through existing petroleum infrastructures. Biodiesel can be used in existing diesel engines either on its own (with perhaps only minor modifications to the engine) or blended with petroleum diesel. Both ethanol and biodiesel are heralded as a sustainable and clean alternative to fossil-fuels in transportation, although there are serious problems associated with such biofuels.

One of the main issues with biofuels, and biorefinery in general, is the large amounts of land suitable for growing food crops which have to be diverted from feeding an ever increasing population to grow the required biomass. While this land could, in principle, come from clearing new land for biomass agriculture, it has been more common for existing farm land to switch from food crops to crops for biofuel production. This is unsustainable, from a humanitarian perspective, because the burning of biofuels in our vehicles at a time when 14% of the world in undernourished and the world's population is rapidly approaching 8 billion is morally obscene. With food shortages, food price inflation, and riots for food occurring in developing countries across the world, it is unethical and unsustainable to propose diverting agricultural land to biofuel production. In addition to switching land use from food to biofuel production, the additional stress that growing biofuels could have in terms of freshwater resources is also problematic. Some estimates place the amount of water required to produce 1 gallon of biofuel as high as 20 000 gallons. To further put this into perspective, it has been estimated that switching half of our transportation fuel to biofuel would require 12 000 cubic kilometers of water per year. This is comparable to the total amount of water which flows down all of the world's rivers. 780 million people in the world do not having access to clean water, and water scarcity is only going to be increasingly exacerbated by global warming. Diverting all of our water to grow biomass for transportation fuel, and polluting the run-off with fertilizers, pesticides, and herbicides, is unsustainable.

Proponents of biofuel production point to how carbon-neutral and environmentally friendly biofuels are, how they create jobs in rural areas and contribute to energy independence from foreign fuel imports. The first point is premised on the fact that carbon dioxide is initially sequestered from the atmosphere during the growth of the biomass, and this offsets the carbon emissions from burning the biofuel. In other words, biofuels have no net contribution to atmospheric carbon dioxide levels and, therefore, by switching from fossil fuels to biofuels we could reduce atmospheric greenhouse gas emissions and reduce global warming trends. However, if biofuels are grown on land then the effects of changing land use has to be taken into consideration. For example, destroying forests and other pristine areas to create land to grow biofuels could release carbon already sequestered in these areas. It could be argued that the effects of changing land use would be temporary and as biofuel crops are typically grown instead of food crops, that land use change is not significant (although if the growth of biofuels is continued then this might no longer be the case).

In terms of overall pollution, it is thought that particulates in the biofuel are more easily oxidized than petrochemical fuels. The combustion of biofuels have been found to have lower emissions than fossil fuel combustion. Replacing fossil fuels with biofuels, therefore, could improve local air quality. In addition, replacing fossil fuels with biofuels will, without doubt, strengthen rural communities, creating rural jobs and infusing much needed wealth into our rural economy. On the face of it, it might appear that proponents of biofuels have good arguments for urging our further use of biofuels (if we ignore the likely food and water shortages).

The largest concern about biofuels is whether it is even worth it. The energy input is very large for the production of biofuels, and there exists serious doubt that there is even a net gain of energy during the process. Is the production of biofuels energetically inefficient? Does it require more energy to produce the fuel than can be obtained from burning the fuel? Energy is required to produce the fertilizers and pesticides, to plow the fields and harvest the crop, to transport the biomass, and to convert the biomass into a usable biofuel. In particular, the oil extraction process can be particularly energy intensive. Recently it was reported that producing ethanol from corn, switchgrass, and wood biomass, or biodiesel using soybeans and sunflower, requires the input of more energy from fossil fuels than would actually be gained from the biofuels that are produced. In contrast studies have found that the production of ethanol is not only energetically efficient but can produce 3 to 5 times the amount of energy than is required to produce the fuel. Both, however, are true. If the biofuel on its own is compared to the energy required to produce it, then the process is energy inefficient, but it is common to not only include the biofuel but also to account for the energy content of the by-products. When by-products are factored in then such land-based biofuels become more economically viable. Still, the high production and processing costs of biofuels mean that it is common for governments to subsidize the growth and production of biofuels, in order to make the biofuels less expensive than fossil fuels. In other words, while blending biofuels with regular fossil fuels can lower the price the customer sees at the gas pump, it can do so by shifting the costs towards higher taxes (after all it is tax dollars which pay for the subsidies). In the US, however, change is hard to achieve. The ethanol industry is closely associated with corn and no American politician would dare stand up to the corn lobby. The US presidential election starts with candidates seeking the nomination from their own party, and an early lead in these primaries can be crucial to the successful campaign of a candidate. The first primary is in Iowa, often referred to as the buckle in the US corn belt, whose economy is significantly impacted by the ethanol industry. It is unlikely, therefore, that politicians in the US would consider not heavily subsidizing and investing in biofuels, particularly ethanol produced from corn.

Algae has emerged as an alternative to land-based biofuels. These aquatic and microscopic (for the most part) organisms have significantly higher growth rates than land-based crops. In some cases, the algae can double its mass in less than 24 hours and contain up to 40% of oils by weight. These are orders of magnitude more productive than land-based crops. However, for optimal growth rates the algae must have access to sufficient nutrients, while to maximize the yield of oil from the algae the algae must be stressed (have a nutrient removed). In other words, the conditions for high growth rate and high oil content are mutually exclusive. There is, therefore, much work to be done in identifying (or genetically engineering) optimum algae species which can grow fast yet still produce large quantities of oil.

In addition to ethanol and biodiesel, solid biofuels still remain popular, and it is possible that algae biomass could make a wonderful solid biofuel, for example, as pellets to be burnt in household pellet stoves for heating. Pellet stoves feed small pellets of solid biofuel (typically made of sawdust) into a small furnace to generate heat. Meanwhile, it has been suggested that the algae could be grown in wastewater, simultaneously treating the wastewater as well as providing nutrients for algae growth. Or that marine algae could be grown in sea water. The competition for water when growing algae biofuels, in contrast to land-based biofuels, is therefore expected to be low. There have also been studies where the rapid growth of algae was stimulated by filtering carbon dioxide emissions from coal-fired power plants directly through the water. This carbon dioxide, of course, will ultimately be released when the algae is burnt, but as with all biofuels means that algal biofuels are essentially carbon neutral (although replacing reflective desert with absorbing algae ponds may warm the planet slightly). The most important aspect of using algae as a biofuel is that the algae can be grown in large open ponds that are constructed on wastelands that are not suitable for agriculture. In other words, there would be no need to divert land from producing food to producing fuel. Algae biofuels, therefore, could provide an end to the food versus fuel debate that hovers over land-based biofuels like the sword of Damocles.

1.6 Hydrogen Production and Fuel Cells

Hydrogen is a way in which energy can be stored and transported. While hydrogen is obviously not a source of energy in its own right (after all it's just an atom, and we're not talking nuclear physics yet), it takes energy to separate hydrogen from molecules (such as water) and some of this energy can be recouped when the hydrogen reforms with other atoms to make different molecules. Therefore, hydrogen can essentially be seen as a way of storing energy. Hydrogen is obtained using electricity or natural gas, which are the real energy sources. The most common way to obtain hydrogen is from natural gas using steam reforming, however upon extracting the hydrogen from the methane gas the reaction produces carbon dioxide and carbon monoxide, which contribute to global warming. Alternatively, water can be split into hydrogen and oxygen atoms using electrolysis (essentially running electricity through the water). The environmental impact of using hydrogen to store energy then depends on how the electricity is produced. For example, burning coal to generate electricity and then using the electricity to obtain hydrogen is a ridiculously inefficient and unsustainable process, which would only lead to the emission of more greenhouse gases and pollution. On the other hand, if solar cells are used to generate the electricity to obtain the hydrogen which can later be converted back to electricity using fuel cells, then hydrogen can be considered to be similar in many respects with the storage of electrical energy using batteries.

The only problem with this scenario is the efficiency of hydrogen production and fuel cells, which convert free hydrogen and oxygen to water and electricity. Around 50% of the electricity required to obtain the hydrogen in the first place is recoverable from the fuel cells (the rest is lost as heat). In other words, more energy is required to liberate hydrogen from the chemical bonds that tether it to methane or water molecules than can ever be recovered by subsequently oxidizing the hydrogen. This raises an important societal question: are there better ways to transport the electrical energy? A hydrogen economy is not premised on simply the electrolysis of water to liberate the hydrogen and fuel cells to recombine the hydrogen with oxygen, the electricity may have to be converted to low voltage direct current (if it isn't already), the water has to be purified and pressurized, and the subsequent hydrogen has to be compressed for transport. Therefore, only around 25% of the electrical energy that is put into a hydrogen economy is recovered at the end, as opposed to the around 90% efficiency of our electric grid. This inefficient process does not particularly lend itself to a future of clean, sustainable, and renewable energy. That said, current internal combustion engines are horribly inefficient and fossil fuels have nevertheless grown to become the dominant transportation fuel. Therefore, it is possible hydrogen fuel cells could still have a role to play within the transportation sector.

A fuel cell is simply a device that reacts air, or rather the oxygen contained within the air, and hydrogen from a tank to produce water vapor and electrical power. A single fuel cell consists of an electrolyte (typically a polymer electrolyte membrane) sandwiched between two electrodes (cathode and anode). The polymer electrolyte membrane is considered the most promising for transportation, although there are alternatives. In a typical fuel cell, hydrogen gas flows through channels to the anode, where a catalyst separates the electrons from the hydrogen ions (protons). The polymer electrolyte membranes are generally saturated with water, and are highly conductive to these protons, but not to the electrons. While the protons are transported across the polymer electrolyte membrane to the other side of the cell, the electrons flow through an external circuit before reaching the cathode. The flow of electrons in the external circuit constitutes the electrical power produced by the fuel cell. Meanwhile, the outside air is drawn towards and flows past the cathode, where the oxygen in the air reacts with the electrons flowing back from the outside circuit and the protons that flow though the membrane to form water. Water management is crucial to the flow of protons in the fuel cell and, hence, one of the major issues in fuel cell technology, especially at low temperatures. (That said, fuel cell vehicles are solving these issues with vehicles able to start at temperatures as low as -37°C and allowing the car's exhaust to drive out moisture from the system upon shutting the engine off.) A single fuel cell produces a voltage of around one volt, but can be stacked with other fuel cells to obtain higher voltages as required.

The electrical power obtained from a fuel cell can be fed to an electric motor and power electric cars, similar to how a battery might power an electric car. Fuel cell vehicles are attractive because the fuel cells do not emit any pollution and are much more efficient that conventional internal combustion engines (which are typically only around 20% efficient at converting chemical energy into useful mechanical energy). Comparing fuel cell vehicles with less efficient vehicles that run on gasoline is perhaps disingenuous, and comparisons with more efficient battery powered electric vehicles might be more appropriate. As the vehicle is essentially an electrical vehicle with a fuel cell instead of a battery, the car can capture mechanical energy during braking, and store it in supplementary batteries, so that the car can use the energy later for acceleration. One advantage of fuel cell vehicles is their ability to store enough hydrogen to allow the vehicles to make long distance journeys (at least comparable to regular vehicles running on gasoline). Electrical vehicles that rely on batteries, on the other hand, are unlikely to ever be able to travel more than a couple of hundred kilometers without recharging. The hydrogen in fuel cell vehicles has to be stored in lightweight carbon-fiber tanks at pressures of several hundred atmospheres, and because of this high pressure filling the tank will not simply be a matter of placing a nozzle into an open tank. The hydrogen must be injected under high pressure into the tank and a tight seal maintained. In contrast, charging electric batteries can be time consuming. However, electric vehicles could simply swap batteries at 'gas' stations rather than have them charged. In other words, owners of electric vehicles could rent the battery, and pay to have an empty battery swapped with a fully charged battery, to save time 'refueling'. Regardless, of whether the transportation industry chooses to primarily invest in electric vehicles or fuel cells, a viable network of service stations would have to be established. This raises some interesting 'chicken or the egg' type questions. Will anyone buy a vehicle with nowhere to refuel it? Will companies install costly refueling infrastructures without any customers thirsty to refuel their vehicles? Presumably only one type of technology will win: either fuel cell vehicles or battery powered vehicles. Whether the electric battery powered vehicles or fuel cell vehicles draws the capital investment and customer adoption, will depend on which technology is best (not necessarily best in terms of sustainability or efficiency, but best for the customer). Cost will play a huge role in deciding which technology wins, and as advances are made (for example, fuel cell vehicles are increasingly using less expensive platinum in the catalyst) the prices of both technologies will likely drop.

Some Potentially Interesting Reading

MacKay, D. (2008). Sustainable Energy-without the hot air. UIT Cambridge.

Peters, G. P., Andrew, R. M., Boden, T., Canadell, J. G., Ciais, P., Le Qur, C., ... & Wilson, C. (2013). The challenge to keep global warming below 2 C. Nature Climate Change, 3(1), 4-6.

Pandolfi, J. M., Connolly, S. R., Marshall, D. J., & Cohen, A. L. (2011). Projecting coral reef futures under global warming and ocean acidification. Science, 333(6041), 418-422.

Dillon, M. E., Wang, G., & Huey, R. B. (2010). Global metabolic impacts of recent climate warming. Nature, 467(7316), 704-706.

Dai, A. (2011). Drought under global warming: a review. Wiley Interdisciplinary Reviews: Climate Change, 2(1), 45-65.

Kosaka, Y., & Xie, S. P. (2013). Recent global-warming hiatus tied to equatorial Pacific surface cooling. Nature, 501(7467), 403-407.

Schiermeier, Q. (2011). Increased flood risk linked to global warming. Nature, 470(7334), 316-316.

Schneider, C. G., Banks, J. M., & Tatsutani, M. (2010). The Toll from Coal: An Updated Assessment of Death and Disease from America's Dirtiest Energy Source. Clean Air Task Force.

Palmer, M. A., Bernhardt, E. S., Schlesinger, W. H., Eshleman, K. N., Foufoula-Georgiou, E., Hendryx, M. S., ... & Wilcock, P. R. (2010). Mountaintop mining consequences. Science, 327(5962), 148-149.

Laden, F., Neas, L. M., Dockery, D. W., & Schwartz, J. (2000). Association of fine particulate matter from different sources with daily mortality in six US cities. Environmental health perspectives, 108(10), 941.

Pope III, C. A., Burnett, R. T., Thun, M. J., Calle, E. E., Krewski, D., Ito, K., & Thurston, G. D. (2002). Lung cancer, cardiopulmonary mortality, and long-term exposure to fine particulate air pollution. Jama, 287(9), 1132-1141.

Dai, L., Zanobetti, A., Koutrakis, P., & Schwartz, J. D. (2014). Associations of Fine Particulate Matter Species with Mortality in the United States: A Multicity Time-Series Analysis. Environmental health perspectives.

Huynh, W. U., Dittmer, J. J., & Alivisatos, A. P. (2002). Hybrid nanorod-polymer solar cells. Science, 295(5564), 2425-2427.

Li, G., Zhu, R., & Yang, Y. (2012). Polymer solar cells. Nature Photonics, 6(3), 153-161.

Snaith, H. J. (2013). Perovskites: the emergence of a new era for low-cost, high-efficiency solar cells. The Journal of Physical Chemistry Letters, 4(21), 3623-3630.

Currie, M. J., Mapel, J. K., Heidel, T. D., Goffri, S., & Baldo, M. A. (2008). High-efficiency organic solar concentrators for photovoltaics. Science, 321(5886), 226-228.

Green, M. A., Emery, K., Hishikawa, Y., Warta, W., & Dunlop, E. D. (2012). Solar cell efficiency tables (version 39). Progress in Photovoltaics: Research and Applications, 20(1), 12-20.

Graham-Rowe, D. (2011). Agriculture: Beyond food versus fuel. Nature, 474(7352), S6-S8.

Wijffels, R. H., & Barbosa, M. J. (2010). An outlook on microalgal biofuels. Science(Washington), 329(5993), 796-799.

Nocera, D. G. (2012). The artificial leaf. Accounts of Chemical Research, 45(5), 767-776.

TerraPower, L. L. C. (2010). Traveling-wave reactors: a truly sustainable and full-scale resource for global energy needs. In Proceedings of ICAPP (Vol. 2010).

2

Introduction to Computer Modeling

CONTENTS

Computer modeling and computer simulation commonly refer to the use of computer programs which are capable of "recreating" a physical system. In other words, physical laws and behaviors can be implemented into computer algorithms such that a computer program mimics a real system. In many physical systems it is not entirely clear how system inputs are related to the outputs. For example, how does the temperature of our planet depend on cloud cover? Or how does the efficiency of a solar cell depend on the nanoscopic morphology of the photovoltaic device? These are questions for which a simple equation is entirely insufficient and a more complicated computational analysis of the physical system is required. Computer simulations, therefore, are required to mimic the behavior and help elucidate the fundamental physics of these systems.

2.1 Brief History of Computer Simulations

The computer era was ushered in by Alan Turing, a mathematician and computer scientist who built the first working computer during the second world war, capable of decrypting German cyphers and finding the setting of the

Enigma, the most advanced German cypher machine. Turing made the most influential contribution in the Allied victory in the war against Nazi Germany, according to Winston Churchill, but in terms of computers and mathematics his influence could have been much greater. His work at Bletchley Park, the home of Britain's code and cypher school, however, was considered top secret and his exploits were not publicly acknowledged until recently. The second world war also led Jon Von Neumann and Stanislaw Ulam to use statistical mathematics (which would have an incredibly profound effect once computers became available) to investigate the behavior of nuclear weapons in Los Alamos and ultimately to help design the nuclear bomb. In 1953 these statistical methods would be harnessed by what might be considered the first computer simulation, capturing the vibration of 64 lattice points, and programmed by Enrico Fermi at Los Alamos. These military applications of computer science would increasingly filter into civilian life after the war and usher in a computer revolution.

In the 1960's computer mainframes became increasingly common, characterized by programs that had to be fed to the computer via punched cards and instantly recognizable due to the large tape drives whirling in the background. These large and cumbersome computers required many people to operate them and had less processing power than a modern refrigerator. However, atomistic computer models involving deterministic molecular dynamics and stochastic Monte Carlo models, continuum climate models (where Lorenz discovered the butterfly effect), and other early simulations evolved quickly. As the computational power and availability of computers have increased, their application and sophistication have also increased. It is currently impossible to find any area of science that has not been touched with computer simulations, with increasingly complex computer simulations not just merely consisting of a mathematical description of a physical system implemented on a computer, but potentially representing a fundamentally new way of doing science.

2.2 Motivation and Applications of Computer Models

Galileo famously said that mathematics is the language of nature. Computer simulations have now taken on that mantle, and are tasked with capturing the fundamental science of experimental systems. As a model, or representation, of the experimental system the simulation is usually considered a low-cost predictive tool: a way in which to change the input variables of the system quickly and efficiently, and predict the potential outcomes. In this way the simulation can guide costly experiments and help tailor the input parameters for the system to optimize performance or system outcomes. However, computer simulations can be much more than a predictive tool.

In some instances an experimental study will throw up peculiar and counter-intuitive results. While objectively true, these experimental results can defy initial intuition and it can fall upon computer simulations to elucidate the experimental system. In particular, capturing a complex system computationally and accounting for multiple facets of the system which interact in complex ways can result in a simulation which captures the experimental system. But, rather than simply capturing and predicting such systems, the computer simulation can be analyzed in detail. Every single variable, every single atom, every point in time can be isolated and investigated in a computer simulation. The entire history of the simulation is available to be scrutinized in detail, to be graphically or statistically represented. Furthermore, different effects on the system can be "turned off" in an entirely unphysical manner. For example, it is possible to turn of effects such as gravity, van der waals interactions or aerodynamic drag to probe the consequences of these effects. Obviously in real situations this would not be possible and it might be difficult to separate the magnitude of different effects in bringing about a given output; however, using computer simulations it is possible to replicate these experimental results and then scrutinize (for example, compare the magnitude of overlapping forces) or turn off different effects (for example, run the simulation is zero gravity) to probe the importance or relevance of different aspects of the system.

It is important to remember that computer simulations are representations only of a real physical system and the mathematical laws or physically-motivated algorithms which constitute the model are important to determining whether or not the simulation is accurate. In other words, if you put garbage into a simulation you're likely to get garbage out of it. A word of caution is offered to experimentalists looking at implementing or interpreting simulation results. In my opinion, any quantitative agreement between experimental systems and a computer simulation are either fortuitous or a consequence of direct fitting of the input parameters. Qualitative agreement is usually the best one can hope for, and it is not uncommon to present your data in dimensionless form (for example, dividing distances by some characteristic length). This does not mean that you can't fit your simulation so that you get quantitative agreement and then quantitatively predict similar systems with different variables, but the further you deviate from the original system that was fitted to the experiment the less accurate you might expect your quantitative prediction.

It is also necessary to consider why you are developing a simulation, and what you expect to obtain from the model, in advance. For instance, prior to implementing a model you might want to ask yourself: Does anyone care about these simulations? Will anything useful come from them?

2.3 Using Spreadsheets for Simulations

In this book we'll be considering spreadsheets for implementing our models and simulations. For the numerical implementation of an analytic or mathematical model spreadsheets are perfectly fine. However, for full simulations with multiple dimensions and lots of variables, spreadsheets are woefully inadequate and inefficient. Anyone who works with computer simulations, termed a modeler, wouldn't dream of implementing the computer model in a slow and inefficient spreadsheet, and would prefer to work in a programing language - of which there are many (C, python, fortran, etc...). Furthermore, using commercial code (black box models) is also sometimes frowned upon, with programmers preferring to implement their computer code from scratch. Some programmers actually take this very seriously, and invest large amounts of time trying to make their codes slightly more efficient and less resource intensive (faster and using less memory). So why a book using spreadsheets to implement models? Well, firstly, as long as the computer model works then it doesn't really matter how efficient it is; although spreadsheets can be prohibitively slow and restrictive, if they work for you, and satisfy your needs, then it's all good. Secondly, this is only an introduction to computer modeling, and the text assumes no prior knowledge of programming. If you want to go further and implement full-blown computer simulations then you will have to learn how to program in a real computer language. That said, some wonderful research can be done with simple models implemented in spreadsheets and if an experimentalist is simply supplementing an experimental study then simple models can be perfectly fine.

The level of detail required in the computer program depends on what question the computer code is being used to answer. However, in general as computers have increased in speed the size and complexity of computer models have also increased, and what minimum system size and model complexity is sufficient to answer basic questions about the system very much depends on the system under investigation. In other words, a successful computer model should further our understanding of the system. Therefore, while a successful model need only include the necessary facets of the system to answer the desired questions, it is sometime necessary to go beyond the simplest models. In some disciplines the nature of the model is considered just as important as, if not more than, the actual system that is being modeled. In other words, if you try to publish work involving a simple model and the referees are in a community advocating much more advanced models, then you can be given a rough time. Therefore, if you decide to get more seriously into computer modeling you may have to go beyond the use of simple spreadsheet models, but as an introduction spreadsheets are a wonderful way to implement simple models, and almost everyone is familiar with spreadsheets.

2.4 Typing Equations into Spreadsheets

When you click on a cell in a spreadsheet, you can either type directly into the cell or you can type into a textbox at the top of the spreadsheet. A snapshot from a spreadsheet is shown in Figure 2.1. If your text extends beyond the width of a cell it can be obscured by information typed in to the next cell along. If you want to increase the width of the cells you can click on the divider between columns at the top and drag it along, or simply double click at the top of the column, at the divider to the right, and it will automatically scale the width to your largest entry.

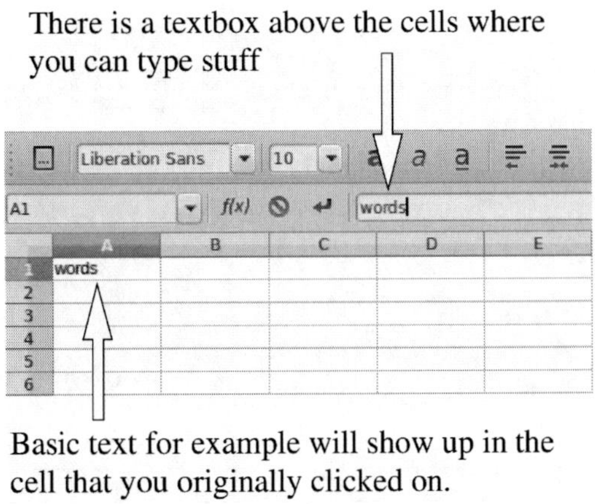

FIGURE 2.1
Typing text into a spreadsheet.

Entering numbers is easy in spreadsheets also (see Figure 2.2), especially if you have a sequence of numbers. Simply type the first two numbers in adjacent cells, highlight those two cells, and then click and drag on the small black box that appear on the bottom right-hand side of the highlighted cells. As you drag over additional cells, these cells will become populated with numbers in the correct series. If only one cell is initially selected then it is assumed that the series simply increments by 1 each time, but if two are initially highlighted then the increment could be any size steps (for example, highlighting and dragging 10 and 20 would result in a series of numbers which increment by 10's). Shift and down arrow on the keyboard usually performs a similar function, but using the mouse is easiest.

The true power behind spreadsheets is in your ability to enter formulas into the spreadsheet cells. When you type an equation always start with an equal sign so that the spreadsheet recognizes the input as a formula and not just some text (as shown in Figure 2.3). The formula can access a wide range of functions (more on this later) and other cells can be referenced by

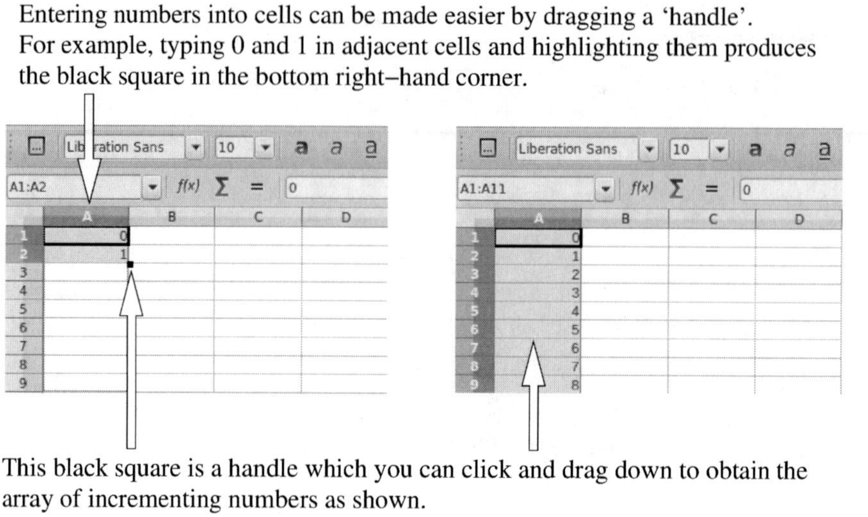

FIGURE 2.2
Entering a series of numbers by clicking and dragging the small black box in the corner.

their column letter and row number. Usually, rather than enter the column letter and row number you can simply click on the cell of the spreadsheet that you would like to reference while you are typing the equation. Figure 2.4 depicts how, similar to how we extended the numbers previously, we can also extend formulas. In other words, if you highlight a cell with a formula in it then you can click and drag the small black box in the bottom right-hand corner of the highlighted box and the formulas will be copied across. If the cell is referenced as simply a column letter and number (say, A1) then as you drag the formulas into adjacent cells the cells they reference will also change (often a highly desirable consequence). However, sometimes the cell you are referencing might be something that will always remain within the cell, and as you drag the formulas you want to continue to reference that cell. In this case, a $ in front of the column letter and row number will help. For example, referencing $A$1 will always remain the same location even if the formula is copied to other locations. Furthermore, you could reference $A1 if you just want the column to remain fixed, or A$1 if you just want the row to remain fixed.

In Figure 2.4 column A is simply a series of numbers starting at 0 and incrementing by 1 as you go down each row. In cell B1 we can type `=A1*2` and the answer would be zero as A1 is zero. However, clicking and dragging the small black box on the bottom right-hand side of cell B1 when it is highlighted and dragging it down will copy this formula into the cell below B1. In this example, the formula is simply multiplying the adjacent cell in column A by

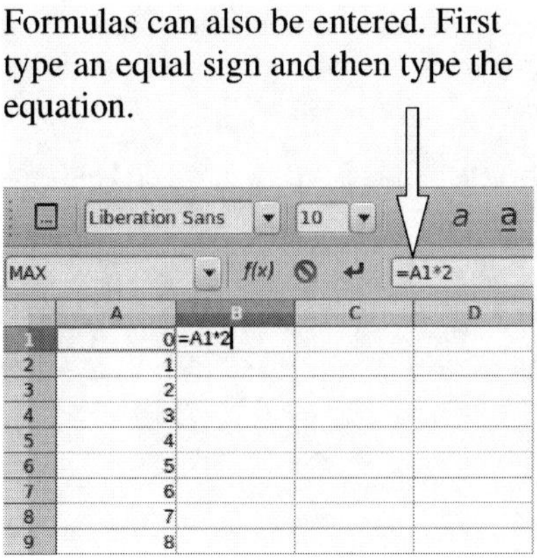

FIGURE 2.3
Typing an equation into a spreadsheet.

a factor of 2. But notice that as we copy it down the row numbers change as well. Alternatively, the formula `=$A$1*2` would make the cells reference always A1 and the answer would always be zero.

In general much more complicated formulas can be considered, and all computers expect the order of operation to be determined by BODMAS. BODMAS stands for brackets, order (or exponents), division, multiplication, addition and subtraction. In other words, `2 + 4 * 5` will result in an answer of 22, `(2 + 4) * 5` will result in an answer of 30, but `2 + (4 * 5)` will result in 22 again because the first thing that occurs is whatever is contained within the brackets, but in the last case the multiplication is bracketed and the multiplication would have occured before addition anyway.

There is a slight quirk to this rule in spreadsheets. Try writing `= -3^2` into a spreadsheet and the result will be 9. In other words, this is interpreted as `= (-3)^2` rather than `= -(3^2)`. When in doubt, therefore, use brackets or parenthesis to make sure operations are performed in the order that you want.

2.5 Functions Available in Spreadsheets

Spreadsheets offer a wide range of functions and tools. Some of the more common formulas which you may encounter in the simulations described here are listed below.

IF()

Formulas can be extended. Type the formula into a box, then click and drag the handle (the black square on the bottom right–hand side of the cell).

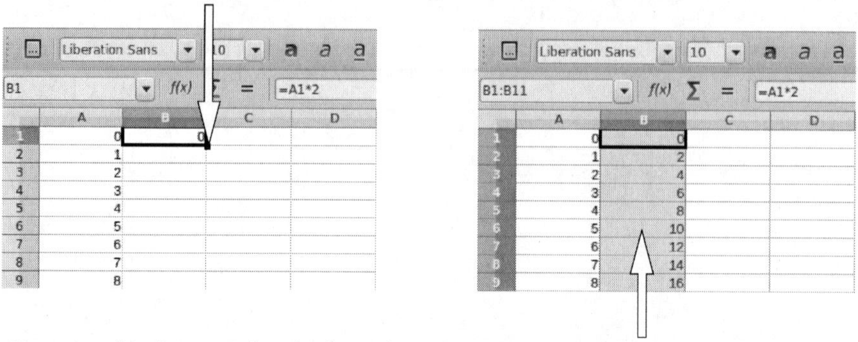

Dragging this down copies the formula to the cells below, and the references move also – the numbers in column A are doubled by the formula in column B.

FIGURE 2.4
Copying formulas across a wide range of cells such that the cells which are referenced change also relative to the cell doing the calculation.

This is a simple if statement. So for example, `=IF(A1>4, 26, 37)` would return the value 26 if the number in A1 is greater then 4 and 37 if it isn't.

AND()

The AND function returns either a TRUE or FALSE flag. So you might have `=AND(A1 > 2, A1 < 5)` which would return a TRUE flag if the number in A1 is both greater than 2 and less than 5. This can also be contained in the IF function: `=IF(AND(A1 > 2, A1 < 5), "Inside", "Outside")` which returns the text "Inside" if the number in cell A1 is between 2 and 5, but will return "Outside" if it isn't.

OR()

The OR function is very similar to AND. `=OR(A1 < 2, A1 > 5)` would return a TRUE flag if either the number in cell A1 is less than 2 or greater than 5.

INDEX()

The function INDEX simply returns the value of an element in a table. For example, `=INDEX(A1:A20, 4)` would return the value in the 4th cell from the array of cells between A1 and A20.

LOOKUP()

The LOOKUP function is useful for looking up a given value in an array and then returning a value based on that lookup. For example `=LOOKUP(37,A1:A20,B1:B20)` would go through the array of cells from A1 to A20 and see if any of them are equal to 37. If one of them is equal to 37 then it returns the corresponding value in the array running from B1 to B20.

PI()

Returns the value of pi (saves you typing in 3.1415927).

SQRT()

SQRT is a function that calculates the square root of a number. `=SQRT(A1)` would simply return the square root of the number contained in cell A1.

SUM()

The function SUM will add up all the values in an array or table of cells. For example, `=SUM(A1:A20)` would return the sum of all the values in the cells from A1 to A20.

AVERAGE()

The function AVERAGE will determine the average of all the values in an array or table of cells. For example, `=AVERAGE(A1:A20)` would return the average or mean of all the values in the cells from A1 to A20.

STDEV()

The function STDEV will return the standard deviation from all the values in an array or table of cells. For example, `=STDEV(A1:A20)` would return the standard deviation for the values in the cells from A1 to A20.

MIN()

The function MIN returns the smallest value from the values in an array or table of cells. For example, `=MIN(A1:A20)` would return the smallest of all the values in the cells from A1 to A20.

MAX()

The function MAX returns the largest value from the values in an array or table of cells. For example, `=MAX(A1:A20)` would return the largest of all the values in the cells from A1 to A20.

There are many more functions available within spreadsheets (especially statistical and economic functions). But here we'll try to limit the number of functions we use. It is worth noting that the same functions will generally work in a variety of spreadsheets (Excel, Libreoffice, Google Drive, etc...)

2.6 Random Numbers

The generation of random numbers was considered something of a contentious issue in the early days of computer simulations, and can still be potentially problematic if you are running a stochastic simulation over many iterations. The problem comes from the pseudo-random nature in which random numbers are generated. Rather than generate a truly random number, which computers being entirely deterministic are incapable of doing, the computer will go through a large array of predetermined random numbers and select different ones each time (sometimes the exact number chosen can depend on the system time of the computer which can introduce a more random nature to the numbers). If too many random numbers are used within a simulation then the computer would recycle through previously used random numbers and potentially introduce artifacts into the simulation. Modern computers can generate a large number of random numbers before you have to worry about this, and as we're running simulations in a spreadsheet we won't need that many random numbers.

The function for generating random numbers in spreadsheets is RAND() which returns a value randomly chosen between 0 and 1. This, of course, could be used within a function to generate a random number between any particular bounds. Note that if you make a change to your spreadsheet then usually (depending perhaps on the spreadsheet environment you are using) all of the random numbers in your spreadsheet are re-generated and all of the cells which are calculated from these values will automatically be recalculated.

2.7 Plotting Data

Visualization and graphics are an important advancement in computer simulations. You can do great research, but unless you can present your results, in a clear and easy to understand way, then your research is essentially worthless. There are many types of plots and software that can be used to generate plots, and all spreadsheets have some plotting capabilities included.

The plots available in most spreadsheets are columns and bar graphs, pie charts, x-y scatter plots or line plots, and bubble and net plots which are perhaps more appropriate for business data. The x-y scatter plot is perhaps the most useful as it can be used, for example, for documenting how the output of a simulation depends on a given input parameter. Excel also contains the capability of generating simple surface plots. However, most of the time data might be taken from the simulation and fed into a separate program that can be used to generate the plots.

There is a large number of software programs that can be used for plotting scientific data, and they all usually require the data to be entered in a specific manner. Some software is commercial and requires the user to purchase a copy. Similar to Excel, which must be purchased from Microsoft, Matlab and Mathematica are available from Mathworks and Wolfram, respectively. While these software packages offer significantly more advanced plotting capabilities (three-dimensional streamlines, vector plots, contour plots on slices through three-dimensional data, etc...) than can be found in spreadsheets, they are also capable of doing much more. In particular, Matlab is capable of running simulations and doing complex matrix computations, and Mathemetica is primarily a symbolic mathematics package for doing complex calculus and algebra. Similar to Libreoffice and Google Drive, which are available freely (Libreoffice is open source), there are many free software applications for plotting data. One of the oldest, and still one of my personal favorites, is Gnuplot which is a simple yet powerful program for visualizing mathematical functions and data. Other software includes Octave and Scilab, which use the same syntax as Matlab, and have similar capabilities except for some of the higher-end plotting functionality. Maxima is an open source symbolic mathematics software package which is similar to Mathematica, and can perform mathematic analysis such as differential and integral calculus, solving linear systems and nonlinear sets of equations, as well as also providing basic plotting capabilities. For more advanced data visualization, OpenDX (IBM's Visualization Data Explorer) is also freely available for performing complex plotting such as three-dimensional surface plots, vector plots, isosurfaces and streamlines. Therefore, there is a large number of freely available open source data visualization software that can be used for plotting your data, and I would urge you to play around with some of these options.

2.8 Macros and Scripts

Macros and scripts are small computer programs which can be run in the background of the spreadsheet environment and perform separate computations. For the purposes of this book, however, the only actions which we will be performing with macros (for the most part) are simple copy and pasting (and we won't be looking at any actual computer code). In particular, most spreadsheets allow you to record a macro and then replay it afterwards. Take the example depicted in Figure 2.5. A number (initially 0) is entered into cell A1 and in cell B1 is the syntax

```
=A1 + 1
```

which simply adds one to the value of the cell A1. Next we copy the contents of cell B1 and paste special (where we only paste the value of the cell, and *not* the syntax contained in the cell) this to cell A1. The number 1 is now occupying cell A1, and the contents of B1 still add one to the value in cell A1, which makes the new value in cell B1 equal to 2. Imagine now that we could record the copy and pasting procedure and save this operation. Each time we replay the copy and paste procedure the numbers increment by 1, and we essentially have a counter which increases by 1 each time we run the procedure. This ability to record procedures in known as either a macro or script (depending on the software) and will help us perform simulations were we need to iterate through time. The only problem is that different software use different programming languages for the macros or scripts and this makes this part of the spreadsheets difficult to transfer between programs (although the rest of the spreadsheet will usually port between software without any trouble).

Cell B1 is simply A1 + 1. Copying the contents of cell B1 and pasting it (using paste special and only pasting the value and not the formula) into cell A1 causes the numbers in cells A1 and B1 to increase by 1. Each time you perform this copy and paste the numbers increment by 1.

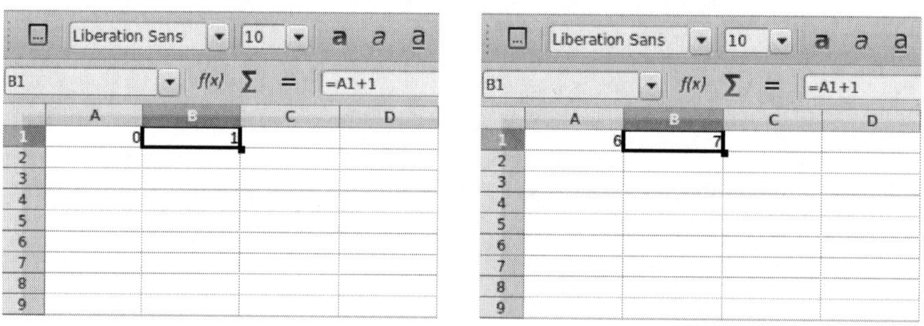

FIGURE 2.5
Example of how a macro can iteratively update a counter.

Excel relies on Visual Basic syntax for the programming language behind the macros contained with the spreadsheet. This is under the development tab (which you may have to enable by right clicking on the ribbon and selecting customize ribbon). Go to the trust center and select run all macros (which is not initially selected and is labeled as not recommended). Now the record macro will appear under the file tab.

In Libreoffice the macros can be written in basic, but are usually written in python. To use macros in Libreoffice it is first necessary to enable this in the software (it isn't by default). Go to Tools - Options - LibreOffice - Advanced, and enable macros in the menu, similar to how this is required to be enabled in Excel. When you are ready to record the macro, go to Tools - Macros - Record Macro. When the macro is recording you can go through the procedure that you would like repeated and press stop when you are finished.

Google Drive also has the potential for macros, or scripts as they are referred to, but this currently requires more programming knowledge to get working. It is not a simple case of recording a macro and then replaying it. Within both Excel and Libreoffice it is also possible to create push buttons, which upon being pushed can then run the macro. This is especially useful when the macro is used to copy the next iteration's information over the previous iterations information, and pressing the button essentially performs an iterative loop in the simulation. Let's not go into the details of the different software here, however, as the pace of the technology is such that the exact procedure for recording and running macros is likely to continue to change periodically.

2.9 Interpolation and Extrapolation

Interpolation and extrapolation essentially means finding the values for a function at any point, when all you to go by are a finite number of discrete points. In other words, say you have a series of points (x_1, y_1), (x_2, y_2), (x_3, y_3), ..., (x_N, y_N), and we want to find the y value that might correspond to some arbitrary x value (not an x-value corresponding to one of the known points). If the desired x, that we are trying to find the corresponding y for is within the range of x's (i.e., greater than the smallest x_i and less than the largest x_i) then we are *interpolating* between the known values. However, if x is outside that range, then we are *extrapolating* beyond the known values. Extrapolation is usually less certain, as the function we are using to represent our values can take off wildly and unexpectedly beyond the known values.

Linear Interpolation and Extrapolation

The simplest method of interpolating between two points is to consider a simple straight line connecting the two points (see Figure 2.6). This linear interpolation is simplest, and for functions with sudden changes in gradient, often the best. If we have two points in space (x_1, y_1) and (x_2, y_2), then the value of y which corresponds to a linear interpolation for the value of x is

$$y = y_1 + (x - x_1)\frac{y_2 - y_1}{x_2 - x_1} \tag{2.1}$$

This is simply a straight line of the form $y = mx + c$, with the slope m being given by $(y_2 - y_1)/(x_2 - x_1)$, and the intercept being at (x_1, y_1). Obviously, if the value of x goes outside the bounds of the data then this function will also extrapolate and estimate values beyond the range of the data. Extrapolation, however, is usually a lot more uncertain than interpretation, especially for functions which can take off suddenly such as polynomial functions.

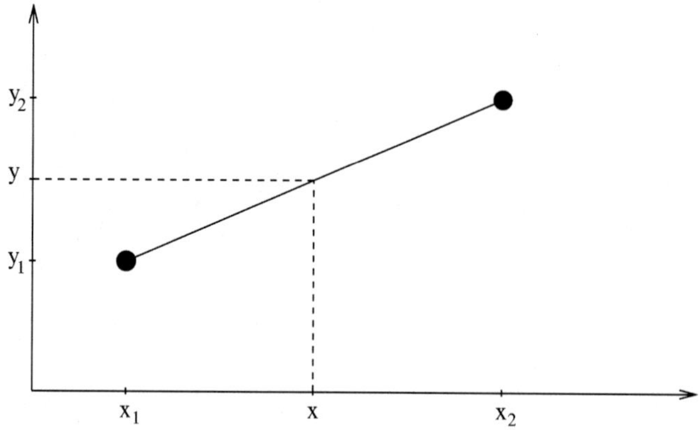

FIGURE 2.6
Example of a simple linear interpolation.

Polynomial Interpolation and Extrapolation

The polynomial interpretation or extrapolation is typically the most common functional form used. Spreadsheets typically have the ability to calculate this form of curve fitting as part of the plotting functions. To do this form of curve fitting by hand, however, can be done quite easily using Lagranges' classical formula

$$\begin{aligned} P(x) &= \frac{(x-x_2)(x-x_3)...(x-x_N)}{(x_1-x_2)(x_1-x_3)...(x_1-x_N)}y_1 \\ &+ \frac{(x-x_1)(x-x_3)...(x-x_N)}{(x_2-x_1)(x_2-x_3)...(x_2-x_N)}y_2 \\ &+ \frac{(x-x_1)(x-x_2)...(x-x_{N-1})}{(x_N-x_1)(x_N-x_2)...(x_N-x_{N-1})}y_N \end{aligned} \tag{2.2}$$

where $P(x)$ is the polynomial function.

A comparison between a simple linear interpolation and the polynomial interpretation obtained using Lagranges' classical formula is depicted in Figure 2.7. The circles represent discrete points in space, where the actual values are known. The solid line represents a linear interpolation between the points (or extrapolation outside of the range of the points). The dashed line is the polynomial interpolation and extrapolation obtained using Lagranges' classical formula. The top example represents a highly curved line, for which the polynomial interpolation is quite smooth. In comparison, the simple linear interpolation is a jagged line, with often sharp discontinuities at the data points. The bottom example represents a curve which consists of two straight lines and a sharp discontinuity in the gradient in the middle of the curve. Here the polynomial fit approximates this curve as a wavy line, and has difficulty handling the sudden discontinuity in the gradient. The appropriate interpolation, therefore, will depend on the nature of the curve being fitted. Also consider the fitted curves outside of the range of the known points. The polynomial fit, in particular, tends to blow up and take off wildly and is probably going to be a poor approximation for the extrapolation of the data.

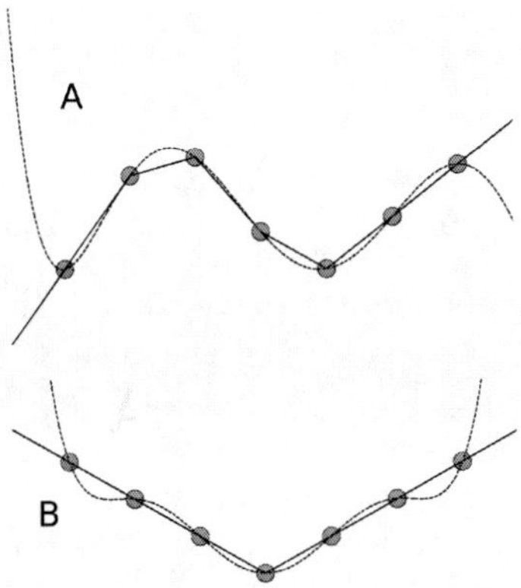

FIGURE 2.7
Comparison between a simple linear interpolation (solid lines) and the polynomial interpretation obtained using Lagranges' classical formula (dashed lines). Two curves are shown a) where the curve is wiggly and b) where the curve is relatively straight but with a sudden discontinuity.

Taylor's Series Expansion

Somewhat related to interpretation and extrapolation is the estimation of values for a function where, rather than have a series of known points, we have just one point and some knowledge of the derivatives about this point. Taylor Series is a way of expanding about this point and obtaining a functional form which can be used to estimate values of the function at some nearby point. In other words, imagine if we had the point (x_1, y_1) along with the derivative of x with respect to y at this point (x_1, y_1') and we wish to find the y value for some arbitrary value of x. We might assume a linear approximation to the function and obtain

$$y = y_1 + y_1'(x - x_1) \tag{2.3}$$

where y'_1 is the derivative of y with respect to x at position x_1. Taylor series expansion takes this further, and considers higher order derivatives. In particular, the value of y is given by

$$y \approx y_1 + \frac{y'_1}{1!}(x - x_1) + \frac{y''_1}{2!}(x - x_1)^2 + \frac{y'''_1}{3!}(x - x_1)^3 + ... \tag{2.4}$$

or more concisely

$$y \approx \sum_{i=0}^{\infty} \frac{d^i y}{dx^i} \frac{(x - x_1)^i}{i!} \tag{2.5}$$

As an example, consider the sine function about the origin. The derivative of sine is cosine (which is 1 at the origin), the second derivative is minus sine (which is zero at the origin), and the third derivative is minus cosine (-1 at origin). We could continue this series forever, but let's just look at the first several terms which would result in a Taylor series approximation of the form

$$y \approx x - \frac{x^3}{3!} + \frac{x^5}{5!} - \frac{x^7}{7!} \tag{2.6}$$

Figure 2.8 compares this approximation with the actual sine function. The solid line is the sine function and the dashed line is the Taylor series expansion with just the first seven terms. Note that over a single cycle of the sine function the approximation is pretty good, but goes off wildly outside of this region. The more terms the more accurate the approximation, and the further from the central point that the approximation is estimated at, the less likely that the approximation is going to be valid.

2.10 Numerical Integration and Differentiation

Throughout this book we will come across derivatives and integrations which we will have to solve numerically. In other words, we will have to discretize the continuous derivatives and integrations onto a finite grid or array so that we can solve these equations within a discrete spreadsheet environment.

Finite Difference Approximations

Finite difference approximations are essentially the application of Taylor series to discrete points in space. However, rather than using known derivatives to estimate values at arbitrary points in space we use the known values at different locations in space in order to estimate the derivatives. Take points in space (x_i, y_i), where the subscript represents the indices for the grid points. The values of the function at these points in space can be used to estimate the derivative of the function. Consider the three points shown in Figure 2.9.

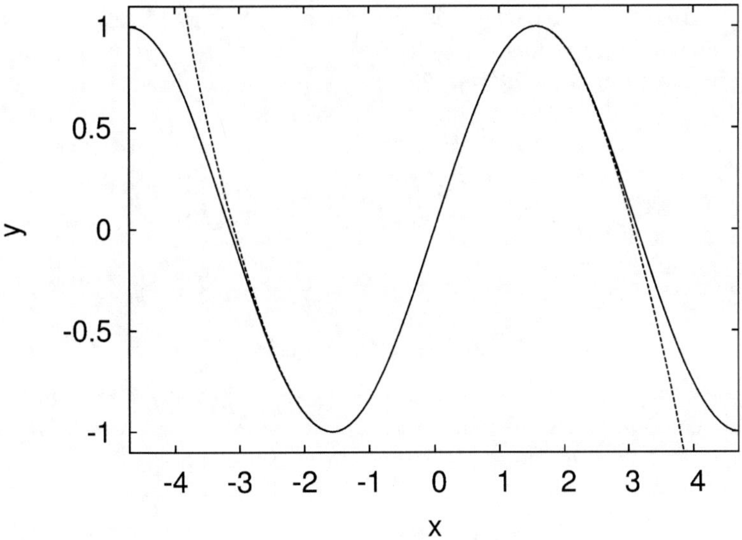

FIGURE 2.8
Comparison between the sine function and a Taylor series approximation (polynomial of degree 7) of the sine function.

The finite difference approximation of the first derivative is shown at the top of Figure 2.9. To derive this we start with Taylor series expansion

$$f \approx f(x) + \frac{f(x)'}{1!}(\Delta x) + \frac{f(x)''}{2!}(\Delta x)^2 + \frac{f(x)'''}{3!}(\Delta x)^3 + ... \tag{2.7}$$

where $f(x)$ is the value of the function at the point at which we are trying to determine the derivative, f is the function at some other neighboring position, and Δx is the distance between these positions (for example, the grid spacing if f and $f(x)$ are on adjacent grid points). For the finite difference approximation, we can write

$$f_{i+1} \approx f_i + \frac{df}{dx}\Delta x \tag{2.8}$$

and similarly,

$$f_{i-1} \approx f_i - \frac{df}{dx}\Delta x \tag{2.9}$$

combining these two equations can result in the following equation:

$$\frac{df}{dx} \approx \frac{f_{i+1} - f_{i-1}}{2\Delta x} \tag{2.10}$$

This is known as the central difference approximation, as is probably the most common way to approximate a derivative. Equations 2.8 and 2.9 are known as forward and backward approximations, respectively, and can be quite useful at boundaries. Note, however, that if we were to include third order terms in Equations 2.8 and 2.9 then the additional terms would cancel, making the central difference approximation more accurate than the forward and backward approximations. In other words, Taylor expansion can give us

$$f_{i+1} \approx f_i + \frac{df}{dx}\Delta x + \frac{d^2 f}{dx^2}\frac{(\Delta x)^2}{2} \tag{2.11}$$

and similarly,

$$f_{i-1} \approx f_i - \frac{df}{dx}\Delta x + \frac{d^2 f}{dx^2}\frac{(\Delta x)^2}{2} \tag{2.12}$$

but in deriving the central difference approximation the additional terms would cancel. We can go further and include more terms in the Taylor series to obtain more accurate approximations. For example, we could consider the following 4 Taylor series

$$f_{i+1} \approx f_i + \frac{df}{dx}\Delta x + \frac{d^2 f}{dx^2}\frac{(\Delta x)^2}{2} + \frac{d^3 f}{dx^3}\frac{(\Delta x)^3}{6} \tag{2.13}$$

$$f_{i-1} \approx f_i - \frac{df}{dx}\Delta x + \frac{d^2 f}{dx^2}\frac{(\Delta x)^2}{2} - \frac{d^3 f}{dx^3}\frac{(\Delta x)^3}{6} \tag{2.14}$$

$$f_{i+2} \approx f_i + \frac{df}{dx}2\Delta x + \frac{d^2 f}{dx^2}\frac{(2\Delta x)^2}{2} + \frac{d^3 f}{dx^3}\frac{(2\Delta x)^3}{6} \tag{2.15}$$

$$f_{i-2} \approx f_i - \frac{df}{dx}2\Delta x + \frac{d^2 f}{dx^2}\frac{(2\Delta x)^2}{2} - \frac{d^3 f}{dx^3}\frac{(2\Delta x)^3}{6} \tag{2.16}$$

Combining these 4 equations it is possible to obtain a more accurate approximation for the finite difference of the form

$$\frac{df}{dx} \approx \frac{8f_{i+1} - 8f_{i-1} - f_{i+2} + f_{i-2}}{12\Delta x} \tag{2.17}$$

While this is of higher order accuracy (including more terms in the Taylor polynomial expansion), it does so by including data points further from the point where the derivative is to be calculated. In terms of stability, therefore, whether the more accurate approximation is more or less stable will depend on the system being simulated. However, when obtaining derivatives for plotting purposes (and not directly included in the iterative loops of a simulation) the higher order approximation will generally yield a slightly smoother function.

In addition to using finite difference to approximate first-order derivatives it is possible to approximate higher orders as well. For example, let's consider the simplest way to obtain the second derivative.

$$f_{i+1} \approx f_i + \frac{df}{dx}\Delta x \tag{2.18}$$

and

$$f_{i-1} \approx f_i - \frac{df}{dx}\Delta x \tag{2.19}$$

were presented earlier as forward and backward finite difference approximations for the derivative at the central point. However, we can consider the derivatives to be central difference approximations for points in space in between grid points $i+1$ and i, and $i-1$ and i, respectively. Then the second derivative is simply the derivative of the derivative, which using another central difference approximation yields

$$\frac{d^2 f}{dx^2} \approx \frac{\frac{f_{i+1}-f_i}{\Delta x} - \frac{f_i - f_{i-1}}{\Delta x}}{\Delta x} = \frac{f_{i-1} + f_{i+1} - 2f_i}{2\Delta x} \tag{2.20}$$

The above procedures can be used to obtain a variety of difference approximations for forward and backward difference approximations, for derivatives of different order and for approximations that use quite a wide range of grid points. While the simpler approximations usually work fine, if problems are encountered with stability then other finite difference approximations can be considered.

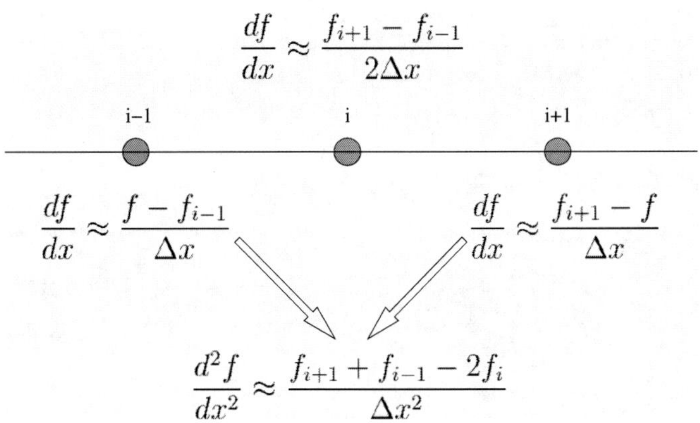

FIGURE 2.9
Finite difference approximation using discrete points in space to represent continuum derivatives.

The calculation of these finite difference approximations is depicted in Figure 2.10. In particular, some data is contained within column B, where here the data is randomly generated as an example. In column C the first derivative is calculated using a finite difference approximation. In particular, the code in cell C3 is

```
=(B4-B2)/2
```

and for this example the grid spacing is simply 1. The second derivative is calculated in column D, with the code in cell D3 being

```
=B2+B4-2*B3
```

and, finally, the first derivative is calculated for a second time using the higher order finite difference approximation. The code in cell E4 is

```
=(8*B5-8*B3+B2-B6)/12
```

If, for example, the distance between points in the x-direction was equal to 0.1 then the above code would change to

```
=(B4-B2)/0.2

=(B2+B4-2*B3)/0.01
```

and

```
=(8*B5-8*B3+B2-B6)/1.2
```

This simple code is copied all the way down and can be used to calculate the derivative for all values of x. Let's look at a simple example of the derivative of the sine function.

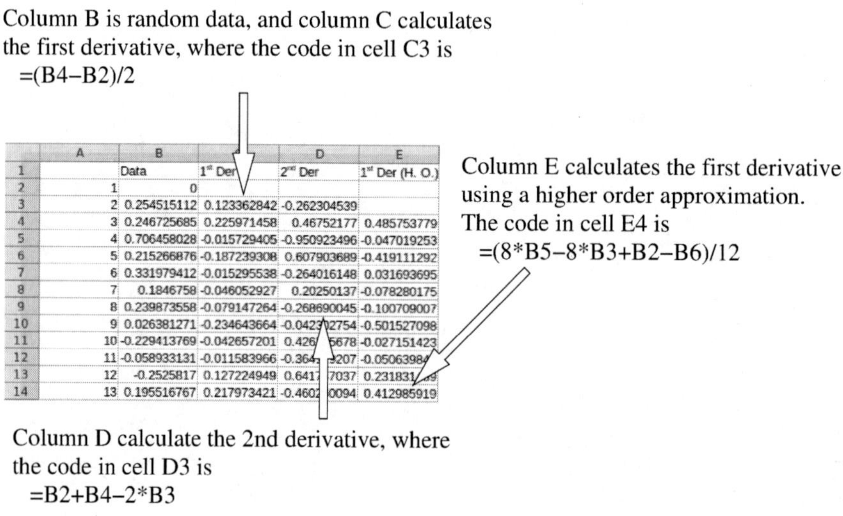

	A	B	C	D	E
1		Data	1st Der	2nd Der	1st Der (H. O.)
2	1	0			
3	2	0.254515112	0.123362842	-0.262304539	
4	3	0.246725685	0.225971458	0.46752177	0.485753779
5	4	0.706458028	-0.015729405	-0.950923496	-0.047019253
6	5	0.215266876	-0.187239308	0.607903689	-0.419111292
7	6	0.331979412	-0.015295538	-0.264016148	0.031693695
8	7	0.1846758	-0.046052927	0.20250137	-0.078280175
9	8	0.239873558	-0.079147264	-0.268690045	-0.100709007
10	9	0.026381271	-0.234643664	-0.042[illegible]02754	-0.501527098
11	10	-0.229413769	-0.042657201	0.426[illegible]5678	-0.027151423
12	11	-0.058933131	-0.011583966	-0.364[illegible]207	-0.0506398[illegible]
13	12	-0.2525817	0.127224949	0.641[illegible]7037	0.231831[illegible]
14	13	0.195516767	0.217973421	-0.460[illegible]0094	0.412985919

FIGURE 2.10
Spreadsheet implementation of finite difference approximations.

The sine function has a first derivative which is equal to the cosine and a second derivative equal to minus the sign. A discrete representation (calculated at discrete data points with a spatial step of 0.1) of the sine wave is given. A plot of the first and second derivative of the sine wave is depicted in Figure 2.11. Note that in this case, the higher order approximation is not shown as it agrees entirely with the lower order approximation. For this simple

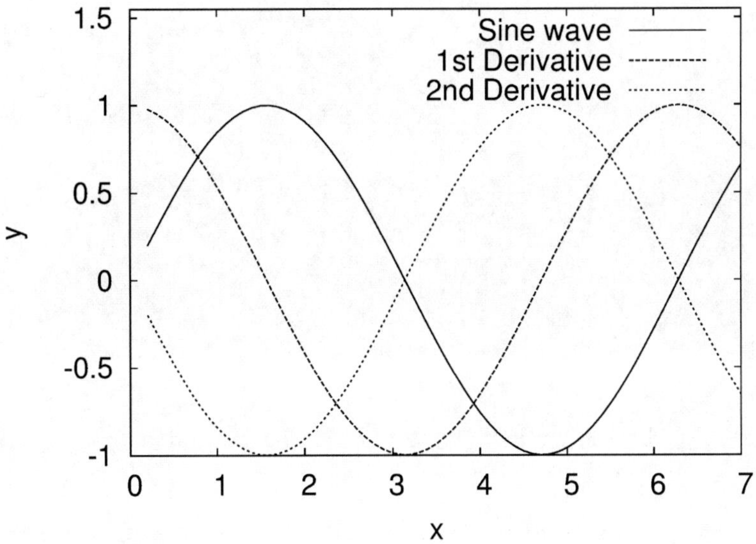

FIGURE 2.11
A sine function, along with its first and second derivatives. The grid spacing is relatively coarse at 0.1.

example the finite difference approximation gives a wonderful approximation of the actual derivative.

Trapezium Rule

For a simple function the integration is essentially the area under a curve. To numerically obtain this area we can break the function into a series of trapeziums as shown in Figure 2.12. The figure shows a curve which might represent the true function and a series of data points which discretize this function. The curve, or rather the area under the curve, is then broken up into trapeziums (4 sided shapes where two of the sides are parallel). This is essentially representing the function as a linear interpolation between known data points (the black circles in Figure 2.12) and then finding the area of the trapeziums underneath this interpolated representation of the function. The area of a trapezium between two data points would be the distance in the x-axis between the two data points multiplied by the average height in the y-axis of the two data points.

Mathematically, we can write the trapezium rule in the following manner

$$\int_{x_1}^{x_N} f(x)dx = \sum_{i=1}^{N-1}(x_{i+1} - x_i)\frac{f_i + f_{i+1}}{2} \tag{2.21}$$

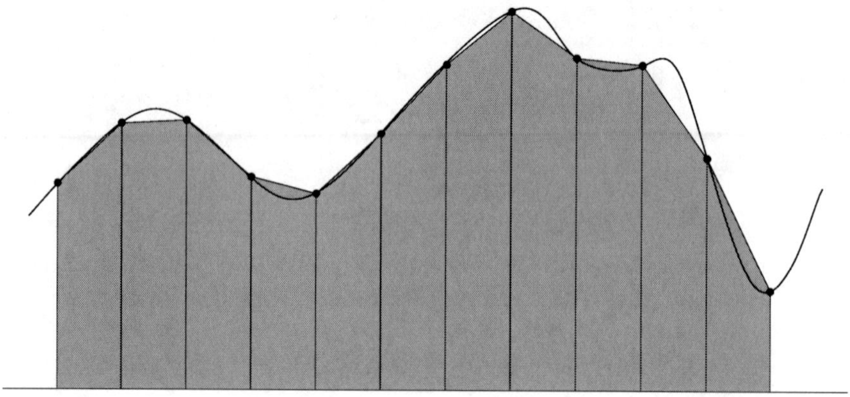

FIGURE 2.12
Illustration of how a continuum function can be integrated by breaking the area under the curve into trapeziums.

In this example we're assuming the integration is over the data range (from x_1 to x_N), although obviously we could integrate over any range. There are alternatives to this rule, such as Simpson's rule, which integrates the function by considering the polynomial interpolation of the data points rather than the linear interpolation. However, the trapezium rule is easy to implement and is usually considered more than adequate.

2.11 Solving Linear Systems

Sometimes a computer model results in a linear system of equations. For example, you might have a bunch of equations of the form

$$\begin{aligned} A_{11}x_1 + A_{21}x_2 + A_{31}x_3 + ...A_{N1}x_N &= b_1 \\ A_{12}x_1 + A_{22}x_2 + A_{32}x_3 + ...A_{N2}x_N &= b_2 \\ A_{13}x_1 + A_{23}x_2 + A_{33}x_3 + ...A_{N3}x_N &= b_3 \\ &\cdots \\ A_{1N}x_1 + A_{2N}x_2 + A_{3N}x_3 + ...A_{NN}x_N &= b_N \end{aligned} \tag{2.22}$$

where the x's are your unknowns, and you have just as many unknowns as you have equations (which means its mathematically solvable). While such systems of equations can be easily solved in other software applications (such as Octave, for example) here we'll look at a couple of ways we might do this using our own computer code (important if we plan to solve such a system of equations at each iteration). Before we do though, it is important to realize that in practice the systems that you solve will invariably be quite sparse. In other words, most of the terms are zero. If we were to write the above system of equations in matrix form

$$\mathbf{A} \cdot \mathbf{x} = \mathbf{b} \tag{2.23}$$

where the bold font represents that $\mathbf{A}$ is a matrix, and both $\mathbf{x}$ and $\mathbf{b}$ are vectors. Then the matrix $\mathbf{A}$ is almost entirely made up of zeros. This arises commonly in computer simulations because we are often dealing with local interactions. In other words, when using the finite difference approximation the derivative is given in terms of local values. Therefore, if the derivative appears in an equation and we discretize the derivative, then in terms of the resulting linear equations we might expect to have an equation whose solution only depends on the local values. In other words, the components of $\mathbf{A}$ which link this site to neighboring sites will have nonzero terms, but all the other terms (the vast majority) will be zero.

Relaxation Method

A simple relaxation method is the simplest to implement, but usually not the most accurate or reliable. Let's consider the most common example, Poisson's equation

$$\nabla^2 \phi = f \tag{2.24}$$

where ∇^2 represents the second derivative in multiple dimensions. For example, in two-dimensions we might write Poisson's equation as

$$\frac{d^2\phi}{dx^2} + \frac{d^2\phi}{dy^2} = f \tag{2.25}$$

which upon using the finite difference approximation mentioned earlier results in an equation of the form

$$\frac{\phi_{i+1,j} + \phi_{i-1,j} - 2\phi_{i,j}}{(\Delta x)^2} + \frac{\phi_{i,j+1} + \phi_{i,j-1} - 2\phi_{i,j}}{(\Delta y)^2} = f_{i.j} \tag{2.26}$$

where $\phi_{i,j}$ represents the value of ϕ at the coordinates i and j, in the x- and y-directions, respectively. Typically the spatial discretizations (the distances between grid points) will be uniform and the same in the x- and y-directions. In other words, $\Delta x = \Delta y$. Therefore, we can re-arrange the above expression to obtain

$$\phi_{i,j} = \frac{1}{4}\left(\phi_{i+1,j} + \phi_{i-1,j} + \phi_{i,j+1} + \phi_{i,j-1} - f(\Delta x)^2\right) \tag{2.27}$$

If we introduce k as an iteration number then we can see how this can be solved iteratively.

$$\phi^{k+1}_{i,j} = \frac{1}{4}\left(\phi^k_{i+1,j} + \phi^k_{i-1,j} + \phi^k_{i,j+1} + \phi^k_{i,j-1} - f(\Delta x)^2\right) \tag{2.28}$$

where we could take the solutions at iteration k and use them to obtain the next iteration's solution at iteration $k+1$. A couple of words of caution. Firstly, when relaxing the above equation make sure the next iteration's values are stored entirely separately from the current iterations. In other words, you don't want to accidentally use the next iteration's values that have just been updated while you're still calculating the next iterations. Secondly, this method of solving the system of linear equations can be quite slow and other methods are considered superior. That said, it does work and there is something to be said for a relaxation method as it enables you to follow the relaxation and if problems occur in your computer code they are generally easier to spot (which helps with debugging).

Conjugate Gradient Method

The conjugate gradient method is included here, because it's one of my favorites. It's elegant and fast. Essentially, the problem is one of finding a minimum and so the technique moves downhill in conjugate steps, and is guaranteed to reach the minimum within a number of steps equal to the number of equations. For our purposes, however, we don't care about the math behind the technique, we just want to use it as a tool to find the solution. Let's go back to our original system of equations

$$\mathbf{Ax} = \mathbf{b} \tag{2.29}$$

If we imagine $\mathbf{x}_0$ to be our initial guess, then the initial residual is defined as

$$\mathbf{d}_0 = \mathbf{r}_0 = \mathbf{b} - \mathbf{A}\mathbf{x}_0 \tag{2.30}$$

We then iterate through the following steps to obtain a solution, defined as when the sum of the residual drops to a sufficiently small value.

$$\alpha_i = \frac{\mathbf{r}_i^T \mathbf{r}_i}{\mathbf{d}_i^T \mathbf{A} \mathbf{d}_i} \tag{2.31}$$

$$\mathbf{x}_{i+1} = \mathbf{x}_i + \alpha_i \mathbf{d}_i \tag{2.32}$$

$$\mathbf{r}_{i+1} = \mathbf{r}_i - \alpha_i \mathbf{A} \mathbf{d}_i \tag{2.33}$$

$$\beta_{i+1} = \frac{\mathbf{r}_{i+1}^T \mathbf{r}_{i+1}}{\mathbf{r}_i^T \mathbf{r}_i} \tag{2.34}$$

$$\mathbf{d}_{i+1} = \mathbf{r}_{i+1} + \beta_{i+1} \mathbf{d}_i \tag{2.35}$$

The above procedure is iterated (where i represents an iteration and $i+1$ represents the next iteration) until the sum of the residuals (the sum of the components of $\mathbf{r}$) is sufficiently small. How small is small enough? Well that all depends on your system. In general the technique will converge to a solution with a small residual, then it will fluctuate around this solution, before blowing up to increasingly worse solutions. Therefore, you should watch the residual closely and try to find when this reaches a minimum. For example, I have run the conjugate gradient twice during a simulation's iteration when necessary: the first time to find the minimum residual, and the second time to reach this minimum. This is probably the most powerful technique for solving a sparse set of linear equations, however, and can reach a solution very quickly.

2.12 Non-linear Equations

Sometimes the equations which have to be solved are not linear. Sometimes the equations that you are trying to solve are nonlinear. Here we'll only consider a single equation (solving large sets of nonlinear equations can be quite tricky) and the Newton-Raphson Method for solving it.

Newton-Raphson Method

The Newton-Raphson method is a method for successively obtaining better solutions to a given equation. Take the following equation,

$$f(x) = 0 \tag{2.36}$$

which is nonlinear and imagine we wish to obtain a solution for x. In other words, find the x values that satisfy the equation. The Newton-Raphson method is simply

$$x_{i+1} = x_i - \frac{f(x_i)}{f(x_i)'} \tag{2.37}$$

where i represents the current iteration, and $f(x_i)'$ is the derivative of the function $f(x_i)$ with respect to x. The above iterative step can be used successively until an adequate solution is obtained. This is a very fast method that can solve the nonlinear equation very efficiently. While it may be extended to systems of equations, we will not consider that here.

2.13 Monte Carlo Simulations

The Monte Carlo model is a stochastic method for optimizing and simulating a wide range of complex systems. Monte Carlo models were the first computer simulations, and remain the most fascinating in my opinion, and therefore it's worth introducing this curious class of models in more detail. The method involves randomly sampling the phase space of the system and usually obtaining the equilibrium state of the system. The random nature of the method led to it being named after the Monte Carlo Casino and the randomness of the games of chance played there. The Monte Carlo method has been applied to many areas associated with alternative energy technologies. In particular, in section 4.2 we will look at a Monte Carlo model for solar cells, where the directed but still random motion of both electrons and holes can be captured inside a photovoltaic device.

However, before we turn our attention to the application of the Monte Carlo method it is worth discussing what the Monte Carlo method is and some simple examples of the Monte Carlo method. There are three main aspects of the Monte Carlo method that I would like to convey. First, the random sampling of the phase space, which is the main aspect of the Monte Carlo method. Second, the extension of the Monte Carlo method to handle dynamics (often referred to as pseudo-dynamics). Finally, I would like to show you the Metropolis algorithm. The Metropolis algorithm allows for the selective sampling of the phase space and stochastic events in the simulation can be either accepted or rejected based on the difference in energy between the system before and after the stochastic event. This allows this random method to capture ordering or organized events. To look at this another way, consider the stochastic nature of evolution with random gene mutations giving rise to biological changes over successive generations. It might seem amazing how evolution can produce such wonders as the duck-billed platypus or 122 tonne dinosaurs from fish that crawled out of the sea. But evolution works on the principle of selecting evolutionary advantageous traits generation after generation such that these amazing developments emerge over long periods of time. The Monte Carlo method is similar, in some aspects, as it selects changes to the system which might, for example, tend to minimize the energy of the system. On short time scales the changes would appear random, and the system doesn't appear to significantly change at all, but over long periods of time incredible order and sophisticated organization can emerge from these simple simulations. This makes the Monte Carlo method both incredibly powerful and incredibly annoying. It's incredibly powerful because it can quite often solve problems that would otherwise be difficult, or too computationally expensive, to tackle with other techniques. However, it can be very annoying to implement because any order or organization doesn't emerge until the simulation has been running for some time. This makes debugging your code a time consuming and complicated endeavor, but well worth the trouble. Let's start with a relatively simple example of the Monte Carlo method.

Simple Monte Carlo Model

This simple example of the Monte Carlo method was introduced to me as an undergraduate, where the experiment consisted of two grids with different colored marbles inside. On one side of the grid all the marbles were blue, while on the other side they were all red. The experiment consisted of randomly selecting two points on the grid and switching the marbles at these locations. In other words, each iteration of the simulation consists of swapping two random marbles such that initially they are separated with all the red on one side and all the blue on the other side, but by the end of the simulation enough marbles have been swapped that the system consists of red and blue marbles randomly dispersed throughout the system. This could represent, for example, the resultant mixing between two separate rooms containing different gases upon opening a door between the two rooms. Note that we will not be simulating the dynamics of this process, only obtaining a final equilibrium dispersion. We will replicate this experiment here using a spreadsheet.

The system will consist of a one-dimensional array (length of 20 cells) where the red and blue marbles are now represented by either a 1 or 0 in the cells. In order to swap the contents of two cells we must first move the contents of one site and then separately move the contents of the other site, and we must stochastically determine which cells we will swap. Because we are only interested in the equilibrium properties, we can swap two sites even if they aren't anywhere near each other. Physically, this might seem a little strange; that two sites (possibly representing particles of a gas) could switch places even though they are not spatially close to one another, but that is the strength of the Monte Carlo method. It allows us to efficiently sample the phase space of the system, the different possible configurations, in an efficient manner without regard to the physical limitations of how such moves might occur in reality.

The spreadsheet is displayed in Figures 2.13 and 2.14, but it is worth noting that there are different ways to organize and implement this model in a spreadsheet and this is just an example. The system consists of 20 sites (we could obviously extend this but small sizes are good for demonstrating the principle). There are two copies of the system: on the left hand side of the spreadsheet is the system after switching one of the sites and on the right hand side of the spreadsheet is the system after switching the second site (effectively swapping the two sites).

The initial conditions are in the cells W6:AP6 and consist of ten ones followed by ten zeroes. The two cells which are to be switched are in the cells U7 and V7, which contain the following code

```
=RANDBETWEEN(1,20)
```

which returns a randomly acquired integer between 1 and 20 (the size of the system considered here). Once we have these two random numbers we switch the sites at these locations. On the left of the spreadsheet (see Figure 2.13) we move the first number to its new location. To do that we insert the following code in cell A7, before copying it across to T7.

```
=IF(A$5=$U7, INDEX($W6:$AP6, $V7), W6)
```

where `A$5` is the cell above A7 that indicates the position in the system, and when this is equal to the random number in U7 then the value will be moved;

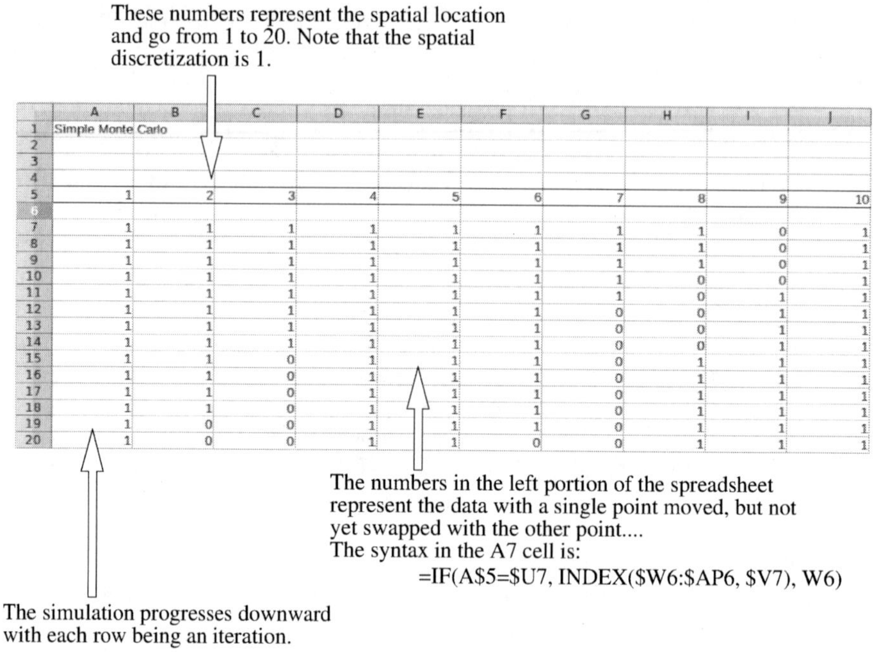

FIGURE 2.13
Spreadsheet implementation of a simple Monte Carlo model.

however if they are not equal then set this equal to the value at the same location in the initial data (cell W6). If the site is identified as the site that is referenced by the first random number then put the value in here from the site referenced by the second random number. To find the value in the site referenced by the second random number, in the above code, we used the following command:

```
INDEX($W6:$AP6, $V7)
```

where `$W6:$AP6` are the cells containing the original data and the cell V7 is the second random number. We do a similar movement, moving the value of the site referenced by the first number into the site referenced by the second number. Once this is done then this will complete the first iteration of the simulation. Because we want to copy the line down and populate the entire spreadsheet with multiple iterations (each row being its own iteration) the data after the swap is directly under the original data (in cells W7:AP7). This way as we copy the data down it is referencing the same sites above it to continue propagating the procedure for each iteration.

The second movement of the swap is achieved using similar code to before

```
=IF(W$5=$V7, INDEX($W6:$AP6, $U7), A7)
```

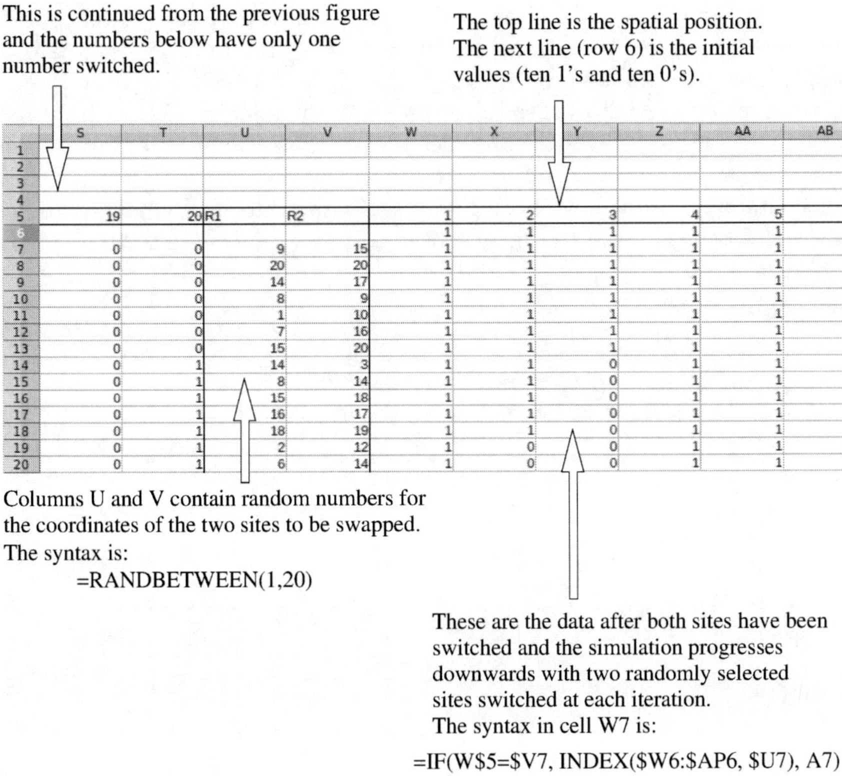

	S	T	U	V	W	X	Y	Z	AA	AB
1										
2										
3										
4										
5	19	20	R1	R2	1	2	3	4	5	6
6					1	1	1	1	1	1
7	0	0	9	15	1	1	1	1	1	1
8	0	0	20	20	1	1	1	1	1	1
9	0	0	14	17	1	1	1	1	1	1
10	0	0	8	9	1	1	1	1	1	1
11	0	0	1	10	1	1	1	1	1	1
12	0	0	7	16	1	1	1	1	1	1
13	0	0	15	20	1	1	1	1	1	1
14	0	1	14	3	1	1	0	1	1	1
15	0	1	8	14	1	1	0	1	1	1
16	0	1	15	18	1	1	0	1	1	1
17	0	1	16	17	1	1	0	1	1	1
18	0	1	18	19	1	1	0	1	1	1
19	0	1	2	12	1	0	0	1	1	1
20	0	1	6	14	1	0	0	1	1	0

FIGURE 2.14
Spreadsheet implementation of a simple Monte Carlo model (continued).

where this code is inserted into cell W7, and then copied across to AP7. If it is the second site that is being swapped, then the first site from the original data is inserted, else the site is obtained from the data where one of the sites was already moved. This completes the swap, with the information from the first site moving to the second, and vice versa the information from the original second sight moving to the first.

Once we have the first line set up we can copy the line down and populate the spreadsheet, with each row switching a randomly chosen pair from the row above. Note that the $ on certain references to cells means that either the row or the column (or both) will persist as the cells are copied. This is often the hardest part of organizing your spreadsheet and special care should be taken to make sure these are correct (although if you mess up then it's usually pretty obvious as the numbers in the simulation won't make any sense).

As the simulation progresses we can measure the iteration number (the number of swaps), but it is customary in Monte Carlo models to refer to the "Monte Carlo steps" (MCS). A MCS is the number of iterations equal to the number of sites (or number of particles) in a particular Monte Carlo Simulation. So in our system a MCS would be 20 iterations, as there are 20 sites. After 10 MCS (200 iterations or rows in the spreadsheet) the system has been mixed enough that the sites with 1 in them are randomly dispersed. This can be seen in Figure 2.15 which shows the system at 0 MCS (initial configuration) and after 10 MCS. Cells that contain 1's are shaded and cells that contain 0's are left white. Initially all of the cells on the left contained ones and all the cells on the right contained zeros, but over the 10 MCS the system has become completely mixed with the distribution of 1's and 0's being completely randomized.

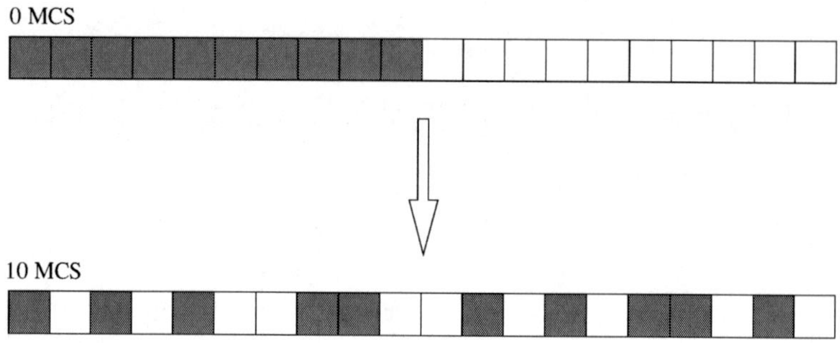

FIGURE 2.15
The transition from ordered phases to random dispersion is depicted in a simple Monte Carlo model.

To further illustrate this trend, Figure 2.16 depicts the number of ones either on the left or right hand side of the system as a function of simulated MCS. Initially all the cells containing ones were on the left hand side of the simulation (there are ten of them) and none were on the right. However, after only about 1 MCS we can see that the distribution of cells containing ones is roughly equal on both sides. Note that the simulation has found this equilibrium (with the cells containing ones equally dispersed between both sides) but that due to the stochastic nature of the simulation there are still fluctuations in the data. Essentially, while the simulation tends towards the equilibrium state, the local moves are not constrained by what the equilibrium state is.

The above model is a useful demonstration of the stochastic nature of the Monte Carlo method, but the Monte Carlo method has other potential attributes that are worth introducing, the first being the implementation of dynamics in the model.

Dynamic Monte Carlo Model

The most powerful aspect of the Monte Carlo method is its ability to efficiently sample the phase space (here by randomly swapping any two sites,

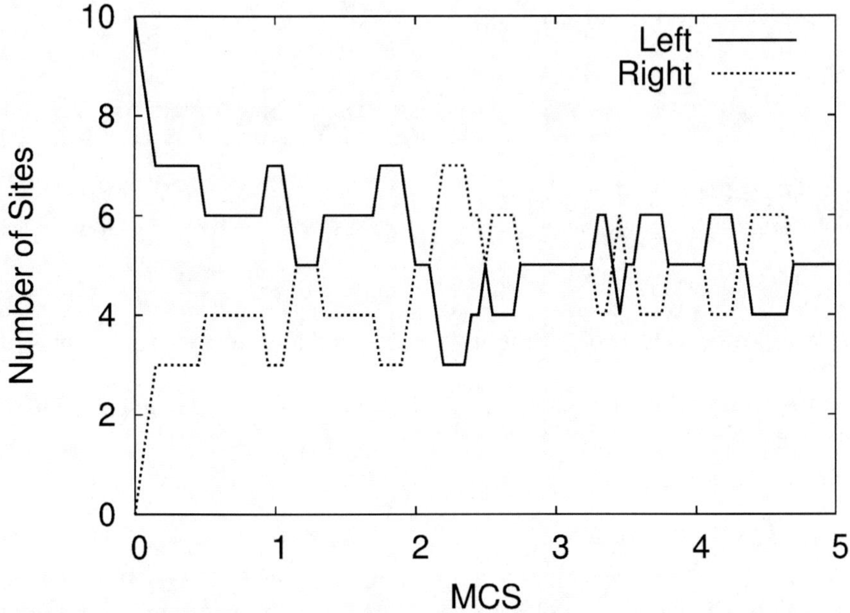

FIGURE 2.16
Plot of the number of sites in the left or right halves of the simulation which contain 1's as a function of the number of Monte Carlo steps. As the system becomes randomized an equal number of 1's will be found in both sides.

regardless as to how far away from each other they happen to be). In the dynamic case, however, we will only swap sites that are adjacent to one another (as shown schematically in Figure 2.17).

This is more representative of reality where, for example, particles of gas only move locally rather than over to the opposite side of the system. So on the positive side of this change we might gain some insights into the dynamics of our system, and how the sites intersperse. However, it will almost certainly take more MCS (or iterations) for the model to reach an equilibrium state. So which is more useful? To reach the equilibrium state in an efficient and computationally expedient manner, or gain interesting insights into the dynamics of how the simulation might reach this equilibrium?

To convert our simple Monte Carlo model from before into a dynamic Monte Carlo model requires only one change. Now the two sites which we are going to swap must be adjacent to each other. This is easy enough to do, right? The relevant section of the spreadsheet is shown in Figure 2.18.

The column U contains the first site which is to be switched and, as before, this is simply a random number between 1 and 20 (the size of the system).

```
=RANDBETWEEN(1,20)
```

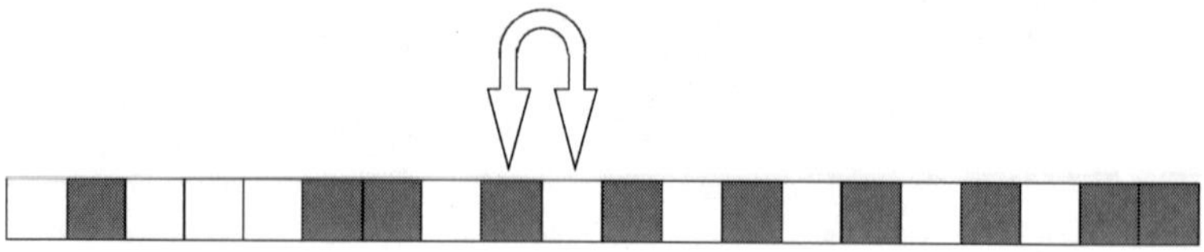

FIGURE 2.17
Illustration of how only local swaps are permitted in a dynamic Monte Carlo model.

However, now instead of picking the second random number to be anything between 1 and 20, we must pick a number either 1 higher or 1 lower than the first random number. Of course, this choice should be random and the relevant code is

```
=IF(RAND()<0.5,U7+1, U7-1)
```

In other words, if a random number between zero and one (`RAND()`) is less than 0.5 (a 50% chance) then the second random number is one greater than the first random number, else it's one less. However, this might cause a problem. What if we choose the first random number to be 20 and then choose to add one to this to get the second random number (i.e., 21)? Then we would be outside of our system bounds (the system size is only 20). There are a couple of ways we might handle this problem. The first might be to prohibit the swap from occurring in this case. In other words, the sites at either end would only be allowed to switch with the adjacent sites towards the center of the system. Alternatively, we can enforce what are referred to as *periodic boundary conditions* onto our system. Under periodic boundary conditions, what leaves the simulation on one side will come around on the opposite side. (Consider the old asteroids video game where when the asteroids left the screen on the left hand side of the screen they reappeared on the right hand side of the screen). In terms of our model, this would mean that if site 20 was to switch with site 21 (which recall doesn't exist) then it would actually switch with site 1 instead. Columns W and X check to see if the second random number is beyond the size of the simulation, and if it is then they enforce the periodic boundary conditions. The code in W7 is

```
=IF((V7<1), 20,V7)
```

and similarly in the cell X7 is the following code

```
=IF((W7>20),1,W7)
```

The rest of the simulation remains unchanged from before.

The results from a typical simulation are shown in Figure 2.19 and 2.20. The graphical representation of the system is shown (as before) with the

The left hand side of the spreadsheet is unchanged
and column U contains random numbers between 1 and 20.
The syntax, as before, is: =RANDBETWEEN(1,20)

Columns V contains a number either plus
or minus 1 from the random number in
column U, using the syntax:

=IF(RAND()<0.5,U7+1, U7−1)

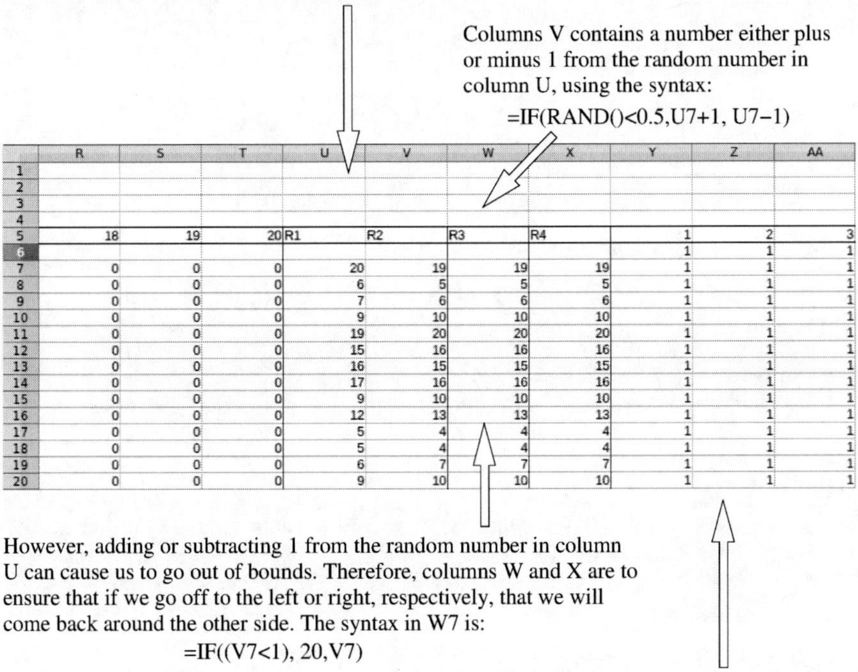

	R	S	T	U	V	W	X	Y	Z	AA
1										
2										
3										
4										
5	18	19	20	R1	R2	R3	R4	1	2	3
6								1	1	1
7	0	0	0	20	19	19	19	1	1	1
8	0	0	0	6	5	5	5	1	1	1
9	0	0	0	7	6	6	6	1	1	1
10	0	0	0	9	10	10	10	1	1	1
11	0	0	0	19	20	20	20	1	1	1
12	0	0	0	15	16	16	16	1	1	1
13	0	0	0	16	15	15	15	1	1	1
14	0	0	0	17	16	16	16	1	1	1
15	0	0	0	9	10	10	10	1	1	1
16	0	0	0	12	13	13	13	1	1	1
17	0	0	0	5	4	4	4	1	1	1
18	0	0	0	5	4	4	4	1	1	1
19	0	0	0	6	7	7	7	1	1	1
20	0	0	0	9	10	10	10	1	1	1

However, adding or subtracting 1 from the random number in column
U can cause us to go out of bounds. Therefore, columns W and X are to
ensure that if we go off to the left or right, respectively, that we will
come back around the other side. The syntax in W7 is:

=IF((V7<1), 20,V7)

To the right remain largely unchanged, and
finishes swapping the two sites. The syntax
in Y7 is

=IF(Y$5=$X7, INDEX($Y6:$AR6, $U7), A7)

FIGURE 2.18
Spreadsheet implementation of a simple dynamic Monte Carlo model.

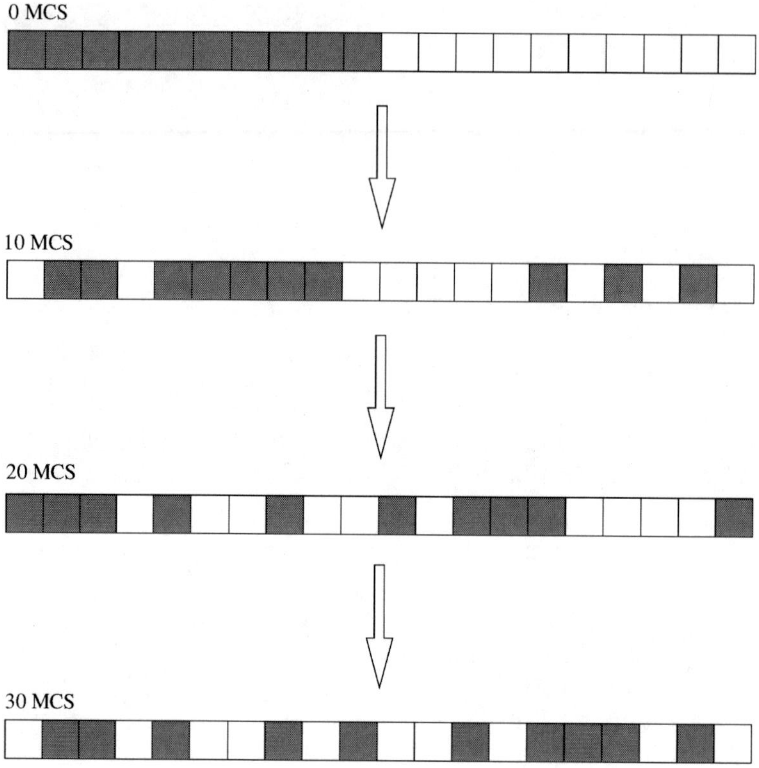

FIGURE 2.19
The transition from ordered phases to random dispersion is depicted in a simple dynamic Monte Carlo model.

shaded squares representing sites containing 1's and the white squares representing sites containing 0's. Initially, at 0 MCS, the system consists of all the 1's located on the left hand side of the system and all the 0's located on the right. However, as the simulation progresses we find that the 1's and 0's become interspersed with one another. Figure 2.20 depicts the number of 1's on the left and right hand side of the system as a function of MCS (which is essentially the time scale in the current model). Note that it takes on the order of 8 MCS before the system reaches equilibrium (equal number of 1's on both left and right sides). However, in the previous model which wasn't dynamic it only took on the order of 1 MCS to reach the same state (see Figure 2.16). Recall that the dynamic Monte Carlo models are less efficient at finding the equilibrium state because they do not sample the phase space as efficiently (being limited to swapping with neighboring sites in our model). It is worth noting that Figure 2.20 represents a variable changing with respect to time, which cannot be said for Figure 2.16. The timescale, however, is measured in

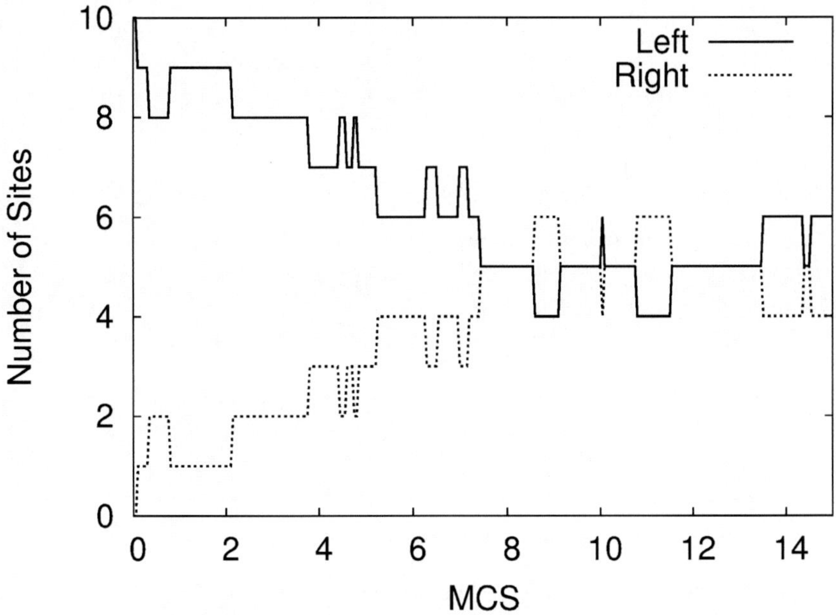

FIGURE 2.20
Plot of the number of sites in the left or right halves of the simulation which contain 1's as a function of the number of Monte Carlo steps in a simple dynamic Monte Carlo model.

terms of MCS and it may be necessary to convert this to a real time scale. To do this will require additional information, such as mapping the simulation results onto either experimental data or theory.

Metropolis Algorithm

The Metropolis algorithm is crucial to the Monte Carlo method. Previously, we looked at systems where the configurations (distribution of 1's and 0's) didn't have different energies. In effect, the system was driven by entropy and, therefore, tended towards a disordered system. The Metropolis algorithm, however, allows us to look at systems where the configurations do possess different energies. In other words, the configurations have probabilities proportional to the Boltzmann factor

$$P = C \exp\left(-\frac{E}{kT}\right)$$

where P is the probability, C is an unknown constant, E is the energy, k is the Boltzmann constant, and T is temperature. We want to explore the phase space of the system, sample different configurations, and preferentially select configurations that are more probable. Consider two configurations, A and B, each of which occurs with probabilities proportional to the Boltzmann factor as above. The ratio of the two probabilities is therefore of the form

$$\frac{P_A}{P_B} = \frac{C\exp\left(-\frac{E_A}{kT}\right)}{C\exp\left(-\frac{E_B}{kT}\right)} = \exp\left(\frac{-(E_A - E_B)}{kT}\right)$$

where the unknown proportionality constant cancels, and we are left with a ratio of the probabilities between the two states in a form that can be easily calculated.

The Metropolis algorithm progresses through the following steps:

1. Starting with a configuration with energy E_A we make a change to the configuration and obtain a new configuration with energy E_B.

2. If E_B is less than E_A then this configuration is more probable and the change is accepted.

3. If E_B is greater than E_A then the new configuration is less probable. The new configuration, however, can still be accepted if

$$\exp\left(\frac{-(E_B - E_A)}{kT}\right) > \epsilon$$

where ϵ is a random number between zero and one. If the difference in energy is small (or the temperature is high) then the ratio in probabilities is close to one and is likely to be greater than ϵ. Therefore, the move is likely to be accepted. In contrast, for large differences in energy (or low temperatures) the move is likely to be rejected.

By following these simple rules, and adequately sampling the phase space of the system under consideration, the resulting configurations obtained by the Monte Carlo method will be the most probable. This allows us to associate energies with different aspects of the configuration and optimize with respect to these energies. For example, in the system considered here we can make the grouping of ones together, or zeros together, energetically favorable and use the Metropolis algorithm to simulate the phase separation of ones and zeros.

Consider a system where the sites are swapped with their nearest neighbors to mimic the dynamics of motion in such as system. However, now we will consider the interactions between neighbors. For example, if we want the sites to group together then we can make the energy of interaction between two sites that are the same (either two ones or two zeros) have a negative energy. Consider the example in Figure 2.21. Here we have a system which undergoes a change in configuration through the swapping in positions between two adjacent sites. The sites before the two that are to be swapped through to

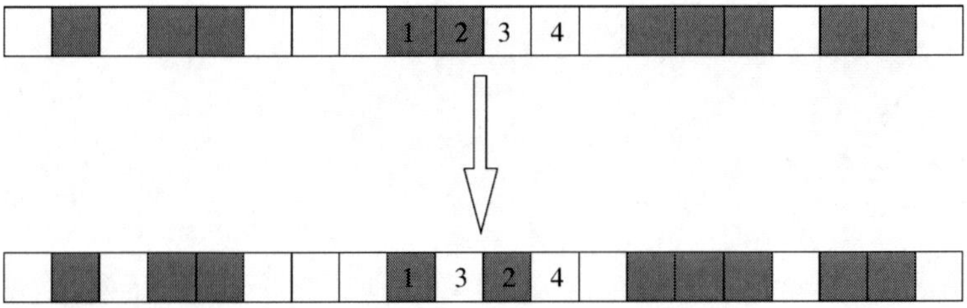

FIGURE 2.21
Illustration of the sites, and neighboring interactions, as two sites are switched. When calculating the change in the energy of the system, upon two sites switching, these sites will be required.

the site afterwards are labeled 1 through 4 (see 2.21). However, sites 2 and 3 are switched. If we consider the energy of interaction (or more precisely the difference in energy between the new and old configurations) then we obtain the following expression for the change in energy

$$\Delta E = E_{13} + E_{24} - E_{12} - E_{34}$$

where E_{ij} is the interaction energy between sites i and j. We use this when calculating the energy difference in the spreadsheet model.

Figures 2.22 and 2.23 display the spreadsheet that is used to implement the Metropolis algorithm. In order to compare the energies we must keep track of the neighbors of the sites which will be switched, in order to calculate the difference in energy. Figure 2.22 shows the left hand side of the spreadsheet and the generation of the random positions to be switched, as before. However, now we also calculate the neighbors to the two sites which are to be switched. Of course this is easily obtained by adding or subtracting 1 from the position of the sites to be switched and making sure that the resultant number is within range. For example, the cell Y7 contains the following code

```
=IF((U7-1) < 1, 20, U7-1)
```

and calculates the site to the left of the first random number to be switched (in cell U7) and makes sure that if we go below 1 then we come back around and the neighboring site is 20 (ensuring periodic boundary conditions).

We can now calculate the energy of interactions associated with the new and old configurations (or rather the aspects which will change between these two configurations). The code in cell AC7 which calculate the old energy is

```
=IF(U7<V7, (IF(INDEX(AG6:AZ6,Y7) = INDEX(AG6:AZ6,U7), -1, 0)
    + IF(INDEX(AG6:AZ6,AB7) = INDEX(AG6:AZ6,X7), -1, 0)),
      (IF(INDEX(AG6:AZ6,AA7) = INDEX(AG6:AZ6,X7), -1, 0)
    +  IF(INDEX(AG6:AZ6,Z7) = INDEX(AG6:AZ6,U7), -1, 0)))
```

The left hand side of the spreadsheet is again the simulation data with one of the sites switched. The syntax in cell T7 is:
=IF(T$5=$U7, INDEX($AG6:$AZ6, $X7), AZ6)

U column contains random numbers between 1 and 20, V column is either 1 greater or lesser than column U. Columns W and X ensure that the data is with range.

	S	T	U	V	W	X	Y	Z	AA	AB
1										
2										
3										
4										
5	19	20	R1	R2	R3	R4	R1-	R1+	R4-	R4+
6										
7	0	1	6	5	5	5	5	7	4	6
8	0	1	16	15	15	15	15	17	14	16
9	0	1	13	12	12	12	12	14	11	13
10	1	1	19	20	20	20	18	20	19	1
11	1	0	13	12	12	12	12	14	11	13
12	1	0	18	19	19	19	17	19	18	20
13	1	0	6	7	7	7	5	7	6	8
14	1	0	13	14	14	14	12	14	13	15
15	1	1	20	19	19	19	19	1	18	20
16	1	0	5	6	6	6	4	6	5	7
17	1	0	5	4	4	4	4	6	3	5
18	1	0	6	7	7	7	5	7	6	8
19	1	0	11	10	10	10	10	12	9	11
20	1	0	11	12	12	12	10	12	11	13
21	1	0	9	8	8	8	8	10	7	9

The columns Y and Z are column U plus and minus 1, respectively. This is because we have to know where the neighboring sites are to determine the energy of interaction. The code in cell Y7 and Z7 is =IF((U7−1) < 1, 20, U7−1) and =IF((U7+1)>20, 1, U7+1)
Similarly, columns AA and AB are column X plus and minus 1.

FIGURE 2.22
Spreadsheet implementation of a Monte Carlo model incorporating the Metropolis algorithm.

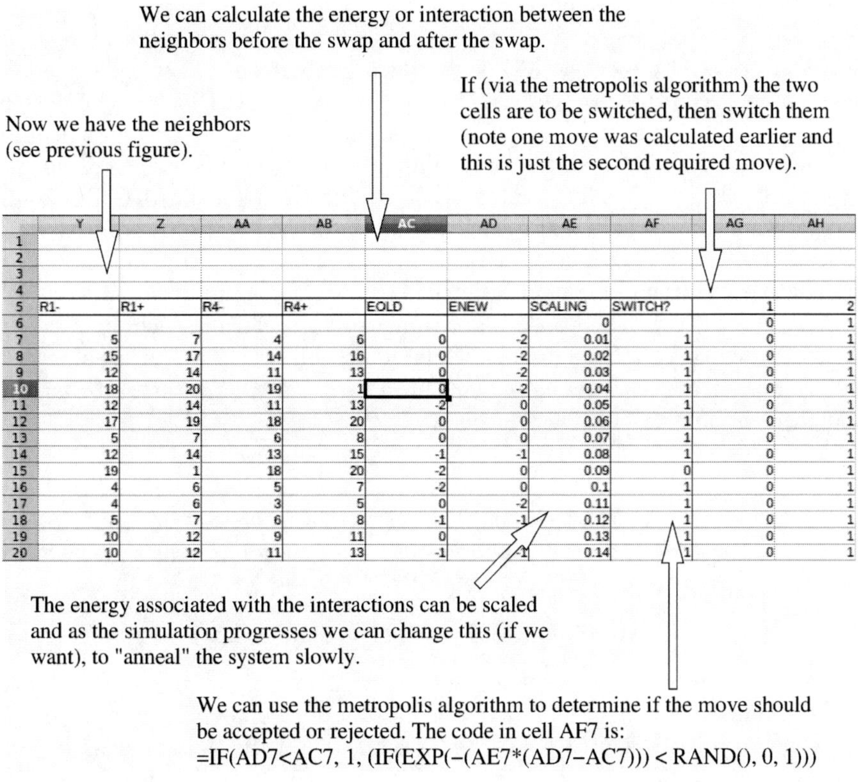

	Y	Z	AA	AB	AC	AD	AE	AF	AG	AH
1										
2										
3										
4										
5	R1-	R1+	R4-	R4+	EOLD	ENEW	SCALING	SWITCH?	1	2
6							0		0	1
7	5	7	4	6	0	-2	0.01	1	0	1
8	15	17	14	16	0	-2	0.02	1	0	1
9	12	14	11	13	0	-2	0.03	1	0	1
10	18	20	19	1	0	-2	0.04	1	0	1
11	12	14	11	13	-2	0	0.05	1	0	1
12	17	19	18	20	0	0	0.06	1	0	1
13	5	7	6	8	0	0	0.07	1	0	1
14	12	14	13	15	-1	-1	0.08	1	0	1
15	19	1	18	20	-2	0	0.09	0	0	1
16	4	6	5	7	-2	0	0.1	1	0	1
17	4	6	3	5	0	-2	0.11	1	0	1
18	5	7	6	8	-1	-1	0.12	1	0	1
19	10	12	9	11	0		0.13	1	0	1
20	10	12	11	13	-1	-1	0.14	1	0	1

FIGURE 2.23
Spreadsheet implementation of a Monte Carlo model incorporating the Metropolis algorithm (continued).

It requires us to know the order that the two sites which are being switched are in (if the first site to the left of the second site). Then we calculate the energy of interaction associated with the old position. For example, the section of code

```
IF(INDEX(AG6:AZ6,Y7) = INDEX(AG6:AZ6,U7), -1, 0)
```

states that if two neighboring sites are the same then the energy associated with this is -1 (more energetically favorable), however if they are different then the energy is 0. The next thing in the model is a scaling factor in column AE. This isn't necessary for the simulation, but it allows us to anneal the system. Essentially we start at a higher temperature and allow the system to move more freely before decreasing the temperature and letting the system settle into a stable equilibrium configuration with little fluctuation. The scaling factor is ramped up slowly to a value of 2 (this value is for the most part arbitrarily chosen, but for some systems it can be determined from experimentally obtained variables). This scaling factor multiplies the difference in energy (the energy associated with the new configuration minus the energy associated with the old configuration).

Now, using the Metropolis algorithm we can determine whether or not to switch the two sites. This is calculated in column AF, and a value of 1 indicates that a switch should occur and a value of 0 indicates that a switch shouldn't be made. The code in cell AF7 is

```
=IF(AD7<AC7, 1, (IF(EXP(-(AE7*(AD7-AC7))) > RAND(), 1, 0)))
```

If the new energy is less than the old energy then the switch should be accepted, else the Metropolis algorithm is used to determine if the switch should occur. This is achieved through the part of the code

```
(IF(EXP(-(AE7*(AD7-AC7))) > RAND(), 1, 0)
```

which quite simply states that if the exponential of minus the scaling, times the difference in energy, is greater than a random number between zero and one then accept the switch. Once we have decided if the switch is going to be accepted or not, it is necessary to modify the code used for the switch. For example, the code in cell AG7 now reads

```
=IF($AF7 = 1, IF(AG$5=$X7, INDEX($AG6:$AZ6, $U7), A7), AG6)
```

which says that if AF7 = 1 (a switch is to occur) then switch as we did in the previous models, but if not then the value in the cell should be the same as the previous iteration and no change should occur.

Sample results from this model are shown in Figure 2.24. The configurations are displayed through the use of shaded squares (ones) and white squares (zeros) and the configurations at 0, 1, 10 and 20 MCS are contrasted. The scaling factor is gradually ramped up until 10 MCS. During the early stages of the simulation, therefore, the sites are swapped much more easily, but as the simulation progresses only swaps which are energetically favorable are likely to be

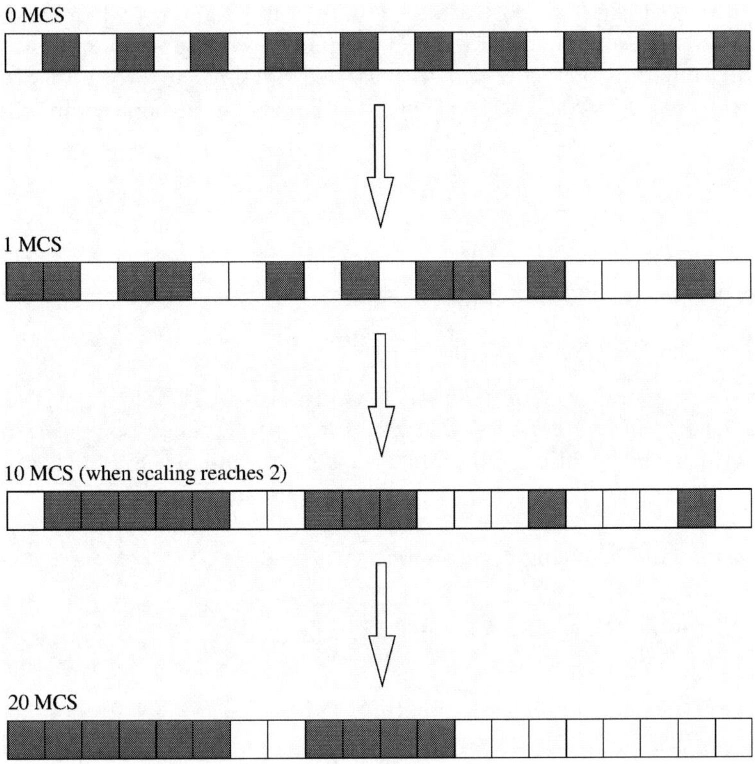

FIGURE 2.24
Illustration of how, using the Metropolis algorithm, a Monte Carlo model can capture the transition from a disordered system to an ordered one; here through phase separation.

accepted. The initial configuration consists of alternating 1's and 0's. This is the most energetically unfavorable configuration as the system energetically favors similar sites being next to each other. As the simulation progresses, because there is an increased likelihood of swaps which cause similar sites to be next to each other, the two phases of ones and zeros separate out. This is not too dissimilar to the process through which oil and water phase separate out into separate phases upon mixing.

This is only a brief introduction to some basic Monte Carlo modeling and we will cover a slightly more advanced model in Chapter 4 when we consider solar cells and solar energy. Monte Carlo models can be really quite complex, and are generally not confined to a lattice but instead are most often off-lattice. In other words, the locations of the particles are not confined to grid points (like the model we implemented) but are allowed to take on any value.

2.14 Exercises

1. Create a spreadsheet program that will iteratively double a given number. Use a macro for the iteration and create a button, that upon being pressed will perform the iteration and double the current number. Successive pressing of the button should continue to double the number.

2. Consider the following data points:

x	y
0	0.2
8	0.6
17	0.4
23	0.2
31	0.1
37	0

Create a spreadsheet program that calculates a polynomial fit to these data points using Lagranges classical formula.

3. Consider the diffusion equation in one-dimension

$$\frac{\partial \phi(x,t)}{\partial t} = \frac{\partial}{\partial x} D(\phi, x) \frac{\partial \phi(x,t)}{\partial x} \tag{2.38}$$

where $\phi(x,t)$ is the concentration of the substance being diffused, and $D(\phi, x)$ is a diffusion coefficient which in principle could depend on position or concentration. For simplicity, let's assume the diffusion coefficient is a constant, and the finite difference approximation is of the form

$$\phi_i^{t+1} = \phi_i^t + \frac{\Delta t D}{(\Delta x)^2}\left[\phi_{i+1}^t + \phi_{i-1}^t - 2\phi_i^t\right] \tag{2.39}$$

where i represents the spatial discretization and t the temporal discretization. Δt is the time step, and Δx the grid spacing. For a system consisting of 40 cells, put ϕ equal to zero everywhere except the central few cells which should be set equal to one. Now solve the diffusion equation for this system. Be careful to choose a sufficiently small Δt to ensure stability.

4. The electric field $\vec{E}$ can be related to electric potential ρ using

$$\nabla \cdot \vec{E} = \frac{\rho}{\epsilon_0} \tag{2.40}$$

or in terms of electric potential, V

$$\nabla^2 V = \frac{\rho}{\epsilon_0} \tag{2.41}$$

which if we were to write in one-dimension using finite difference approximations might look like

$$\frac{V_{i+1} + V_{i-1} - 2V_i}{(\Delta x)^2} = \frac{\rho}{\epsilon_0} \tag{2.42}$$

Simulate a system consisting of a series of 20 cells, with a positive charge density at one end and a negative charge density at the other end. a) Use the relaxation method to solve this system. b) Use the conjugate gradient method to solve this same system.

5. Consider a spherical ball of density ρ floating in water (which for simplicity we assume has a density of 1). The ball is only partially submerged, with the volume underneath the water given by

$$V = \int_0^H \pi r^2(h)\, dh = \int_0^H \pi(1 - (1-h)^2)\, dh = H^2 - \frac{H^3}{3} \tag{2.43}$$

assuming the radius of the ball is also unity (i.e., 1). To find how far the ball will be submerged we simply balance the weight of the ball with the buoyancy

$$V = \rho_{ball}\frac{4}{3}\pi \tag{2.44}$$

where recall the density of water and the radius of the ball are both taken to be 1. This results in the following nonlinear equation

$$H^3 - 3H^2 + 4\rho_{ball} = 0 \tag{2.45}$$

Use the Newton-Raphson method to solve this problem and find the height,

H, as a function of the density of the ball ρ_{ball}.

6. In a program above, utilizing the Metropolis algorithm, different sites that were swapped were constrained to always be adjacent to each other. This mimics the dynamics of particle movement in real systems (resulting in a dynamic Monte Carlo model). Remove this constraint from the model and allow any two sites to be swapped. What difference does this make to the dynamics and final equilibrated state of the system?

7. The scaling factor in the Metropolis algorithm for the Monte Carlo simulation above was gradually increased to a value of 2. But why this value? Are the simulation's results sensitive to this value? How about how quickly the system ramps up to this value? Code the spreadsheet such that the step over which the scaling factor is increased each iteration can be changed along with the maximum value of the scaling factor (these two quantities, the step size and the maximum scaling, should be placed at the top of the simulations so that they can be changed easily). Change the step size to 0.02 and 0.05, and change the maximum scaling to 1 and 5. How does changing these variables affect the final configurations and the dynamics of the system?

Additional Reading

Press, W. H. (2007). Numerical recipes 3rd edition: The art of scientific computing. Cambridge University Press, Cambridge.

Allen, M. P., & Tildesley, D. J. (1987). Computer simulation of liquids. Oxford University Press, Oxford.

3

Global Warming and Pollution

CONTENTS

Before discussing the alternative technologies that can be used to generate energy, and exploring the simple computer models which elucidate the function of these technologies, it is worth turning our attention to some simple models which explore the reasons *why* we need alternative energy technologies. In particular, in this chapter we will explore some simple models related to global warming and the dispersion of pollutants. Alternative energy technologies are usually more expensive than traditional fossil fuel based energy sources, and this has limited our transition to cleaner energy sources. However, there are good reasons to make the change to cleaner energy technologies, and we will look at some of the models which exemplify the consequences of anthropogenic greenhouse gases and the dispersion of other pollutants.

While the computer models used to predict future climate change are generally incredibly complicated (taking vast amounts of time to run on very large supercomputers), we can still gain invaluable insights into the basic science of global warming through much simpler models. Here we will look at models which balance the energy into our planet from the sun with the energy out from our planet in the form of thermal radiation. Furthermore, we will look at a simple heat transfer model to look at the role of clouds in dictating our climate. Note, however, that these simple models, while exemplifying the basic science behind these phenomena, lack the complexity and multiple feedback mechanisms which make predicting the future climate of our planet so incredibly difficult.

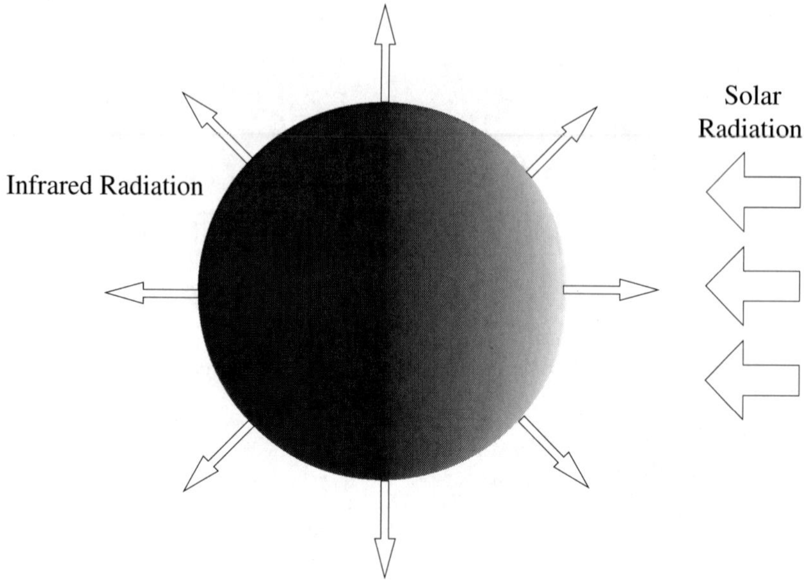

FIGURE 3.1
Schematic of the Global Energy Balance Model.

3.1 Global Warming: Global Energy Balance Model

The simplest model of global warming is the global energy balance model. A schematic of the global energy model is shown in Figure 3.1. The global energy balance model is a zero-dimensional model and simply consists of an equation which balances the energy in with the energy out of our planet. In this model the planet is assumed to be in equilibrium, with the temperature being held constant by the balance between solar energy incident on the Earth and thermal radiation emitted from the Earth. In particular, this model is called the global energy balance models because the properties of the entire planet are considered to be averaged over the entire planet. The only free variable in the global energy balance model is the temperature of the Earth.

The global energy balance model does not require a spreadsheet, as it simply consists of a single equation which balances the energy in with the energy out and a single free variable, the temperature of the planet. However, implementing the global energy balance model in a spreadsheet allows variables to be changed easily, with the results of changing the variables instantaneously calculated in the spreadsheet.

As mentioned, the global energy balance model balances (i.e., sets equal to each other) the energy absorbed by our planet from the sun with the energy emitted in thermal radiation from our planet (which increases with temperature).

It is worth noting that the energy absorbed by our planet is emitted by the sun and only comes from a single direction, and interacts with the side of our planet facing the sun. However, the thermal radiation emitted by the Earth is emitted in all directions, from the entire surface area of our spherical planet. The amount of incoming solar radiation, or insolation, is roughly $S = 1370Wm^{-2}$ and this is known as the "Solar Constant" (although it should be noted that this value varies slightly over time). It is essentially the energy per second per unit area (hence the units Wm^{-2}), or intensity of sunlight, incident on the Earth. Note, however, the further you move away from the sun the lower the intensity of sunlight as the power from the sun is spread out over a greater area. (One extreme science fiction solution to global warming is to move the Earth further from the Sun to reduce solar insolation!)

The intensity of sunlight incident on Earth is $S = 1370Wm^{-2}$, only if the sun is directly above us. Near the equator of our planet we might expect to observe a relatively high energy density, whereas in other regions of our planet the normal to the surface of the planet makes an increasingly high angle with respect to the direction of incident sunlight. In other words, as we move away from the equator we see the sun at an angle and, therefore, the intensity of light is less. We'll consider this in more detail later when implementing the zonal energy balance model. For now, the amount of energy (per second) from the sun that is incident on our planet depends on the area of our planet that faces the sun. In other words, as the sun looks down on Earth it sees a circle whose radius is that of our planet, rather than seeing a three-dimensional sphere. For our planet, therefore, we consider the area illuminated to be the area of a circle whose radius is that of our planet. The amount of energy incident per second is, therefore

$$E_{incident} = \pi R^2 S \tag{3.1}$$

where R is the radius of the Earth and S is the solar constant. However, not all of this incident sunlight will be absorbed by our planet, and a fraction will be reflected back out to space

$$E_{reflected} = \alpha \pi R^2 S \tag{3.2}$$

where α is the albedo, or reflectivity, of the Earth and is expressed as a fraction. The albedo of the planet depends on the composition of the Earth's surface and its cloud cover. For example, snow and ice are considered to have a relatively high albedo (on the order of 0.9) whereas oceans, which absorb a lot of energy, have a low albedo (on the order of 0.1). This is important when considering global warming. For example, as the planet heats up some ice will melt, revealing darker oceans underneath, and this will result in an increase in absorption further raising our planet's temperature. Or as the planet heats up, an increase in water vapor and cloud cover is often predicted. The clouds both increase the albedo, reflecting sunlight back into space (reducing the Earth's temperature), and absorb more outgoing thermal radiation (heating up the planet). Our Earth's combined albedo (averaging over the entire planet) is roughly 0.3. In other words, on average our planet reflects about 30% of incoming solar radiation back into space. The total energy input (per second) which is absorbed by our planet, therefore, is the energy incident minus the energy reflected

$$E_{in} = (1 - \alpha)\pi R^2 S \tag{3.3}$$

As the planet warms up from the solar radiation, it also emits thermal radiation back out into space. All warm bodies emit thermal radiation (think about the images you might see from an infrared thermal camera) and as a body gets hotter it radiates more. The amount of energy emitted by a body is described by the Stefan-Boltzmann law, and increases with temperature to the power of 4. In particular, the amount of radiation emitted in a second from the Earth's surface is of the form

$$E_{emitted} = \sigma T^4 4\pi R^2 \tag{3.4}$$

where σ is the Boltzmann constant ($5.67 \times 10^{-8} W m^{-2} K^{-4}$) and T is the temperature in Kelvins. However, our planet has an atmosphere that absorbs and re-emits some of this outgoing radiation back to Earth (the greenhouse effect). Therefore, only a fraction of this energy is allowed to escape our planet. This fraction ϵ is known as the transparency of our atmosphere. Putting this to one would mimic a planet without an atmosphere or any greenhouse effect. A very low value of ϵ, approaching zero, on the other hand would mimic a planet whose atmosphere is opaque to thermal radiation like Venus, with a run-away greenhouse effect. The energy leaving our planet is, therefore, of the form

$$E_{out} = \epsilon\sigma T^4 4\pi R^2 \tag{3.5}$$

It is worth noting that our continuing emission of carbon dioxide is very slightly decreasing the transparency of our atmosphere, but the approximate amount of outgoing thermal radiation which escapes our planet is about 60% of the thermally emitted radiation. Therefore, the transparency, $\epsilon = 0.6$. At equilibrium the energy in will balance with the energy out, and we can set $E_{in} = E_{out}$.

$$(1 - \alpha)\pi R^2 S = \epsilon\sigma T^4 4\pi R^2 \tag{3.6}$$

Notice that the radius of the planet cancels out and, upon rearranging this equation for temperature, we can obtain an expression for our planet's temperature of the form

$$T = \left[\frac{S(1 - \alpha)}{4\sigma\epsilon}\right]^{\frac{1}{4}} \tag{3.7}$$

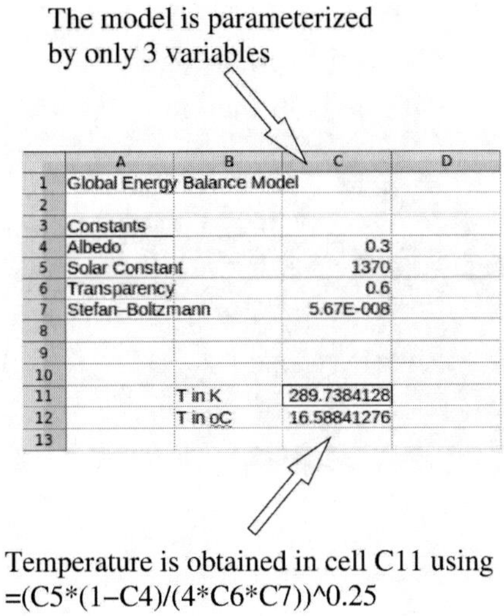

FIGURE 3.2
A spreadsheet implementation of the Global Energy Balance Model.

The global energy balance model described above can be easily implemented in a spreadsheet (as shown in Figure 3.1). In particular, the variables in the model (solar constant, albedo and transparency) can be changed and the temperature instantly calculated. While this model is very simple, it is worth noting that using this model we actually get reasonable agreement with the actual temperature of our planet. Recall that $S = 1370Wm^{-2}$ is the solar constant, $\alpha = 0.3$ is the albedo (or reflectivity) of our planet, $\epsilon = 0.6$ is the transparency of our atmosphere to outgoing radiation, and $\sigma = 5.67 \times 10^{-8}Wm^{-2}K^{-4}$ is the Boltzmann constant. Putting these numbers into our spreadsheet model we get a temperature of 289 K or 16.6 ^{o}C. Given the simplicity of the above model, this is reasonably close to the experimentally measured value of around 15 ^{o}C.

3.2 Global Warming: Zonal Energy Balance Model

We will now extend the global energy balance model in a single dimension. In particular, the planet is considered to be discretized into zones, where the planet is broken up depending on latitude. The properties of the planet are considered to be averaged over large regions, rather than be averaged over the entire planet. In particular, these quantities are averaged over latitudinal zones, instead of globally as in the global energy balance model. In other

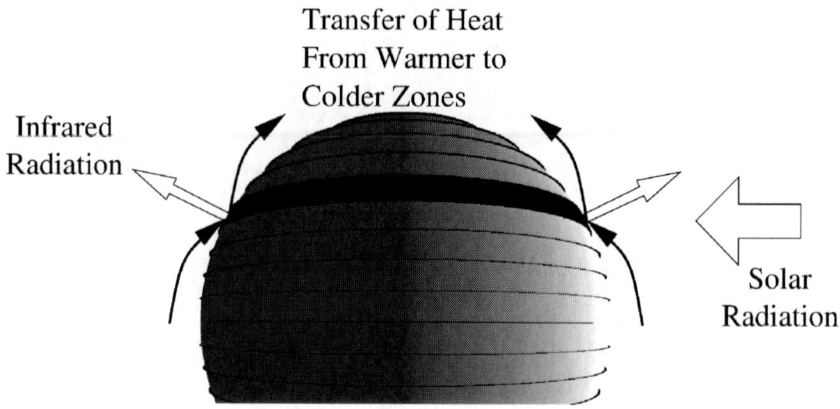

FIGURE 3.3
Schematic of the Zonal Energy Balance Model.

words, we have similar equations to the global energy balance model considered previously, but this time the model must account for the flow of energy from latitudinal zones to their neighboring zones. In this model, the expression describing the incoming energy from the sun must be adjusted to account for the amount of light incident at different latitudes. Recall that the spherical geometry means that light is incident at more of an angle, relative to the normal direction, as one increases latitude. (At the equator the sun rises to be directly above around noon, but as you travel north, for example, the sun arcs through the sky to the south and is never directly over head.) Each zone also has a different albedo, or reflectivity, which depends on the amount of land and sea within the range of latitude which defines the zone. This might also depend on the type of land covering (snow, desert, forests, etc...) and the fraction and type of cloud covering. At each latitudinal zone, the surface temperature can be calculated, taking into consideration the different albedos and balancing the energy in with the energy out (similar to how we solved the global energy balance model). It is worth noting that the zonal energy balance model can also be extended to include the effects of atmospheric CO_2 concentrations. In particular, a variable atmospheric concentration of CO_2 can be incorporated into the Stefan-Boltzmann law. Recall that the Stefan-Boltzmann law was already modified in the global energy balance model to account for the absorption of outgoing infrared radiation by the atmosphere (i.e., the greenhouse effect) and so this term can now be further modified to account for atmospheric CO_2 concentrations.

Recall that the zonal energy balance model differs from the global energy balance model, in that the planet is split up into different zones (see Figure 3.3). In particular, the planet is discretized latitudinally into a given number of zones (here 20) and the properties of these zones are taken to be the averaged properties over the entire zone. The energy absorbed by a given zone is of the form

$$E_{in} = \frac{S}{4}\mu_x(1 - \alpha_x) \tag{3.8}$$

where, as before, $S = 1370 Wm^{-2}$, the albedo α_x depends on latitude and position x (where x is obtained from the sine of the latitude)

$$x = \sin(\phi) \tag{3.9}$$

where ϕ is the latitude and x runs from -1 (South Pole) to 1 (North Pole). μ is an approximate correction factor to account for the varying sunlight received at different latitudes and can be given by

$$\mu_x = 1 - 0.477\frac{3x^2 - 1}{2} \tag{3.10}$$

Interestingly, while the albedo is given an initial value in the model, depending on the average reflectivity in a given latitudinal zone, if a particular zone is covered in ice and snow (considered to occur when the temperature dips below -10 ^oC) then the albedo will increase to account for the increased reflection of the ice and snow now covering that particular zone. In particular, the albedo is set to 0.62 if the temperature drops below 10 ^oC. This introduces some interesting nonlinearity into the model, and the final temperatures are highly dependent on the initial conditions of this model.

The energy which leaves a given latitudinal zone is of the form

$$E_{out} = A + BT_x - 5.397 \ln\frac{C}{C_0} \tag{3.11}$$

where the usual T^4 term from the Stefan-Boltzmann law is now approximated by a linear function with the variables A and B being fitted constants from the Stefan-Boltzmann law (modified for the greenhouse effect). A plot of a linear approximation to Stefan-Boltzmann law over a range of temperatures between -20 ^oC and 40 ^oC is shown in Figure 3.4.

Over the temperature range of interest, the Stefan-Boltzmann law is reasonably approximated by a linear function, and this approximation is commonly made when implementing this model. Note that the the greenhouse effect is further modified to account for changes in CO_2 concentration, C, from the the pre-industrial CO_2 concentration, C_0. In particular, C is the concentration of CO_2 in the atmosphere and C_0 = 280 ppm is the pre-industrial concentration of CO_2, before anthropogenic greenhouse gas emissions.

To allow for the transport of heat between zones we can introduce a term that essentially evens out the temperature distribution:

$$K\left(T_x - T_s\right) \tag{3.12}$$

where $K = 3.80 Wm^{-2}C^{-1}$ is a constant and T_s is the average temperature over all the zones.

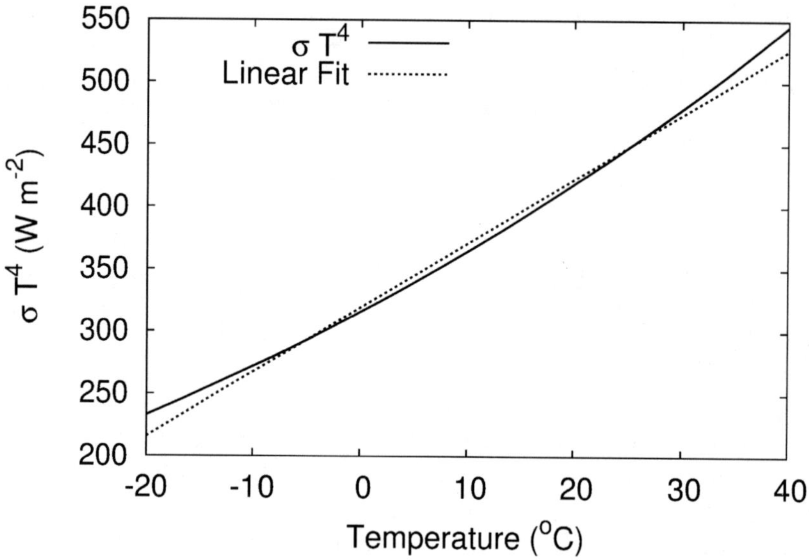

FIGURE 3.4
A linear approximation to Stefan-Boltzmann's law is often used in this model. The Stefan-Boltzmann law is plotted using a solid line and the linear fit is the dashed line.

While the above model can be solved for the equilibrium (by setting the energy into a given zone equal to the energy out) energy balance, here we evolve the system towards this equilibrium using a dynamic model. In particular, the dynamics of the zonal energy balance model can be captured using the following equation

$$D\frac{dT_x}{dt} = \frac{S}{4}\mu_x(1-\alpha_x) - \left[A + BT_x - 5.397\ln\frac{C}{C_0}\right] - K\left(T_x - T_s\right) \tag{3.13}$$

where D dictates the dynamics of the system and can be thought of as the heat capacity or the thermal inertia of the system, t is time, and T_x is the temperature at the latitudinal position x.

A screen shot from a spreadsheet implementing the zonal energy balance model is shown in Figure 3.5. The constant values for the parameters required in the model are included at the top of the spreadsheet, which makes it easy to change these variables. Also at the top of the spreadsheet is the average global temperature predicted for the planet after 60 iterations in the model. (Essentially, this is the averaged temperature over all zones after the system has usually reached a steady state.) This is useful because, upon changing one of the constants at the top (or the initial temperature or albedo), the results from the model are instantaneously apparent, at least in terms of the global temperature. For example, the concentration of CO_2 can be varied from 280 ppm (the preindustrial concentration) to 400 ppm (the current concentration). The effect of this elevation in CO_2 concentrations is to increase the temperature of the planet from 13.55 ^oC to 13.93 ^oC.

In the spreadsheet, the first column is the parameter x (which recall is obtained from the latitude, ϕ, using $\sin\phi$) The second column is the correction factor μ_x which depends on x. The third column is the initial albedo, α_x, which is the albedo before considering whether or not the poles will be covered with ice and snow. Note that even with the same parameters the planet could be entirely covered with ice and snow (snowball Earth) or have no significant amounts of ice and snow, depending on the initial conditions. The initial albedo is somewhat arbitrarily prescribed here to be larger at the poles and less at the equator (although there are values for these quantities in the literature), as this roughly reflects experimental observations. The initial temperature is included in the fourth column and is also somewhat arbitrarily set to be colder at the pole and warmer at the equator. However, these conditions can be changed easily enough in the model. Given the temperature in the zones, the albedo is recalculated (to see if any zones have dropped below -10 ^oC and should be covered with ice and snow) at every time step. With the new albedo and the previous temperatures, the new temperatures are calculated using a finite difference approximation

$$T_x^{t+1} = T_x^t + D\Delta t\left\{\frac{S}{4}\mu_x(1-\alpha_x) - \left[A + BT_x - 5.397\ln\frac{C}{C_0}\right] - K\left(T_x - T_s\right)\right\} \tag{3.14}$$

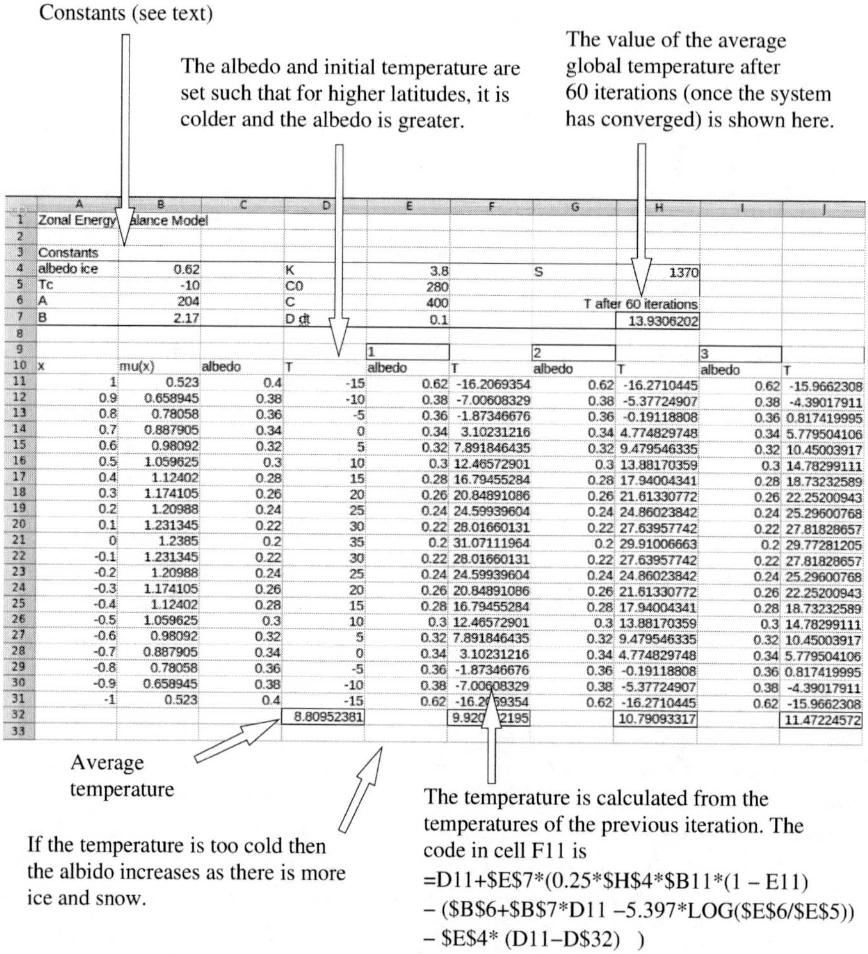

	A	B	C	D	E	F	G	H	I	J
1	Zonal Energy alance Model									
2										
3	Constants									
4	albedo ice	0.62		K	3.8		S	1370		
5	Tc	-10		C0	280					
6	A	204		C	400			T after 60 iterations		
7	B	2.17		D dt	0.1			13.9306202		
8										
9					1		2		3	
10	x	mu(x)	albedo	T	albedo	T	albedo	T	albedo	T
11	1	0.523	0.4	-15	0.62	-16.2069354	0.62	-16.2710445	0.62	-15.9662308
12	0.9	0.658945	0.38	-10	0.38	-7.00608329	0.38	-5.37724907	0.38	-4.39017911
13	0.8	0.78058	0.36	-5	0.36	-1.87346676	0.36	-0.19118808	0.36	0.817419995
14	0.7	0.887905	0.34	0	0.34	3.10231216	0.34	4.774829748	0.34	5.779504106
15	0.6	0.98092	0.32	5	0.32	7.891846435	0.32	9.479546335	0.32	10.45003917
16	0.5	1.059625	0.3	10	0.3	12.46572901	0.3	13.88170359	0.3	14.78299111
17	0.4	1.12402	0.28	15	0.28	16.79455284	0.28	17.94004341	0.28	18.73232589
18	0.3	1.174105	0.26	20	0.26	20.84891086	0.26	21.61330772	0.26	22.25200943
19	0.2	1.20988	0.24	25	0.24	24.59939604	0.24	24.86023842	0.24	25.29600768
20	0.1	1.231345	0.22	30	0.22	28.01660131	0.22	27.63957742	0.22	27.81828657
21	0	1.2385	0.2	35	0.2	31.07111964	0.2	29.91006663	0.2	29.77281205
22	-0.1	1.231345	0.22	30	0.22	28.01660131	0.22	27.63957742	0.22	27.81828657
23	-0.2	1.20988	0.24	25	0.24	24.59939604	0.24	24.86023842	0.24	25.29600768
24	-0.3	1.174105	0.26	20	0.26	20.84891086	0.26	21.61330772	0.26	22.25200943
25	-0.4	1.12402	0.28	15	0.28	16.79455284	0.28	17.94004341	0.28	18.73232589
26	-0.5	1.059625	0.3	10	0.3	12.46572901	0.3	13.88170359	0.3	14.78299111
27	-0.6	0.98092	0.32	5	0.32	7.891846435	0.32	9.479546335	0.32	10.45003917
28	-0.7	0.887905	0.34	0	0.34	3.10231216	0.34	4.774829748	0.34	5.779504106
29	-0.8	0.78058	0.36	-5	0.36	-1.87346676	0.36	-0.19118808	0.36	0.817419995
30	-0.9	0.658945	0.38	-10	0.38	-7.00608329	0.38	-5.37724907	0.38	-4.39017911
31	-1	0.523	0.4	-15	0.62	-16.2[illegible]69354	0.62	-16.2710445	0.62	-15.9662308
32				8.80952381		9.920[illegible]2195		10.79093317		11.47224572
33										

FIGURE 3.5
Spreadsheet implementation of the Zonal Energy Balance Model.

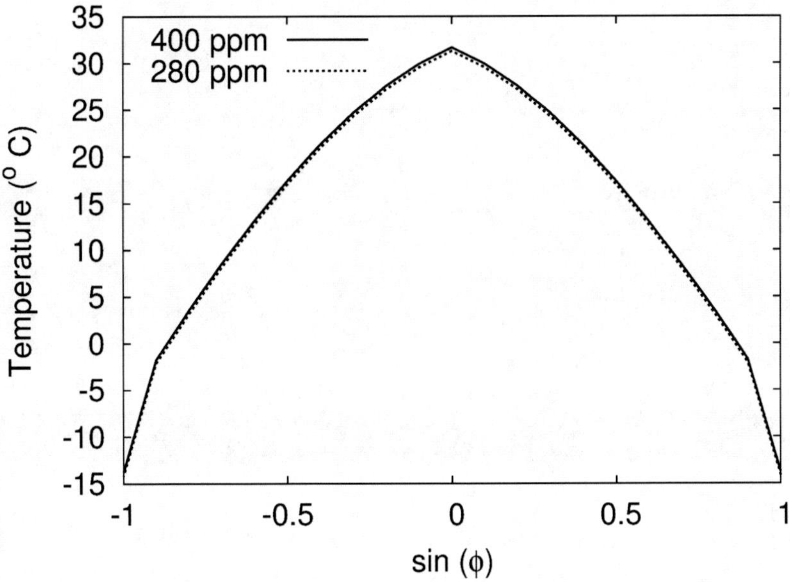

FIGURE 3.6
Temperature as a function of the sine of the latitude for current (400 ppm) and pre-industrial (280 ppm) CO_2 concentrations.

where the temperature at the new time, $t+1$, is calculated in terms of the temperature at time t. The system is evolved over a number of time steps, moving horizontally across the spreadsheet, until the system settles down to a steady state solution. Here the system appears to settle down quite quickly with the quantity $D\Delta t$ being set to 0.1 (if this quantity is too large then the system will become unstable).

The steady state solution (that is when the system becomes equilibrated and the system no longer changes) is given in Figure 3.6. In particular, the temperature as a function of x is plotted for two steady state solutions: one when the concentration of CO_2 is 280 ppm and another when the concentration of CO_2 is 400 ppm. Notice that the difference between the two curves is very small. However, there are a couple of points that need to be considered when interpreting this model. First, the planet is discretized in one-dimension in the current model (ignoring longitudinal variations or variations with altitude) and some of the initial conditions or parameters could be off. Secondly, there are no complicated feedback mechanisms included in the model (for example, the release of further CO_2 from the oceans as the temperatures rise or changes in cloud cover). Even though the model is really still very simple, the predicted global increase in temperature as a consequence of global warming

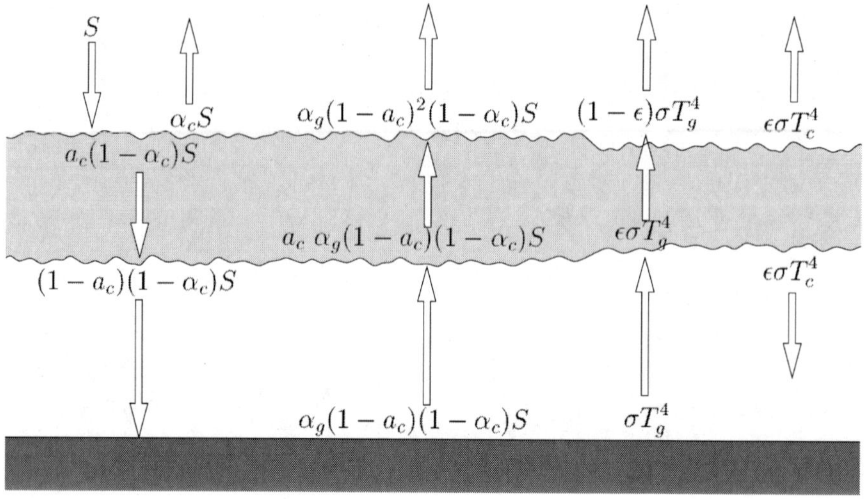

FIGURE 3.7
Schematic of a radiative model of a cloud.

is comparable to what has been seen experimentally (0.8 ^{o}C) for the initial values considered here.

3.3 Global Warming: Radiative Model of a Cloud

To consider the effects of cloud cover on the temperature of our planet we turn our attention to a simple radiative model of a cloud. In particular, the radiative model simply balances the radiative energy coming into the cloud with the radiative energy coming out of the cloud. When these balance we can determine the temperature of the ground beneath. Recall that cloud cover is a crucial component to determining the effects of global warming. For example, during the day the clouds can reflect back sunlight and help cool the planet, but they can also absorb outgoing infrared radiation and warm the planet. The presence of more clouds as a consequence of global warming, therefore, could lead to either a positive feedback or negative feedback mechanism which could further warm the planet or help stabilize its temperature, respectively. This is one of the greatest uncertainties in climate science at the moment and is crucial to our predictions for future warming trends. While the model presented here is relatively crude and simple, it does offer interesting insights into the effects of cloud cover on the Earth's temperature.

A schematic of the model is depicted in Figure 3.7. The cloud is depicted as a band of gray through the center of the image and the ground is dark gray and at the bottom. The arrows represent the direction of heat flow (through either solar or thermal radiation). It is worth noting that the model is only a simple quasi-one-dimensional model, in that we do not include convection effects, or any spatial effects in either altitude, latitude, or longitude. The far left of Figure 3.7 depicts the incoming solar radiation, S, as it enters the atmosphere (where S is simply the solar constant as before). Some of the incoming solar energy, α_c, will be reflected by the cloud back into space. Of the fraction not reflected, $1-\alpha_c$, some of the light, $a_c(1-\alpha_c)S$, will be absorbed by the cloud. The constant α_c represents the albedo of the cloud, and a_c is the absorptivity, or absorptance, of the cloud to solar radiation. Whatever solar energy not absorbed by the cloud will be transmitted through the cloud and reach the Earth Some of this energy will be reflected back up to the clouds, $\alpha_g(1-a_c)(1-\alpha_c)S$, where α_g is the albedo of the ground. As this reflected light is again incident on the cloud, some of the energy will be absorbed, $a_c\alpha_g(1-a_c)(1-\alpha_c)S$, and some of the energy will be transmitted through the clouds, $\alpha_g(1-a_c)^2(1-\alpha_c)S$.

However, the energy from the sun is not the only radiative energy in the system; there is also the thermal radiation from the ground and thermal radiation from the cloud itself. The Earth is considered a black body radiator and the emission of thermal radiation from the Earth is σT_g^4. where σ is the Stefan-Boltzmann constant and T_g is the temperature of the ground (which we are interested in). Some of this thermal radiation will be absorbed by the cloud, $\epsilon\sigma T_g^4$ and some of this thermal radiation will be transmitted by the cloud, $(1-\epsilon)\sigma T_g^4$ and make its way out into space. Here, ϵ is the emissivity of the clouds (the ability of the cloud to emit thermal radiation relative to a perfect black body) and the absorptivity of thermal radiation by the cloud is considered to be the same as its emissivity. The emission of thermal radiation from the cloud (both towards the Earth and out into space) is $\epsilon\sigma T_g^4$.

If we consider the energy flowing into the cloud from the top, with the energy flowing out of the cloud from the top, we obtain the following radiative energy balance

$$S=\alpha_c S+\alpha_g(1-a_c)^2(1-\alpha_c)S+\epsilon\sigma T_c^4+(1-\epsilon)\sigma T_g^4 \tag{3.15}$$

In addition, we can write down an expression for the energy balance between the energy absorbed by the cloud and the energy emitted by the cloud

$$a_c(1-\alpha_c)S+a_c\alpha_g(1-a_c)(1-\alpha_c)S+\epsilon\sigma T_g^4=2\epsilon\sigma T_c^4 \tag{3.16}$$

Finally, we could write down a third equation for the energy balance at the ground

$$(1-\alpha_g)(1-a_c)(1-\alpha_c)S+\epsilon\sigma T_c^4=\sigma T_g^4 \tag{3.17}$$

This gives us three equations describing the radiative energy balance in the system. However, in the model there is only 2 unknowns, the temperature of the cloud and the temperature of the ground. Therefore, there is some redundancy in the system of equations and we only need two equations. For

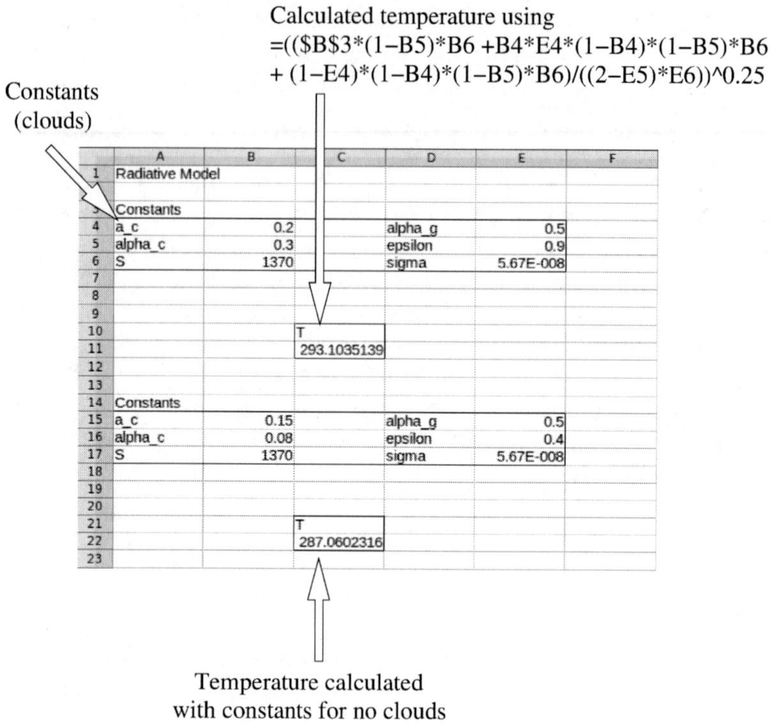

FIGURE 3.8
Spreadsheet implementation of the radiative model of a cloud.

example, combining the first two equations and re-arranging for T_g yields the following expression

$$T_g = \left[\frac{a_c(1-\alpha_c)S + a_c\alpha_g(1-a_c)(1-\alpha_c)S + (1-\alpha_g)(1-a_c)(1-\alpha_c)S}{(2-\epsilon)\sigma}\right]^{\frac{1}{4}} \tag{3.18}$$

This is the equation which we will solve in a spreadsheet. Note that this is simply a 'plug and chug' problem now and we don't necessarily require a spreadsheet (or any computer program) to solve the problem, but a spreadsheet will make it easier to change the parameters in the model and see, instantaneously, the resulting temperature of the planet as a consequence of these changes.

The implementation of this model in a spreadsheet is depicted in Figure 3.8. In particular, the constants used in the model are written at the top of the spreadsheet and the code written in cell C11 is

```
=(($B$3*(1-B5)*B6 +B4*E4*(1-B4)*(1-B5)*B6 +
(1-E4)*(1-B4)*(1-B5)*B6)/((2-E5)*E6))^0.25
```

which is just the equation above. The main aspect of solving this model, therefore, is not the implementation which is relatively easy, but the choice of constants. Recall that different types of clouds will have different values for these constants. For example, even if there are no clouds present then the atmosphere without clouds will have some influence and the values for the albedo, absorptivity, and emissivity of the clouds will not be zero. In particular, without clouds the constants are $\alpha_c = 0.08$ (scattering alone), $a_c = 0.15$, and $\epsilon = 0.4$. However, when we consider the sky to possess clouds, then the albedo can vary from around 0.1 to 0.9 depending on the size and concentration of water (or ice) droplets, and the thickness and density of the clouds. In general, the smaller the drops and the greater the water content, the greater the cloud albedo. Low, thick clouds (such as stratocumulus with hight water content) have a relatively high albedo, whereas high, thin clouds (such as Cirrus which due to their altitude are made up entirely of ice particles) tend to have a low albedo.

The absorption of incident solar energy by clouds, a_c, is roughly 0.15 when the sky is clear of clouds, but can be as high as 0.35 under heavy overcast conditions. For relatively low altitude clouds, an increase in a_c from around 0.2 to 0.35 is might be accompanied by an increase in albedo from 0.1 to, say, 0.6 or 0.7 as both of these parameters vary with cloud thickness. For high altitude clouds, the absorption of solar radiation is relatively small (no different to thin air). However, high altitude clouds do absorb infrared radiation and tend to enhance atmospheric greenhouse warming. The average emissivity (and absorption of infrared radiation by the clouds in the current model) is on the order of 0.967 for low altitude clouds, 0.781 for middle altitude clouds, and 0.616 for high altitude clouds.

Global warming is not the only negative consequence of our current processes of generating energy. For example, pollution from power plants and traffic results in the release of fine particulate matter (which can cause premature death in people with heart or lung disease, and cause respiratory problems in the general population). Sulfur dioxide, besides causing acid rain, causes thousands of premature deaths, heart attacks, and incidents of respiratory disease each year. Nitrogen oxides cause ground level ozone which exacerbates asthma and increases susceptibility to chronic respiratory disease. Add in the detrimental health effects from other pollutants such as mercury, lead, and arsenic, and it becomes easy to see that the role of pollution is very important when contrasting the impacts of traditional and alternative energy technologies. Here we will be looking at some of the simple models which can be used to track the dispersal of pollutants over time.

3.4 Pollution: Gradient Transport Theory

The simplest way to predict the dispersion of a pollutant is through the use of the gradient transport theory. Here, the concentration, C, diffuses out and away from the pollution source, but also the wind velocity, U, can also result in the spread of the pollution. If we consider the wind to be blowing in the x direction, then the following equation can be used update the concentration:

$$\frac{\partial C}{\partial t} + U\frac{\partial C}{\partial x} = D\left(\frac{\partial^2 C}{\partial x^2} + \frac{\partial^2 C}{\partial y^2} + \frac{\partial^2 C}{\partial z^2}\right) + R + S \tag{3.19}$$

where C is the concentration, t is time, U is the air speed in the x-direction (similar terms for the y and z directions are omitted here), and the x-direction is defined as the direction of the wind. D is the diffusion coefficient, R is a term which represents the source of the pollution, and S represents possible pollution sinks. However, we can simplify this expression slightly. Imagine that the velocity of the wind above is constant (and does not depend on x) then we can take the velocity to be zero and imagine that the system is moving along with this velocity. In other words, we ignore the velocity and as the concentration diffuses out, we imagine that it is also moving across with the wind (and the frame of reference for the model is moving). In this case we have a simpler expression

$$\frac{\partial C}{\partial t} = D\left(\frac{\partial^2 C}{\partial x^2} + \frac{\partial^2 C}{\partial y^2} + \frac{\partial^2 C}{\partial z^2}\right) \tag{3.20}$$

where the source and sink terms are also omitted to capture the dynamics of only a solitary puff of pollution. The pollutant is assumed to have been released in a sudden burst, and there is no source or sink terms subsequent to the initial release of the pollution. What we are left with is essentially a Gaussian diffusion of the pollutant. In one dimension this model reduces to

$$\frac{\partial C}{\partial t} = D\left(\frac{\partial^2 C}{\partial x^2}\right) \tag{3.21}$$

or, in terms of the discrete cells of the spreadsheet used to solve this,

$$\frac{C_i^{t+1} - C_i^t}{\Delta t} = D\frac{C_{i-1}^t - 2C_i^t + C_{i+1}^t}{\Delta x^2} \tag{3.22}$$

where we have used the finite difference approximation (discussed in section 2.10) to replace the continuous second derivative with a discrete approximation that we can solve in a spreadsheet. In particular, C_i^t is the concentration at position i and time t, Δt is the time step, and Δx is the spatial step. In the model formulation here it might seem obvious to absorb the time step and spatial step into the diffusion constant (as all three quantities are constants and multiply each other). However, let's keep the time step separate, as this is usually related to stability (and we want small values to ensure stability) and just absorb the spatial step into the diffusion constant. In effect, we can take the spatial step, and time step if we wanted to, to be equal to one so we don't have to worry about how big the system is or how long the time is in the simulation. This is often referred to as making the model "dimensionless", where the dimensionless diffusion coefficient is all that is required. In other words, the diffusion of a large puff of pollution from a smoke stack is exactly the same as the diffusion of the aroma from a spray of perfume. The length and time scales are completely different, but from a simulation point of view we might be able to view them in the same way.

To solve this equation we use the spreadsheet shown in Figure 3.9. The simulation progresses in time in subsequent rows going down, and moves across spatially going across the columns horizontally. The diffusion coefficient (including spatial step) and time are given at the top, with the simulation domain given below. Equation 3.22 above requires the concentration at sites to the left, C_{i-1}^{t}, and right, C_{i+1}^{t}, of a site to update the concentration at that site, and obtain C_{i}^{t+1}. However, at the boundaries of the simulation there might not be any concentration to the left or right. For example, in row B which represents the furthest site to the left of the simulation domain there would be no concentrations to the left of this point. But we need a concentration for equation 3.22 to solve the concentrations in row B. Therefore, we need boundary conditions.

Here, we apply what are called Neumann boundary conditions where we assign what the derivative is at the boundary. In particular, we set the derivative at the boundary to be zero and make the values of the concentrations in row A equal to the values of the concentrations in row B (the values being the same makes the derivative equal to zero). Alternatively, we might have set Dirichlet boundary conditions, where instead of setting the values in row A equal to the values in row B, we instead set the values in row A equal to a constant. For example, we could have set the concentration in row A (and the last row to the right of the simulation) equal to zero. In essence, the concentration at a certain distance away from where the puff is expanding from is set to zero and considered to be negligible (even if this isn't really the case). Furthermore, we could have set periodic boundary conditions such that what diffuses out from the left comes back around to the right of the simulations, and vice versa. In other words, periodic boundary conditions would capture the diffusion from an infinite series of expanding puffs of smoke all separated by a distance equal to the simulation size. The Neumann boundary conditions used here simulate the diffusion in a closed box that does not allow the concentration to leave (hence the spatial gradient of the concentration is equal to zero). An alternative way to look at these boundary conditions is to consider that they mimic an infinite array of domains, but rather than each domain being the same in an infinite series here it would mimic each domain boundary being a reflection of the domain next to it. In other words, the symmetry at

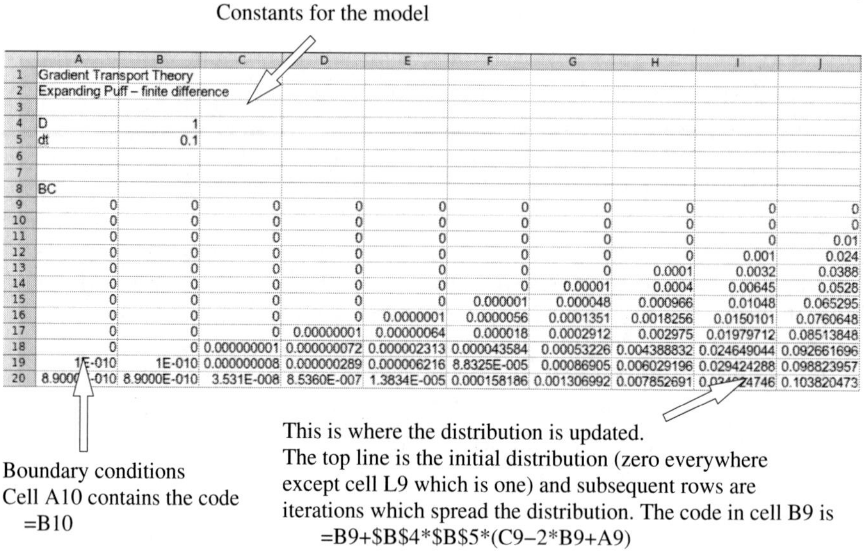

	A	B	C	D	E	F	G	H	I	J
1	Gradient Transport Theory									
2	Expanding Puff – finite difference									
3										
4	D	1								
5	dt	0.1								
6										
7										
8	BC									
9	0	0	0	0	0	0	0	0	0	0
10	0	0	0	0	0	0	0	0	0	0
11	0	0	0	0	0	0	0	0	0	0.01
12	0	0	0	0	0	0	0	0	0.001	0.024
13	0	0	0	0	0	0	0	0.0001	0.0032	0.0388
14	0	0	0	0	0	0	0.00001	0.0004	0.00645	0.0528
15	0	0	0	0	0	0.000001	0.000048	0.000966	0.01048	0.065295
16	0	0	0	0	0.0000001	0.0000056	0.0001351	0.0018256	0.0150101	0.0760648
17	0	0	0	0.00000001	0.00000064	0.000018	0.0002912	0.002975	0.01979712	0.08513848
18	0	0	0.000000001	0.000000072	0.000002313	0.000043584	0.00053226	0.004388832	0.024649044	0.092661696
19	1E-010	1E-010	0.000000008	0.000000289	0.000006216	8.8325E-005	0.00086905	0.006029196	0.029424288	0.098823957
20	8.9000E-010	8.9000E-010	3.531E-008	8.5360E-007	1.3834E-005	0.000158186	0.001306992	0.007852691	0.0[illegible]24746	0.103820473

FIGURE 3.9
Spreadsheet implementation of a simple model using the Gradient Transport Theory.

the boundary conditions could arise if the domain to the left or right of the simulated domains were identical symmetric images. Although I imagine this is more confusing to imagine than simply a closed box!

The concentration as a function of distance (in arbitrary units, or a.u., because we took this to be dimensionless) is plotted for the above simulation in Figure 3.10. The initial concentration is one in the center and this is then evolved such that the concentration diffuses out from the center. Recall, in this model we assume the frame of reference of the model moves with the wind (constant velocity) and so even though the puff of smoke expands out from a given site, in reality we might imagine this puff of smoke expanding out and convecting along with the wind. The concentrations after 5, 20, and 40 iterations are plotted (the 5th, 20th, and 40th rows of the region of the spreadsheet containing the simulation domain).

In the above analysis, we looked at concentration of a pollutant diffusing out as a smooth quantity. However, in reality we might expect to see some turbulence in the system. The effects of turbulence, however, can be quite difficult to predict because turbulence, by definition, is characterized by the chaotic motion of the fluid. One way to get around this is to express the concentrations as mean concentrations, rather than actual concentrations. In this manner, we can write the following equation

$$\frac{\partial \bar{C}}{\partial t} + \bar{U}\frac{\partial \bar{C}}{\partial x} = K_x \frac{\partial^2 \bar{C}}{\partial x^2} + K_y \frac{\partial^2 \bar{C}}{\partial y^2} + K_z \frac{\partial^2 \bar{C}}{\partial z^2} \tag{3.23}$$

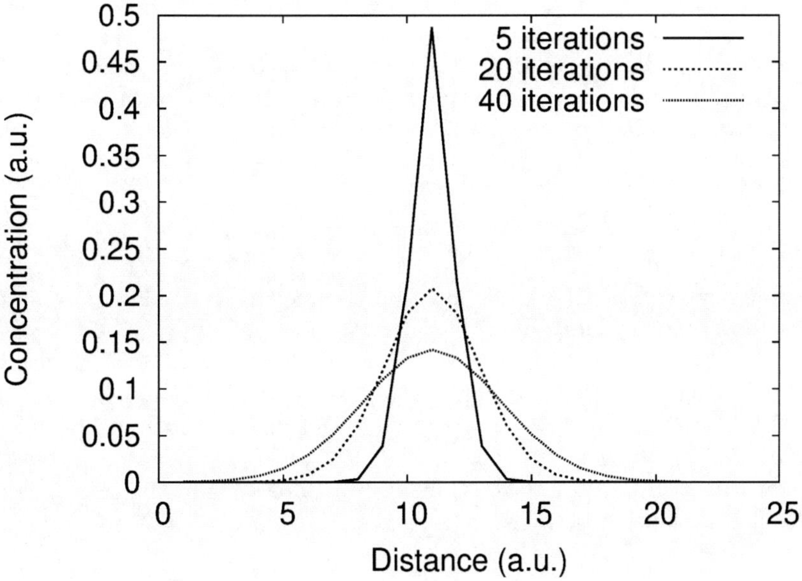

FIGURE 3.10
Plot of concentration as a function of distance generated using the Gradient Transport Theory. Pollutants spread from the localized source over time.

where the bar on top of the different quantities represents that this is the mean value (rather than the turbulently varying local value). Furthermore, the molecular diffusion can be neglected (hence, no diffusion coefficient, D in the above equation) as the turbulent motion of the fluid dictates the diffusion process. This equation is very similar to the model we just solved, with the only difference being the eddy diffusivities, K's, instead of the molecular diffusivities, D's. Let's solve this model in two dimensions (three dimensions is hard to capture in a two-dimensional spreadsheet, especially when you're solving the simulation as a function of time). First, we'll assume Neumann boundary conditions as before

$$K_z \frac{\partial \bar{C}}{\partial z} = 0 \tag{3.24}$$

which mimics perfectly reflecting (or symmetric) boundary conditions at the floor. Note that this might not be entirely correct, as some of the pollutant is likely to be deposited on the flow but we'll consider this effect a little later. The two-dimensional equation that we must solve is

$$\frac{\partial \bar{C}}{\partial t} = K_x \frac{\partial^2 \bar{C}}{\partial x^2} + K_z z \frac{\partial^2 \bar{C}}{\partial z^2} - \bar{u} \frac{\partial \bar{C}}{\partial x} \tag{3.25}$$

where now we ignore the y-direction, and assume the wind blows in only the x-direction. Furthermore, the eddy diffusivity in the z-direction, K_z is taken to be linearly dependent on height and replaced with, $K_z z$. We can use the finite difference method to solve this system as before, but now we will assume that the concentration of the source is no longer just a simple puff but could, in principle, be anything we want (in practice, we'll just keep it at a constant value of 1 for the duration of the simulation). In particular, the discretization of the above equation becomes

$$\frac{\bar{C}^{t+1}_{i,j} - \bar{C}^{t}_{i,j}}{\Delta t} = K_x \frac{\bar{C}^{t}_{i-1,j} - 2\bar{C}^{t}_{i,j} + \bar{C}^{t}_{i+1,j}}{\Delta x^2} + K_z z \frac{\bar{C}^{t}_{i,j-1} - 2\bar{C}^{t}_{i,j} + \bar{C}^{t}_{i,j+1}}{\Delta z^2} - \bar{U} \frac{\bar{C}^{t}_{i+1,j} - \bar{C}^{t}_{i-1,j}}{2\Delta x} \tag{3.26}$$

However, as we are trying to solve a two-dimensional system (in x- and z-directions) as a function of time in a spreadsheet simulation, we will have to use a macro to transfer data. Recall, because a spreadsheet is inherently two-dimensional, solving a two-dimensional system as a function of time requires us to handle the time dimension using copies of the simulation domain in different sections of the spreadsheet (here, the two dimensions in the spreadsheet correspond to the two spatial dimensions and we don't have a extra third dimension for time).

Figure 3.11 depicts the strategy that we will use to solve this system. There will be three renditions of the simulation domain (with rows and columns representing the x- and z-direction). The top section of the spreadsheet will contain the initial concentrations (at time t = 0). We keep a version of the simulation domain with the initial concentrations in case we decide to start the simulation again, in which case we can simply copy and paste the values from this domain down to the middle section of the spreadsheet. Below the top section of the spreadsheet (with initial concentrations) is the middle section, which consists of the current concentration values in the simulation. This middle section consists of constant values and there are no equations here. The bottom section of the spreadsheet is below the middle section, and here we calculate the concentrations in the next time step based on the values in the previous time step (stored in the middle section of the spreadsheet).

Initially, we might copy the initial concentration from the top section to the middle section, and the concentrations after the first time step would be calculated in the bottom section. However, to evolve the simulation forward we need to iteratively calculate the concentrations over many time steps. Rather than have many different realizations of the simulation domain, one after the other, it is better to simply copy the values from the bottom section up to the middle section. Then the middle section will contain the values from the first iteration, and the bottom section will calculate the next iteration (the second iteration) from the new data in the middle section. Note that we must only copy and paste the values, however, rather than the equations as we want the constant values to serve as the input to the calculation of the next iteration. If we can simply copy and paste (paste special with values only) multiple times, then each time we copy and paste will constitute an iteration in the simulation.

To perform the copy and pasting required for the simulation's iteration we use a macro. Recall that a macro is simply some stored commands that can be run at some later time by "running" the macro (see the section on macros and scripts in Chapter 2). In particular, we can "record" a macro first where we copy the bottom section and paste (paste special with values only) the values into the second section. Once we have recorded this macro, we can create a button on our spreadsheet and tell the spreadsheet to perform the macro (and redo the copy and paste) every time we press the button. Therefore, every time we press the button the simulation iterates a single time step and we have the current concentrations in the spreadsheet for this given iteration. Note that all data from previous iterations is lost as we evolve the simulation, but this is not important as we can copy and paste the values at regular intervals into a separate spreadsheet if we want to track the concentrations as a function of time (which is how the data plotted in Figure 3.15 is obtained).

The spreadsheet implementation of this model is depicted in Figures 3.12 to 3.14. In particular, Figure 3.12 shows the top of the simulation. Again, we assume Δx is equal to one, and just allow the time step to be varied to ensure stability of the model. The constant velocity u, and the eddy diffusivities, k_x and k_z, are defined at the top (although, strictly speaking, in this model the eddy diffusivity in the z-direction depends on height and given by $k_z z$).

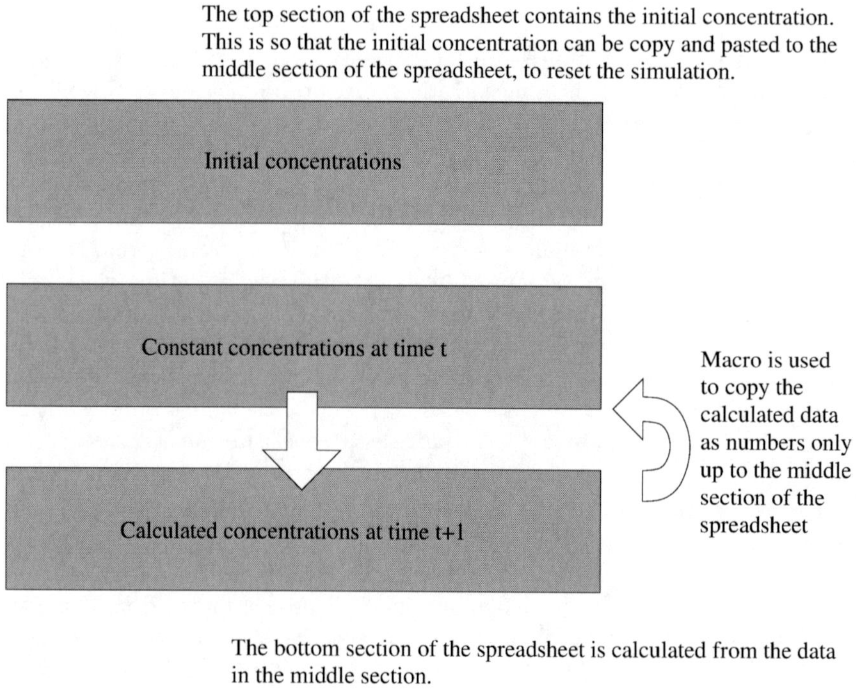

FIGURE 3.11
Schematic showing a possible strategy for implementing a two-dimensional model of the Gradient Transport Theory in a two-dimensional spreadsheet.

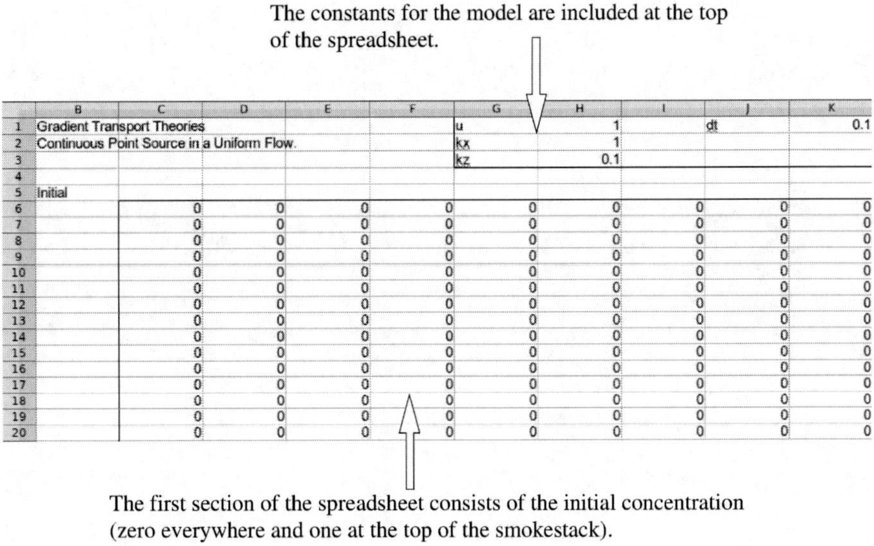

	B	C	D	E	F	G	H	I	J	K
1	Gradient Transport Theories					u	1		dt	0.1
2	Continuous Point Source in a Uniform Flow.					kx	1			
3						kz	0.1			
4										
5	Initial									
6		0	0	0	0	0	0	0	0	0
7		0	0	0	0	0	0	0	0	0
8		0	0	0	0	0	0	0	0	0
9		0	0	0	0	0	0	0	0	0
10		0	0	0	0	0	0	0	0	0
11		0	0	0	0	0	0	0	0	0
12		0	0	0	0	0	0	0	0	0
13		0	0	0	0	0	0	0	0	0
14		0	0	0	0	0	0	0	0	0
15		0	0	0	0	0	0	0	0	0
16		0	0	0	0	0	0	0	0	0
17		0	0	0	0	0	0	0	0	0
18		0	0	0	0	0	0	0	0	0
19		0	0	0	0	0	0	0	0	0
20		0	0	0	0	0	0	0	0	0

FIGURE 3.12
Spreadsheet implementation of a two-dimensional model of the Gradient Transport Theory.

Next is the top section of the spreadsheet, a section of the spreadsheet which contains the initial concentration of pollution (zero everywhere except at the top of the smokestack).

Figure 3.13 depicts the transition in the spreadsheet from the middle section, containing the current concentrations (as constant values), to the bottom section which contains the concentrations at the next time step (calculated). The middle section contains constant values of concentrations which are copied and pasted (only values) from the bottom section. Note that this snapshot of the spreadsheet is a few iterations into the simulation, and the values in the middle section around the smock stack can be seen. The top of the smokestack always has a concentration of 1 (cell F64), and we can see slightly elevated concentrations around the top of the smokestack as the concentration diffuses out.

The button in the center simply runs a macro which copies and pastes the values from the bottom section to this middle section. The macro, which constitutes a single iteration in the model, is created by "recording" the actions of you copying and pasting the values from the bottom section to the middle section once. Every time you run the macro it will "play back" those same commands, copying and pasting the latest calculated concentrations from the bottom section as constant values in the middle section.

It is the bottom section of the spreadsheet which calculates the concentrations at the next iteration from the current concentrations (in the middle

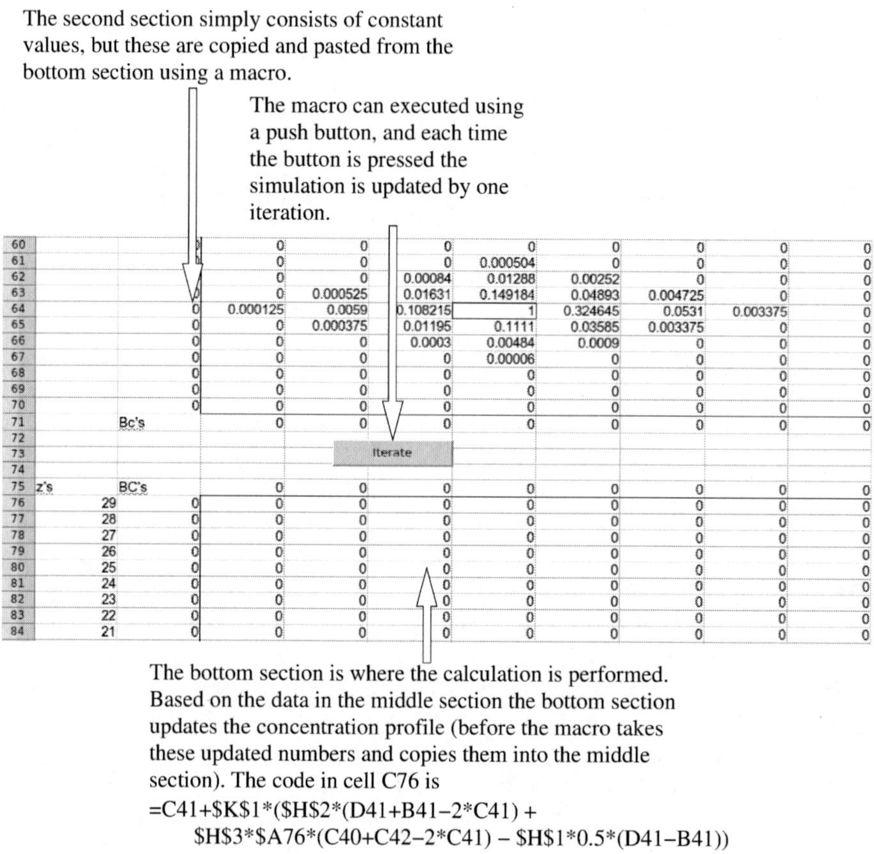

60			0	0	0	0	0	0	0	0	0
61			0	0	0	0	0.000504	0	0	0	0
62			0	0	0	0.00084	0.01288	0.00252	0	0	0
63			0	0	0.000525	0.01631	0.149184	0.04893	0.004725	0	0
64			0	0.000125	0.0059	0.108215	1	0.324645	0.0531	0.003375	0
65			0	0	0.000375	0.01195	0.1111	0.03585	0.003375	0	0
66			0	0	0	0.0003	0.00484	0.0009	0	0	0
67			0	0	0	0	0.00006	0	0	0	0
68			0	0	0	0	0	0	0	0	0
69			0	0	0	0	0	0	0	0	0
70			0	0	0	0	0	0	0	0	0
71		Bc's		0	0	0	0	0	0	0	0
72											
73											
74											
75	z's	BC's		0	0	0	0	0	0	0	0
76		29	0	0	0	0	0	0	0	0	0
77		28	0	0	0	0	0	0	0	0	0
78		27	0	0	0	0	0	0	0	0	0
79		26	0	0	0	0	0	0	0	0	0
80		25	0	0	0	0	0	0	0	0	0
81		24	0	0	0	0	0	0	0	0	0
82		23	0	0	0	0	0	0	0	0	0
83		22	0	0	0	0	0	0	0	0	0
84		21	0	0	0	0	0	0	0	0	0

FIGURE 3.13
Spreadsheet implementation of a two-dimensional model of the Gradient Transport Theory (continued).

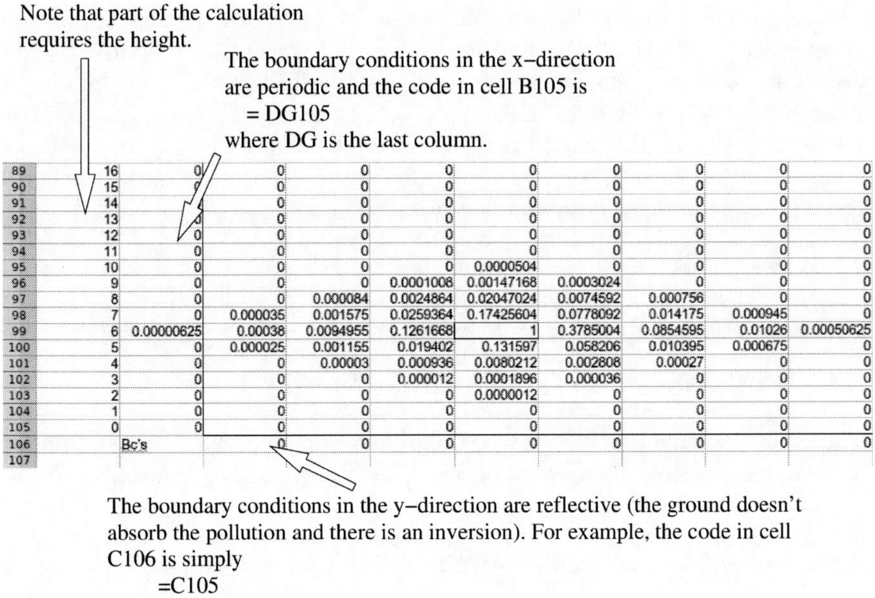

89	16	0	0	0	0	0	0	0	0	0
90	15	0	0	0	0	0	0	0	0	0
91	14	[illegible]	0	0	0	0	0	0	0	0
92	13	0	0	0	0	0	0	0	0	0
93	12	0	0	0	0	0	0	0	0	0
94	11	0	0	0	0	0	0	0	0	0
95	10	0	0	0	0	0.0000504	0	0	0	0
96	9	0	0	0	0.0001008	0.00147168	0.0003024	0	0	0
97	8	0	0	0.000084	0.0024864	0.02047024	0.0074592	0.000756	0	0
98	7	0	0.000035	0.001575	0.0259364	0.17425604	0.0778092	0.014175	0.000945	0
99	6	0.00000625	0.00038	0.0094955	0.1261668	1	0.3785004	0.0854595	0.01026	0.00050625
100	5	0	0.000025	0.001155	0.019402	0.131597	0.058206	0.010395	0.000675	0
101	4	0	0	0.00003	0.000936	0.0080212	0.002808	0.00027	0	0
102	3	0	0	0	0.000012	0.0001896	0.000036	0	0	0
103	2	0	0	0	0	0.0000012	0	0	0	0
104	1	0	0	0	0	0	0	0	0	0
105	0	0	0	0	0	0	0	0	0	0
106		Bc's	0	0	0	0	0	0	0	0
107										

FIGURE 3.14
Spreadsheet implementation of a two-dimensional model of the Gradient Transport Theory (continued).

section). In other words, the discrete equation above is used in the bottom section. The syntax in cell C76 is

```
=C41+$K$1*($H$2*(D41+B412*C41) +

$H$3*$A76*(C40+C422*C41) - $H$1*0.5*(D41B41))
```

where `C41` is the cell containing the current concentration (middle section), `$K$1` is the time step, `$H$2` is the eddy diffusivity k_x, `$H$3*$A76` represents $k_z z$ and `$H$1` is the velocity u. The syntax in cell C76 is dragged across and down to fill the entire simulation domain in the bottom section, and calculate the updated concentrations everywhere.

Figure 3.14 shows the bottom section, and these newly calculated concentrations. Note that the cell F99 is not calculated like all the other cells, but is kept at a constant value of 1, to mimic the continual emittance of pollution from the smokestack. Column A contains the z values (required for calculating eddy diffusivity in the z-direction). Note that in the z-direction, at the floor, we implement Neumann boundary conditions and make the gradient of concentration equal to zero (row 106 is simply equal to row 105). In the x-direction we implement periodic boundary conditions, with column B equal to column DG (end of simulation domain) and, similarly, column DH equal to column C.

The concentrations after 10 iterations, 50 iterations, 100 iterations, and 200 iterations of the simulation are shown in Figure 3.15. The areas which are darker represent higher concentrations of pollution. As you can see, the pollution moves to the right with the wind, as the simulation progresses. Also, because the eddy diffusivity in the z-direction depends on height, the pollution spreads out more as it rises (the eddy diffusivity higher in the air is considered to be greater than the eddy diffusivity near the ground).

3.5 Pollution: Gaussian Plume Model

The Gaussian plume model was originally developed in 1932 (by O.G. Sutton in a paper entitled "A Theory of Eddy Diffusion in the Atmosphere"). As we have seen, treating eddy diffusion similarly to molecular diffusion (albeit taking the diffusion to be over larger spatial scales), the concentration of pollutants from a smoke stack might be expected to be Gaussian in profile. If we assume that the time averaged pollutant concentration downwind from a source can be captured using a Gaussian (normal) distribution curve then we can create a simple expression for the concentrations as a function of position and time. The Gaussian plume model assumes that the plume concentration is Gaussian perpendicular to the direction of the wind, but becomes increasingly spread out as we move downwind. The main assumption behind this model is that over relatively short periods of time (such as, say, a few hours) we can assume the air pollutant emissions and meteorological conditions are steady state (they do not vary as a function of time). The Gaussian plume model, therefore, might typically be applied to sources of pollution which are fixed in location and release a steady stream of pollutants, such as coal fired power plants.

To summarize, the basic Gaussian plume model assumes that the velocity profile does not vary with respect to either time or space, the point source is constant, and there are no loss mechanisms (either through chemical reactions or deposition). The concentration of pollutants only diffuses in the cross-wind and vertical directions (and not in the direction of wind velocity), and this diffusion produces pollutant concentration profiles which can be mathematically captured using a Gaussian function. One further addition must also be added to this simple model. When the concentration of pollutants diffuses down to the ground then, generally, the basic model assumes that there is no absorption or deposition of the pollutants to the ground. Therefore, as the pollutants diffuse away from the centerline of the plume (traveling downwind) the concentrations of the pollutants would mathematically diffuse into the ground (and below the ground) as expressing the concentration of pollution in the vertical direction as a simple Gaussian function would not differentiate between the ground and the air. To overcome this problem, the pollution is

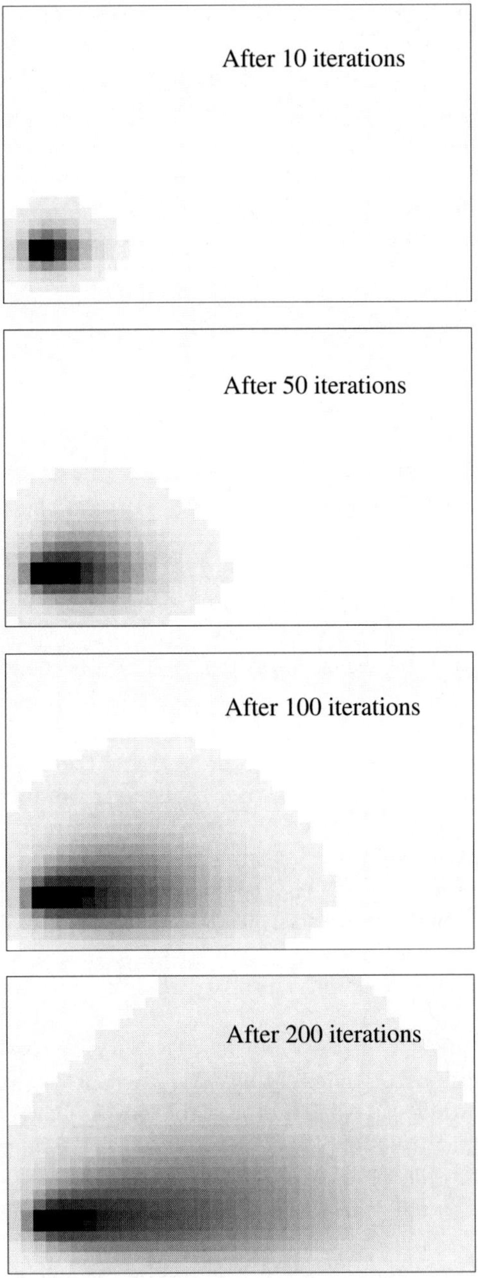

FIGURE 3.15
Pollution concentrations in the x-z plane from a continuous pollution source are shown at various times after the source of the pollution is switched on. The concentrations are calculated using a two-dimensional implementation of the Gradient Transport Theory.

assumed to 'reflect' from the ground and a mirroring effect is assumed to take place. The upward dispersion of pollutants from this reflection can be captured very easily by placing a second smoke stack at a depth of H *below* the ground. In other words, the diffusion of pollutants from this imaginary smoke stack below ground would mimic the reflection of pollutants at the ground. This is believed to be a good approximation for some pollutants (such as CO, SO_2, and NO_2) but might not be expected to apply to other pollutants (such as fine particulate matter).

The concentration of pollutants is given by the following equation

$$\bar{C}(x, y, z, H) = \frac{Q}{2\pi \bar{U} \sigma_y \sigma_z} \exp\left(-\frac{y^2}{2\sigma_y^2}\right)$$
$$\left\{\exp\left[-\frac{(z-H)^2}{2\sigma_z}\right] + \exp\left[-\frac{(z+H)^2}{2\sigma_z}\right]\right\} \tag{3.27}$$

where $\bar{c}(x, y, z, H)$ is the average concentration as a function of position (x, y and z) and the effective stack height, H. Note that here H is the effective stack height, which is not the actual height of the smoke stack. Instead we must take into consideration that the emissions will be warmer than the surrounding air (and hence buoyant) and have an initial upward velocity. Therefore, the effective stack height is greater than the actual height of the smoke stack. Q is the source strength or emission rate, and $\bar{U}$ is the mean transport velocity across the plume.

Note that the equation contains a term for a smoke stack above ground and a term for a smoke stack below ground (where the difference between vertical location and effective plume height is $z - H$ and $z + H$, respectively). Furthermore, the amount that the concentrations spreads out (the width of the plume) is controlled by the parameters σ_y and σ_z. σ_y is the cross wind dispersion parameter and σ_z is the vertical dispersion parameter, which both depend on the distance downwind from the smoke stack, x. In particular, these dispersion parameters are typically defined by stability classes, developed by Pasquill in 1961, and commonly used expressions for these parameters are given in Tables 3.1 and 3.2.

The parameters A through F represent stability classifications based on qualitative descriptions of the weather. A represents an unstable, or turbulent, atmosphere while F represents a stable atmosphere. In particular, the stability can depend upon the surface wind speed (smaller wind speeds are more unstable), the daytime incoming solar radiation (stronger radiation is more unstable), or night time cloud cover (with less cloud cover being more stable).

A spreadsheet implementation of this model is depicted in Fig. 3.16. The constants in the model are defined at the top, where the dispersion parameters are calculated from the Pasquill type C conditions in the urban environment. To make the calculations in the cells a little simpler, we combine the constants that are at the beginning of Equation 3.27 into a single constant. The code in cell F5 is of the form

TABLE 3.1
Open country.

Pasquill type	σ_y	σ_z
A	$0.22x(1+0.0001x)^{-\frac{1}{2}}$	0.2 x
B	$0.16x(1+0.0001x)^{-\frac{1}{2}}$	0.12 x
C	$0.11x(1+0.0001x)^{-\frac{1}{2}}$	$0.08x(1+0.0002x)^{-\frac{1}{2}}$
D	$0.08x(1+0.0001x)^{-\frac{1}{2}}$	$0.06x(1+0.0015x)^{-\frac{1}{2}}$
E	$0.06x(1+0.0001x)^{-\frac{1}{2}}$	$0.03x(1+0.0003x)^{-1}$
F	$0.04x(1+0.0001x)^{-\frac{1}{2}}$	$0.016x(1+0.0003x)^{-1}$

TABLE 3.2
Urban areas.

Pasquill type	σ_y	σ_z
A-B	$0.32x(1+0.0004x)^{-\frac{1}{2}}$	$0.24x(1+0.001x)^{-\frac{1}{2}}$
C	$0.22x(1+0.0004x)^{-\frac{1}{2}}$	$0.2x$
D	$0.16x(1+0.0004x)^{-\frac{1}{2}}$	$0.14x(1+0.0003x)^{-\frac{1}{2}}$
E-F	$0.11x(1+0.0004x)^{-\frac{1}{2}}$	$0.08x(1+0.0015x)^{-\frac{1}{2}}$

```
=H1/(2*PI()*E1*H2*H3)
```

which calculate this constant. Next we calculate the concentration as a function of z and y, for a given value of the downwind direction x. Only positive values of y and z are considered because the system is symmetric about $y = 0$ and below $z = 0$ is the ground, where we do not need to calculate airborne concentrations. The code in cell B9 (which is subsequently dragged across and down to fill the entire simulation domain) calculates the concentration as a function of y and z using Equation 3.27. The code in cell B9 is

```
=$F$5 * EXP(-B$8*B$8/(2*$H$2*$H$2)) *

=(EXP(-($A9-$E$2)*($A9-$E$2)/(2*$H$3*$H$3)) +

=EXP(-($A9+$E$2)*($A9+$E$2)/(2*$H$3*$H$3) ) )
```

Note that when the code in cell B9 references the y and z coordinates it references the cells as `B$8` and `$A9`, respectively. The first one keeps the row constantly at 8 (as this row contains the y values) while the second keeps the column constantly at A (as this column contains the z values).

A contour of the concentration of pollution in the plane $x = 100$ is shown in Fig. 3.17. The concentration is plotted as a function of distance in the y and z directions (where in the spreadsheet above, $\Delta y = \Delta z = 10$). Only the positive y and z values are plotted, as negative z is below ground and

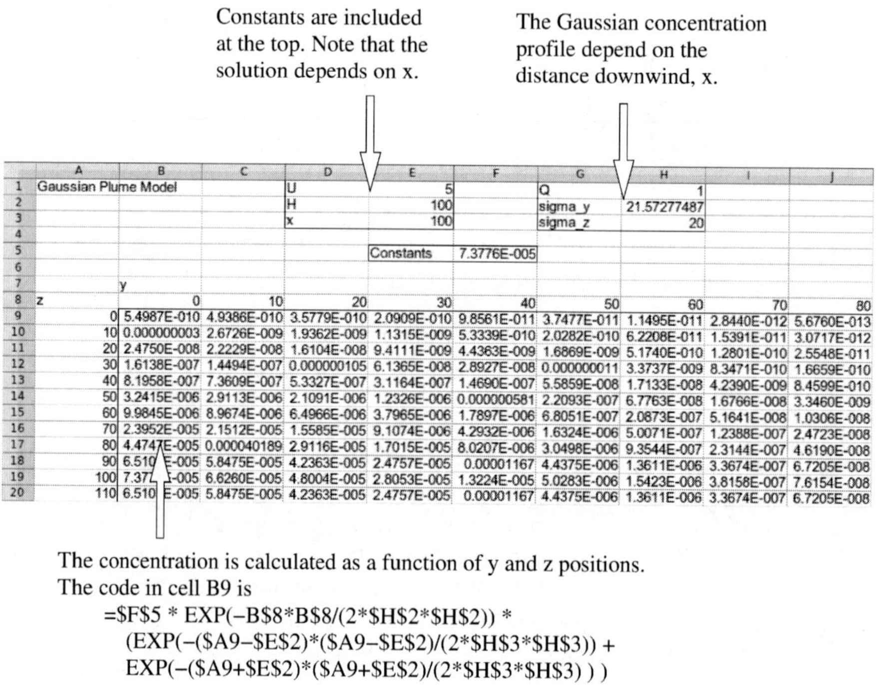

	A	B	C	D	E	F	G	H	I	J
1	Gaussian Plume Model			U	5		Q	1		
2				H	100		sigma_y	21.57277487		
3				x	100		sigma_z	20		
4										
5					Constants	7.3776E-005				
6										
7		y								
8	z	0	10	20	30	40	50	60	70	80
9	0	5.4987E-010	4.9386E-010	3.5779E-010	2.0909E-010	9.8561E-011	3.7477E-011	1.1495E-011	2.8440E-012	5.6760E-013
10	10	0.000000003	2.6726E-009	1.9362E-009	1.1315E-009	5.3339E-010	2.0282E-010	6.2208E-011	1.5391E-011	3.0717E-012
11	20	2.4750E-008	2.2229E-008	1.6104E-008	9.4111E-009	4.4363E-009	1.6869E-009	5.1740E-010	1.2801E-010	2.5548E-011
12	30	1.6138E-007	1.4494E-007	0.000000105	6.1365E-008	2.8927E-008	0.000000011	3.3737E-009	8.3471E-010	1.6659E-010
13	40	8.1958E-007	7.3609E-007	5.3327E-007	3.1164E-007	1.4690E-007	5.5859E-008	1.7133E-008	4.2390E-009	8.4599E-010
14	50	3.2415E-006	2.9113E-006	2.1091E-006	1.2326E-006	0.000000581	2.2093E-007	6.7763E-008	1.6766E-008	3.3460E-009
15	60	9.9845E-006	8.9674E-006	6.4966E-006	3.7965E-006	1.7897E-006	6.8051E-007	2.0873E-007	5.1641E-008	1.0306E-008
16	70	2.3952E-005	2.1512E-005	1.5585E-005	9.1074E-006	4.2932E-006	1.6324E-006	5.0071E-007	1.2388E-007	2.4723E-008
17	80	4.4747E-005	0.000040189	2.9116E-005	1.7015E-005	8.0207E-006	3.0498E-006	9.3544E-007	2.3144E-007	4.6190E-008
18	90	6.510[illegible]E-005	5.8475E-005	4.2363E-005	2.4757E-005	0.00001167	4.4375E-006	1.3611E-006	3.3674E-007	6.7205E-008
19	100	7.37[illegible]-005	6.6260E-005	4.8004E-005	2.8053E-005	1.3224E-005	5.0283E-006	1.5423E-006	3.8158E-007	7.6154E-008
20	110	6.510[illegible]E-005	5.8475E-005	4.2363E-005	2.4757E-005	0.00001167	4.4375E-006	1.3611E-006	3.3674E-007	6.7205E-008

FIGURE 3.16
Spreadsheet implementation of the Gaussian Plume Model.

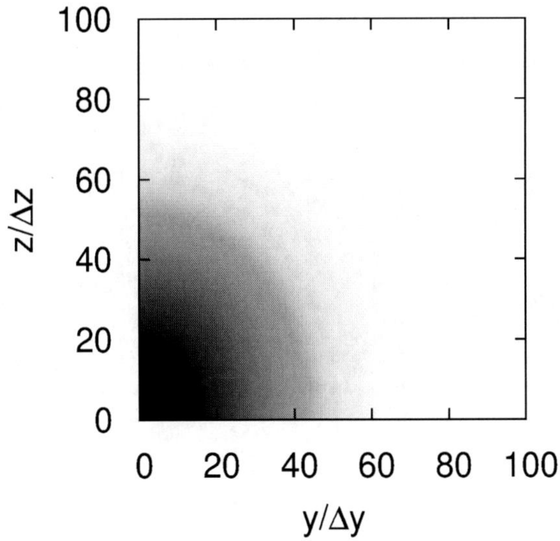

FIGURE 3.17
Contour plot of pollution concentrations calculated using the Gaussian Plume Model.

the system is symmetric about $y = 0$. This gives us a two-dimensional plane of concentrations cutting through the three-dimensional system. We can also solve this system for $z = 0$ and obtain the pollutant concentrations at ground level.

If we put $z = 0$ in Equation 3.27 then we obtain the following equation

$$\bar{C}(x, y, 0, H) = \frac{Q}{\pi \bar{U} \sigma_y \sigma_z} \exp\left(-\frac{y^2}{2\sigma_y^2}\right) \exp\left(-\frac{H^2}{2\sigma_z^2}\right) \tag{3.28}$$

where the two source terms (at $z = H$, and the mirrored source at $z = -H$) are identical. This equation can be evaluated as a function of x and y to obtain the concentration profile on the ground down wind from a fixed steady-state pollution source. The spreadsheet evaluation of Equation 3.28 is depicted in Figure 3.18.

At the top of the spreadsheet is the constant velocity, the effective plume height, and the source strength (here taken to be unity). As we go across the columns of the spreadsheet the cross-wind position, y, increases while as we go down the rows of the spreadsheet the distance down wind, x, increases. The first column is the distance downwind. The second and third columns are the dispersion parameters in the cross-wind and vertical directions, σ_y and σ_z, respectively. Recall that these dispersion parameters depend on the distances traveled in the downwind direction, x, and result in the predicted pollution

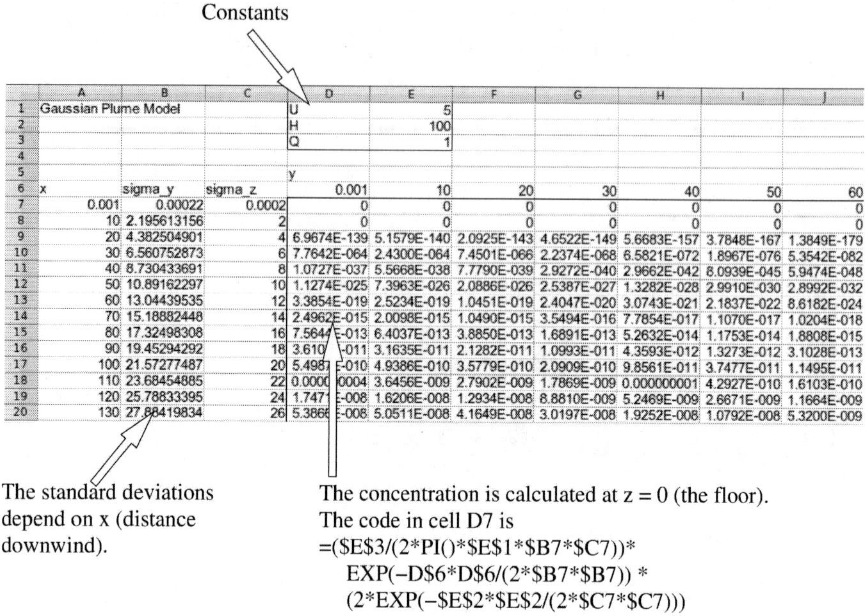

	A	B	C	D	E	F	G	H	I	J
1	Gaussian Plume Model			U	5					
2				H	100					
3				Q	1					
4										
5				y						
6	x	sigma_y	sigma_z	0.001	10	20	30	40	50	60
7	0.001	0.00022	0.0002	0	0	0	0	0	0	0
8	10	2.195613156	2	0	0	0	0	0	0	0
9	20	4.382504901	4	6.9674E-139	5.1579E-140	2.0925E-143	4.6522E-149	5.6683E-157	3.7848E-167	1.3849E-179
10	30	6.560752873	6	7.7642E-064	2.4300E-064	7.4501E-066	2.2374E-068	6.5821E-072	1.8967E-076	5.3542E-082
11	40	8.730433691	8	1.0727E-037	5.5668E-038	7.7790E-039	2.9272E-040	2.9662E-042	8.0939E-045	5.9474E-048
12	50	10.89162297	10	1.1274E-025	7.3963E-026	2.0886E-026	2.5387E-027	1.3282E-028	2.9910E-030	2.8992E-032
13	60	13.04439535	12	3.3854E-019	2.5234E-019	1.0451E-019	2.4047E-020	3.0743E-021	2.1837E-022	8.6182E-024
14	70	15.18882448	14	2.4962E-015	2.0098E-015	1.0490E-015	3.5494E-016	7.7854E-017	1.1070E-017	1.0204E-018
15	80	17.32498308	16	7.5644E-013	6.4037E-013	3.8850E-013	1.6891E-013	5.2632E-014	1.1753E-014	1.8808E-015
16	90	19.45294292	18	3.610[illegible]-011	3.1635E-011	2.1282E-011	1.0993E-011	4.3593E-012	1.3273E-012	3.1028E-013
17	100	21.57277487	20	5.498[illegible]-010	4.9386E-010	3.5779E-010	2.0909E-010	9.8561E-011	3.7477E-011	1.1495E-011
18	110	23.68454885	22	0.000[illegible]0004	3.6456E-009	2.7902E-009	1.7869E-009	0.000000001	4.2927E-010	1.6103E-010
19	120	25.78833395	24	1.747[illegible]-008	1.6206E-008	1.2934E-008	8.8810E-009	5.2469E-009	2.6671E-009	1.1664E-009
20	130	27.88419834	26	5.386[illegible]-008	5.0511E-008	4.1649E-008	3.0197E-008	1.9252E-008	1.0792E-008	5.3200E-009

FIGURE 3.18
Spreadsheet implementation of the Gaussian Plume Model to calculate ground level pollution concentrations.

concentration being more spread out as the plume moves downwind. The code in cell D7, which simply calculates the concentration using Equation 3.28, is

```
=($E$3/(PI()*$E$1*$B7*$C7)) *
 EXP(-D$6*D$6/(2*$B7*$B7)) *
(EXP(-$E$2*$E$2/(2*$C7*$C7)))
```

This code is then dragged both across and down the spreadsheet to populate the spreadsheet with the calculated concentrations of pollution as a function of cross-wind and vertical directions.

The concentration profile calculated using this spreadsheet model is depicted in 3.19. Only the concentration as a function of positive y values is calculated as the system is symmetric about $y = 0$. Furthermore, in this model there is considered to be no eddy diffusivity in the wind direction and, therefore, the pollutant concentrations on the ground upwind from the pollutant source are zero. As expected, the concentration of pollutants is more concentrated and less spread out on the ground near the smoke stack (lower values of x). However, as we move downwind to higher values of x we find that the concentration becomes more spread out. This model, therefore, pre-

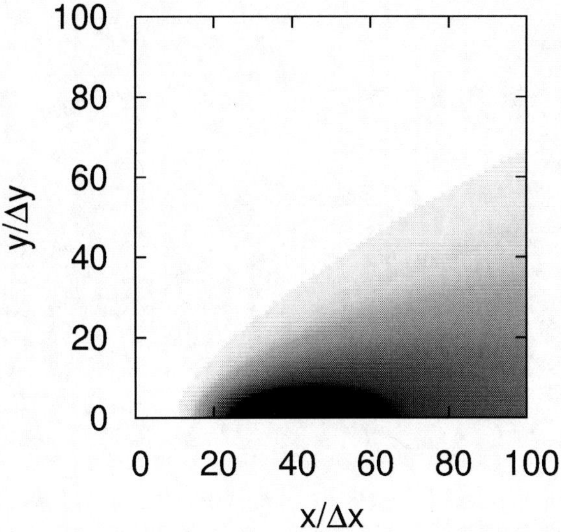

FIGURE 3.19
Ground level pollution concentrations calculated using the Gaussian Plume Model.

dicts the concentration of pollution in the air down wind from a point source, assuming steady state conditions.

3.6 Pollution: Gaussian Plume Model with Settling of Pollutants

The Gaussian plume model can be extended to account for the settling of pollutants. In other words, the pollution might fall down from the air as a consequence of gravity as the plume is both blown downwind and diffuses out. One method to account for this settling is to simply assume that the effective stack height of the plume (the center line about which the plume disperses) is decreased as the plume moves downwind. This can be achieved using the following equation

$$\acute{H} = H - \frac{v_s x}{\bar{U}} \tag{3.29}$$

where $\acute{H}$ is the modified effective stack height which we use in our Gaussian plume model, H is the effective stack height which we used previously, v_s is the settling velocity of the pollutants (for example, the terminal velocity

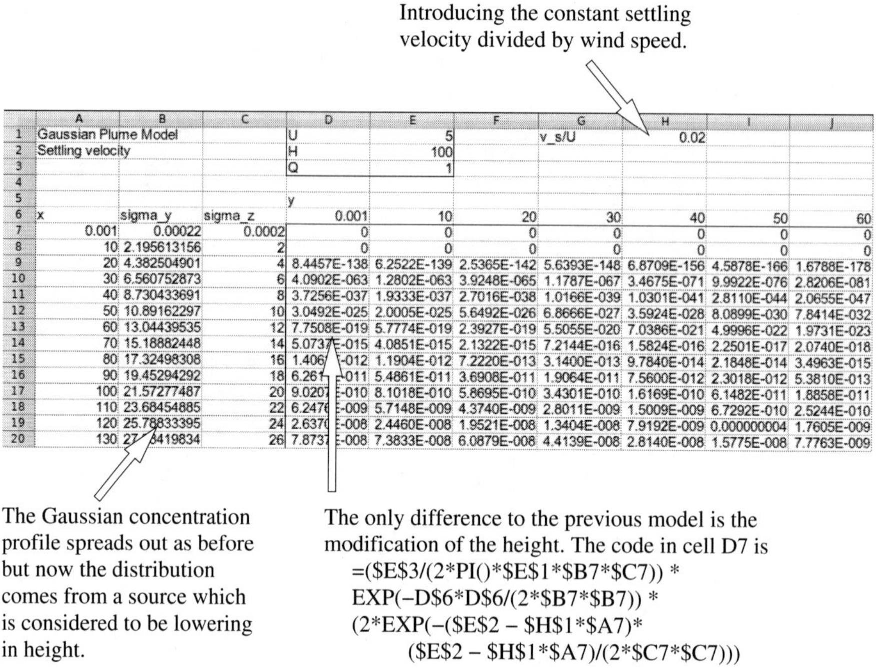

	A	B	C	D	E	F	G	H	I	J
1	Gaussian Plume Model			U	5		v_s/U	0.02		
2	Settling velocity			H	100					
3				Q	1					
4										
5				y						
6	x	sigma_y	sigma_z	0.001	10	20	30	40	50	60
7	0.001	0.00022	0.0002	0	0	0	0	0	0	0
8	10	2.195613156	2	0	0	0	0	0	0	0
9	20	4.382504901	4	8.4457E-138	6.2522E-139	2.5365E-142	5.6393E-148	6.8709E-156	4.5878E-166	1.6788E-178
10	30	6.560752873	6	4.0902E-063	1.2802E-063	3.9248E-065	1.1787E-067	3.4675E-071	9.9922E-076	2.8206E-081
11	40	8.730433691	8	3.7256E-037	1.9333E-037	2.7016E-038	1.0166E-039	1.0301E-041	2.8110E-044	2.0655E-047
12	50	10.89162297	10	3.0492E-025	2.0005E-025	5.6492E-026	6.8666E-027	3.5924E-028	8.0899E-030	7.8414E-032
13	60	13.04439535	12	7.7508E-019	5.7774E-019	2.3927E-019	5.5055E-020	7.0386E-021	4.9996E-022	1.9731E-023
14	70	15.18882448	14	5.0737E-015	4.0851E-015	2.1322E-015	7.2144E-016	1.5824E-016	2.2501E-017	2.0740E-018
15	80	17.32498308	16	1.406[illegible]-012	1.1904E-012	7.2220E-013	3.1400E-013	9.7840E-014	2.1848E-014	3.4963E-015
16	90	19.45294292	18	6.261[illegible]-011	5.4861E-011	3.6908E-011	1.9064E-011	7.5600E-012	2.3018E-012	5.3810E-013
17	100	21.57277487	20	9.0207[illegible]-010	8.1018E-010	5.8695E-010	3.4301E-010	1.6169E-010	6.1482E-011	1.8858E-011
18	110	23.68454885	22	6.247[illegible]-009	5.7148E-009	4.3740E-009	2.8011E-009	1.5009E-009	6.7292E-010	2.5244E-010
19	120	25.78833395	24	2.637[illegible]-008	2.4460E-008	1.9521E-008	1.3404E-008	7.9192E-009	0.000000004	1.7605E-009
20	130	27[illegible]8419834	26	7.873[illegible]-008	7.3833E-008	6.0879E-008	4.4139E-008	2.8140E-008	1.5775E-008	7.7763E-009

FIGURE 3.20
Spreadsheet implementation of Gaussian Plume Model with settling.

of falling particulates), and $\bar{U}$ is the average velocity in the downwind direction, x. Incorporating this modified effective stack height into the Gaussian plume model for ground level concentrations (Equation 3.28) results in a new expression of the form

$$\bar{C}(x, y, 0, H) = \frac{Q}{\pi \bar{U} \sigma_y \sigma_z} \exp\left(-\frac{y^2}{2\sigma_y^2}\right) \exp\left[-\frac{\left(H - \frac{v_s x}{U}\right)^2}{2\sigma_z^2}\right] \tag{3.30}$$

where the effective height of the smoke stack now decreases as the plume moves downwind. The implementation of this new equation into a spreadsheet is depicted in Figure 3.20. The only difference between this spreadsheet and the one depicted in Figure 3.18 is the constant ratio of settling velocity to downwind velocity, included at the top of the spreadsheet, and the modification of the effective stack height. In particular, the effective stack height is now reduced as a function of downwind direction when calculating the concentration of pollutants. This is considered a reasonable approximation to the problem of settling pollutants as long as the modified effective stack height does not go below the ground.

The effect of including the settling of pollutants is depicted in Figure 3.21. In particular, along the line $y = z = 0$ the concentration of pollution is plotted

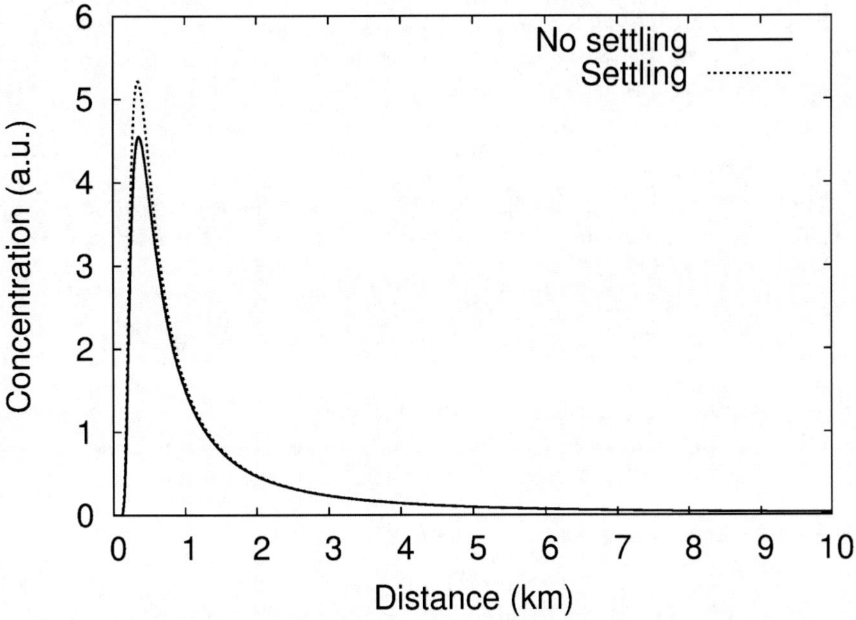

FIGURE 3.21
Concentration as a function of distance calculated using the Gaussian Plume Model downwind from the pollution source. Pollution which settles and pollution which does not settle are contrasted.

as a function of downwind direction, x. The effect of settling is predicted to increase ground-level concentrations, especially near to the source of the pollution. However, while this model captures the settling of pollutants as they fall from the air, it does not capture the deposition of pollutants on the ground.

3.7 Pollution: Gaussian Plume Model with Deposition of Pollutants

The capture the deposition of pollutants on the ground can be more difficult as pollutants must be removed from the system, and no longer propagate downwind from the smoke stack. The simplest way to achieve this effect is to imagine that the source strength is reduced as a function of downwind distance (similar to how the effective stack height was reduced in the settling model). Furthermore, the mirrored source at an effective height of $-H$ can also be reduced to account for the reduction in pollutants 'reflecting' off of the ground and back into the air. However, a more rigorous modification of the Gaussian plume model to account for the deposition of pollutants on the ground consists of a deposition flux of the form

$$F_D(x, y) = v_D \bar{C}(x, y) \tag{3.31}$$

where v_D is the deposition velocity and $\bar{C}(x, y)$ is the average concentration of pollution at the ground. In the deposition model, the concentration in the $y = 0$ plane is given by

$$\bar{C}(x, z) = Q_o D(x, z, H) - \int_0^x v_D \bar{C}(\acute{x}, 0) D(x - \acute{x}, z, 0) d\acute{x} \tag{3.32}$$

where the first term accounts for the diffusion of pollution from the smoke stack and the second term accounts for the loss of pollutants through deposition upwind of the location where the concentration is being calculated. Q_o is the source strength and $D(x, z, H)$ is the diffusion term in the vertical direction. In other words,

$$D(x, z, H) = \frac{1}{\sqrt{2\pi}\bar{U}\sigma_z} \left\{ \exp\left[-\frac{(H - z)^2}{2\sigma_z^2} \right] + \exp\left[-\frac{(H + z)^2}{2\sigma_z^2} \right] \right\} \tag{3.33}$$

and, as before, the dispersion parameter σ_z, depends on the distance downwind from the source.

The first term in Equation 3.32 is the concentration of pollution with no deposition, and the second term (with the integration) takes into consideration the loss of pollution concentrations due to deposition on the ground. The integration in Equation 3.32 is over positions downwind from the smokestack, up to the position x where we are calculating the concentration. In other words, the pollution concentration at x depends on the concentrations of pollutants on the ground prior to x, and to solve for the concentration at x we must first solve for ground level concentrations from the smoke stack up to position x. The concentration at position x, therefore, is the usual concentration due to the smoke stack minus a string of negative "pollution sources" all along the floor leading from the smoke stack to x, where the negative "pollution sources" account for the removal of pollutants due to deposition. Before we solve this system, however, consider the same equations but taking $z = 0$ (as well as $y = 0$).

$$\bar{C}(x) = Q_o D(x, H) - \int_0^x v_D \bar{C}(\acute{x}) D(x - \acute{x}, 0) d\acute{x} \tag{3.34}$$

and

$$D(x, H) = \frac{1}{\sqrt{2\pi}\bar{U}\sigma_z} 2 \exp\left[-\frac{H^2}{2\sigma_z^2} \right] \tag{3.35}$$

This system of equations predict the one-dimensional concentration profile along the floor downwind from a smoke stack, while taking into consideration the deposition of pollutants. The one-dimensional form is much easier to solve using a spreadsheet.

The prediction of pollutant concentrations using the Gaussian plume model with deposition is made difficult because of the integration which must be numerically estimated. In particular, Figure 3.22 depicts an iterative strategy for solving Equation 3.34 which spatially maps on to the spreadsheet model. In other words, the spreadsheet consists of the same seven sections and they are spatially organized in the spreadsheet in exactly the same way as they are laid out in Figure 3.22. As we go along the columns in the spreadsheet we increase our distance downstream and at the top of the spreadsheet the concentration of pollutants with no deposition, $Q_oD(x,H)$, is calculated. Next we include the concentration of pollutants at iteration i, given by $\bar{C}_i(x,H)$. Initially, the concentration at the 1st iteration can be set equal to the concentration that might be expected from a system with no deposition. In other words,

$$\bar{C}_1(x,H) = Q_oD(x,H) \tag{3.36}$$

From this concentration, which recall is along the floor downwind from the pollution source, we can estimate the amount of pollution which might be deposited on to the floor. In order to do this we must numerically evaluate the integration in Equation 3.34.

The third section of the spreadsheet, as we move down from top to bottom in Figure 3.22, calculates the dispersion parameter at a position x due to a pollutant source at $\acute{x}$. In other words, the removal of pollution at $\acute{x}$ will influence the concentration of pollution calculated at x, and recall that in this model the removal of pollution is treated as a negative pollution source. Therefore, the first part of evaluating the integration in Equation 3.34 is to calculate the dispersion parameter at discrete values of x and $\acute{x}$.

Next we calculate the argument of the integration, $\bar{C}_i(\acute{x})D(x-\acute{x},0)$, as a function of the discrete values of x and $\acute{x}$. Note that in order to calculate $D(x-\acute{x},0)$ we need the dispersion parameter σ_z calculated in the above section. The integration that we are evaluating is

$$\int_0^x \bar{C}_i(\acute{x})D(x-\acute{x},0)d\acute{x} \tag{3.37}$$

where the effective height of the stack is zero as the ‘stack’ in this term is the deposition of pollutants on the floor. The constant deposition velocity can be pulled out of the integration. The discretized version of this integration is

$$\sum_0^x \bar{C}_i(\acute{x})D(x-\acute{x},0)\Delta\acute{x} \tag{3.38}$$

where we now discretize the integration using small steps of size $\Delta\acute{x}$. Note, however, that the summation is only over distances where the source, $\acute{x}$, is closer to the stack than where we are calculating the concentration, x. In other words, we only need to sum over values of $\acute{x}$ that are less than x. Therefore, in the fifth section of the spreadsheet (going down the sections in Figure 3.22), the values of $\bar{C}_i(\acute{x})D(x-\acute{x},0)$ are copied from the fourth section, except for when $\acute{x} > x$ where we put the values in the cells to zero.

In the sixth section (second from bottom) we can sum over the cells from the fifth section in a given column to give us the integration at a given distance from the stack, x. In particular, we sum over the values of $\bar{C}_i(\acute{x})D(x-\acute{x},0)$ in the fifth section and then multiply this summation with the deposition velocity, v_D and the discretization of $\acute{x}$, $\Delta\acute{x}$. Finally, this integration is subtracted from the concentrations $\bar{C}_i(x)$ to give updated concentrations, $\bar{C}_{i+1}(x)$. These updated concentrations can be copied and pasted back up to the second section of the spreadsheet and replace $\bar{C}_i(x)$, such that the spreadsheet calculates the next iteration. In this manner, the model iteratively converges to a point where subsequent iterations in our procedure do not significantly change the concentrations and we can assume that this represents the solution to our system.

The spreadsheet implementation of this strategy is depicted in Figures 3.23 through 3.27. Figures 3.23 shows the top of the spreadsheet and the concentration at the ith iteration. The top of the spreadsheet consists of the constants required in the model (including the spatial discretization that we will use in the integration). Row 7 contains the x values (increasingly further away from the smoke stack as we move across the spreadsheet to the right), row 9 calculates the dispersion parameter σ_z as a function of this distance, and row 11 calculates the pollution concentration with no deposition, $Q_oD(x,H)$.

Row 13 contains the current iteration of the concentration as a function of x (going across the spreadsheet), which initially is set equal to concentration with no deposition. Just below this row of concentrations is a button labeled "Iterate". This button simply runs a macro which copies and pastes the concentrations from the $i+1^{th}$ iteration to the concentrations at the ith iteration (using paste special with values only). Once the concentrations have been copied then the spreadsheet will automatically repopulate with new values, calculating the next iteration's concentrations, which can, again, be copied to the top of the spreadsheet by pressing the "Iterate" button. At the bottom of Figure 3.23 is the calculation of σ_z, which can be obtained using one of the expressions in Tables 3.1 or 3.2.

Figure 3.24 depicts the next region of the spreadsheet; the calculation of $\bar{C}_i(\acute{x})D(x-\acute{x},0)$, as a function of the discrete values of x and $\acute{x}$. The code in cell B126, at the top left of the region calculating $\bar{C}_i(\acute{x})D(x-\acute{x},0)$, is

```
=B$13*(2/(SQRT(2*PI())*$E$3*B19)) *
        EXP(-$E$2*$E$2/(2*B19*B19))
```

and this code can be dragged across and down cells to calculate $\bar{C}_i(\acute{x})D(x-\acute{x},0)$ for all values of x and $\acute{x}$. For values of $\acute{x}$ less than x the cells have

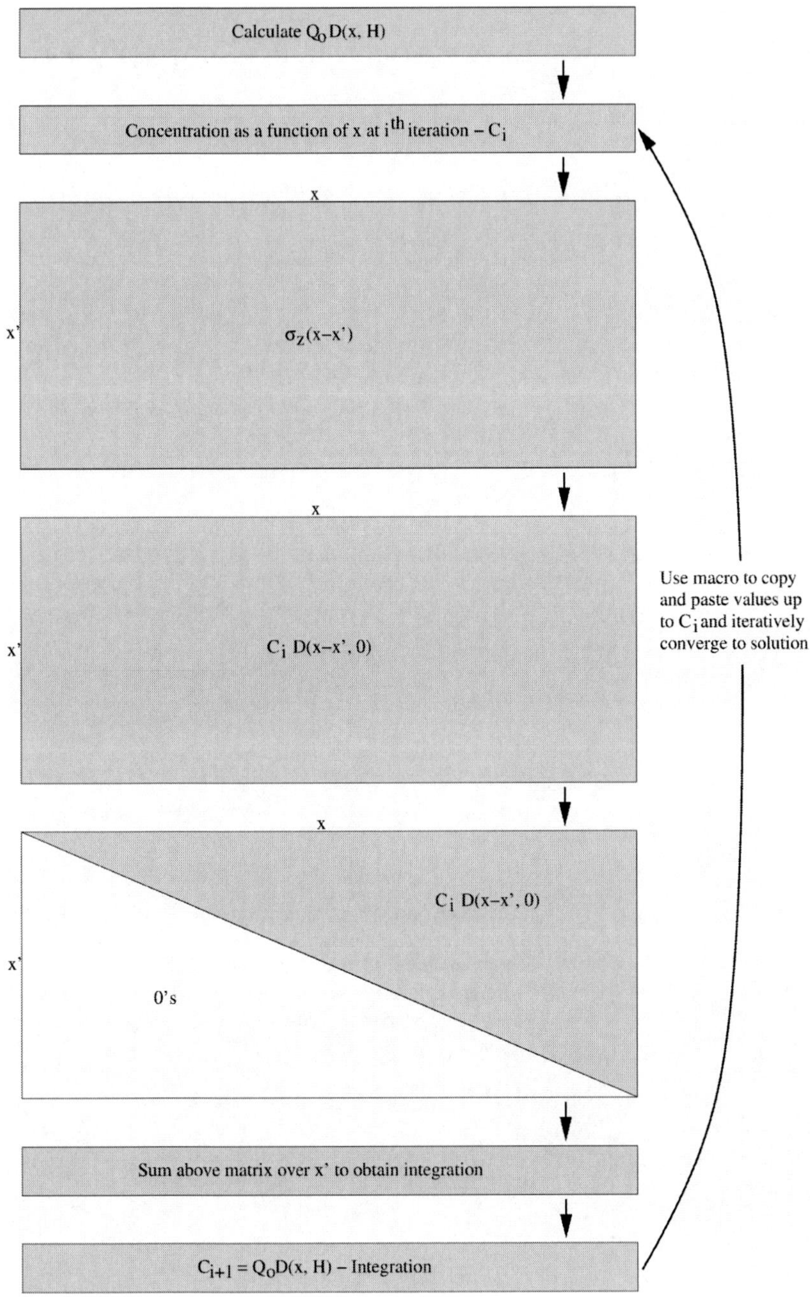

FIGURE 3.22
Schematic showing the implementation of the Gaussian Plume Model with deposition of the pollutants in a spreadsheet.

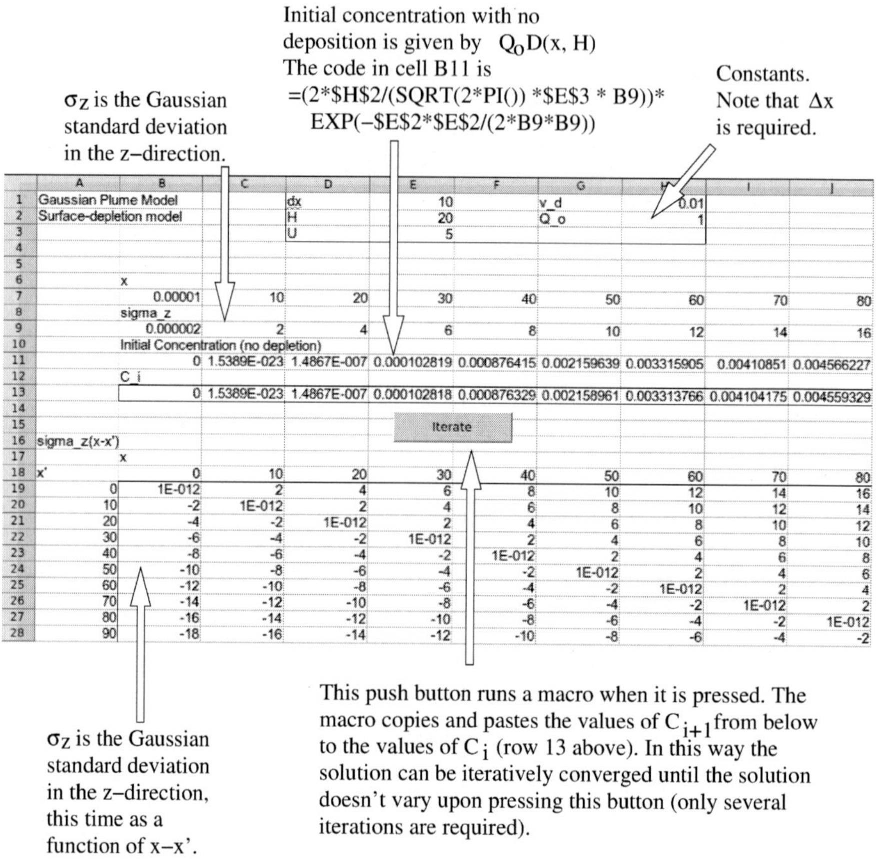

	A	B	C	D	E	F	G	H	I	J
1	Gaussian Plume Model			dx	10		v_d	0.01		
2	Surface-depletion model			H	20		Q_o	1		
3				U	5					
4										
5										
6		x								
7		0.00001	10	20	30	40	50	60	70	80
8		sigma_z								
9		0.000002	2	4	6	8	10	12	14	16
10		Initial Concentration (no depletion)								
11		0	1.5389E-023	1.4867E-007	0.000102819	0.000876415	0.002159639	0.003315905	0.00410851	0.004566227
12		C_i								
13		0	1.5389E-023	1.4867E-007	0.000102818	0.000876329	0.002158961	0.003313766	0.004104175	0.004559329
14										
15					Iterate					
16	sigma_z(x-x')									
17		x								
18	x'	0	10	20	30	40	50	60	70	80
19	0	1E-012	2	4	6	8	10	12	14	16
20	10	-2	1E-012	2	4	6	8	10	12	14
21	20	-4	-2	1E-012	2	4	6	8	10	12
22	30	-6	-4	-2	1E-012	2	4	6	8	10
23	40	-8	-6	-4	-2	1E-012	2	4	6	8
24	50	-10	-8	-6	-4	-2	1E-012	2	4	6
25	60	-12	-10	-8	-6	-4	-2	1E-012	2	4
26	70	-14	-12	-10	-8	-6	-4	-2	1E-012	2
27	80	-16	-14	-12	-10	-8	-6	-4	-2	1E-012
28	90	-18	-16	-14	-12	-10	-8	-6	-4	-2

FIGURE 3.23
Spreadsheet implementation of Gaussian Plume Model with deposition of the pollutants.

Gaussian standard deviation in the z–direction as a function of x–x'. Note the negative values are not important as these will be ignored.

115	960	-192	-190	-188	-186	-184	-182	-180	-178	-176
116	970	-194	-192	-190	-188	-186	-184	-182	-180	-178
117	980	-196	-194	-192	-190	-188	-186	-184	-182	-180
118	990	-198	-196	-194	-192	-190	-188	-186	-184	-182
119	1000	-200	-198	-196	-194	-192	-190	-188	-186	-184
120										
121										
122										
123	C_i D(x-x',0)									
124		x								
125	x'	0	10	20	30	40	50	60	70	80
126	0	0	2.3683E-046	2.2103E-014	1.0572E-008	0.000000768	4.6626E-006	1.0988E-005	0.000016862	2.0819E-005
127	10	0	0	2.2879E-030	1.5286E-011	9.0103E-008	1.8921E-006	7.1565E-006	1.3609E-005	1.8732E-005
128	20	0	-2.3683E-046	0	1.5823E-027	1.3029E-010	0.000000222	2.9042E-006	8.8635E-006	1.5118E-005
129	30	0	-2.2879E-030	-2.2879E-030	0	1.3486E-026	3.2098E-010	3.4072E-007	0.000003597	9.8465E-006
130	40	0	-1.5823E-027	-2.2103E-014	-1.5823E-027	0	3.3225E-026	4.9266E-010	0.000000422	3.9959E-006
131	50	0	-1.3487E-026	-1.5286E-011	-1.5286E-011	-1.3486E-026	0	5.0996E-026	6.1018E-010	4.6878E-007
132	60	0	-3.3235E-026	-1.3030E-010	-1.0572E-008	-1.3029E-010	-3.3225E-026	0	6.3160E-026	6.7784E-010
133	70	0	-5.1029E-026	-3.2108E-010	-9.0111E-008	-9.0103E-008	-3.2098E-010	-5.0996E-026	0	7.0164E-026
134	80	0	-6.3227E-026	-4.9298E-010	-2.2205E-007	-7.6803E-007	-2.2198E-007	-4.9266E-010	-6.3160E-026	0
135	90	0	-7.0271E-026	-6.1082E-010	-3.4093E-007	-1.8926E-006	-1.8921E-006	-3.4072E-007	-6.1018E-010	-7.0164E-026

As part of the integration it is required to solve C_i D(x–x', 0).
We will sum this quantity in a little while, but for now we calculate the values as a function of x–x' (again negative values are not important).
The code in cell B126 is

```
=B$13*(2/(SQRT(2*PI())*$E$3*B19) )* EXP(–$E$2*$E$2/(2*B19*B19))
```

FIGURE 3.24
Spreadsheet implementation of Gaussian Plume Model with deposition of the pollutants (continued).

negative numbers (these are not important as we will remove them from the calculation), and values of $\acute{x}$ equal to x the cells are equal to zero.

When evaluating the integral, which is integrating from the smoke stack up to the point where the pollution concentration is being calculated, it is only necessary to vary $\acute{x}$ up to the value of x. In other words, the values of $\bar{C}_i(\acute{x})D(x-\acute{x},0)$ for values of $\acute{x}$ greater than x are irrelevant to our calculations. Therefore, we can remove these values and replace them with zeros. Figure 3.25 depicts the region of the spreadsheet where this is calculated. In particular, the code contained in cell B233 is

```
=IF($A233<B$232, B126, 0)
```

which compares the values of $\acute{x}$ and x, and if $\acute{x}$ is less than x then the calculated value for $\bar{C}_i(\acute{x})D(x-\acute{x},0)$ above is included in the cell, otherwise the cell is set to equal zero.

Finally, the bottom of the spreadsheet is depicted in Figure 3.26. In particular, the calculated values of $\bar{C}_i(\acute{x})D(x-\acute{x},0)$ from above (only when $\acute{x}$ is less than x) are summed over $\acute{x}$. The code in cell B336 is

```
=SUM(B233:B333)*$H$1*$E$1
```

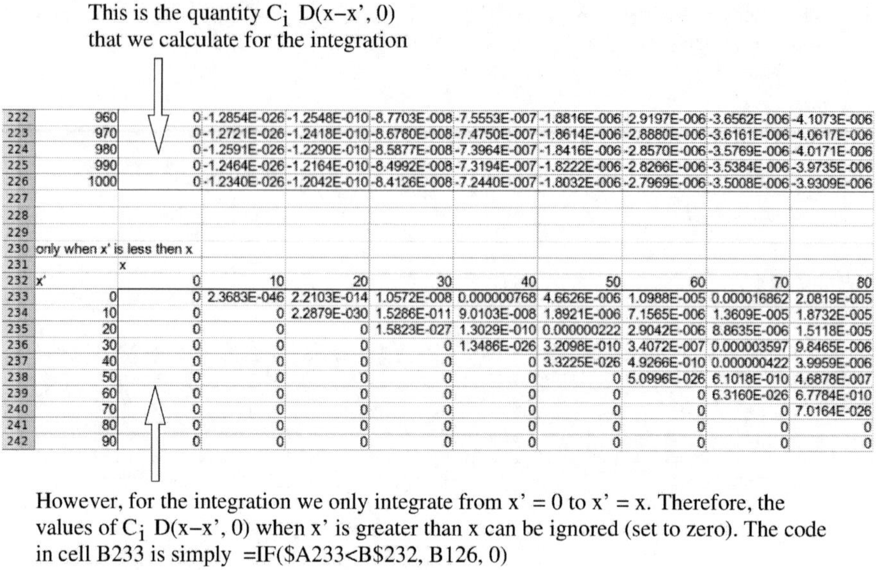

222	960	0	-1.2854E-026	-1.2548E-010	-8.7703E-008	-7.5553E-007	-1.8816E-006	-2.9197E-006	-3.6562E-006	-4.1073E-006
223	970	0	-1.2721E-026	-1.2418E-010	-8.6780E-008	-7.4750E-007	-1.8614E-006	-2.8880E-006	-3.6161E-006	-4.0617E-006
224	980	0	-1.2591E-026	-1.2290E-010	-8.5877E-008	-7.3964E-007	-1.8416E-006	-2.8570E-006	-3.5769E-006	-4.0171E-006
225	990	0	-1.2464E-026	-1.2164E-010	-8.4992E-008	-7.3194E-007	-1.8222E-006	-2.8266E-006	-3.5384E-006	-3.9735E-006
226	1000	0	-1.2340E-026	-1.2042E-010	-8.4126E-008	-7.2440E-007	-1.8032E-006	-2.7969E-006	-3.5008E-006	-3.9309E-006
227										
228										
229										
230	only when x' is less then x									
231		x								
232	x'	0	10	20	30	40	50	60	70	80
233	0	0	2.3683E-046	2.2103E-014	1.0572E-008	0.000000768	4.6626E-006	1.0988E-005	0.000016862	2.0819E-005
234	10	0	0	2.2879E-030	1.5286E-011	9.0103E-008	1.8921E-006	7.1565E-006	1.3609E-005	1.8732E-005
235	20	0	0	0	1.5823E-027	1.3029E-010	0.000000222	2.9042E-006	8.8635E-006	1.5118E-005
236	30	0	0	0	0	1.3486E-026	3.2098E-010	3.4072E-007	0.000003597	9.8465E-006
237	40	0	0	0	0	0	3.3225E-026	4.9266E-010	0.000000422	3.9959E-006
238	50	0	0	0	0	0	0	5.0996E-026	6.1018E-010	4.6878E-007
239	60	0	0	0	0	0	0	0	6.3160E-026	6.7784E-010
240	70	0	0	0	0	0	0	0	0	7.0164E-026
241	80	0	0	0	0	0	0	0	0	0
242	90	0	0	0	0	0	0	0	0	0

FIGURE 3.25
Spreadsheet implementation of Gaussian Plume Model with deposition of the pollutants (continued).

where the cells H1 and E1 contain the constants v_D and Δx, respectively. In other words, row 336 contains the latest evaluation (at the ith iteration) of the integral

$$\int_0^x v_D \bar{C}(\acute{x}) D(x - \acute{x}, 0) d\acute{x} \tag{3.39}$$

from Equation 3.34. Note that the calculation of the integration requires the evaluation of the concentration profile, $\bar{C}(\acute{x})$. However, upon the calculation of this integration we will use this calculation to in turn update the concentration profile, which means that we now have to recalculate the integration! This is the reason why we have to iteratively solve the system. Upon updating the concentration in row 338, by subtracting the integration from the initial concentration where no deposition occurs, this updated concentration is copied to the top of the spreadsheet in row 13 to use this updated concentration to calculate the integral and the next iteration of the concentration profile. After only several iterations, however, the system converges such that the difference in concentration profiles between iterations is very small (at least for the parameters considered here). At this point the simulation is complete and the concentration profile captures the effects of deposition.

The concentration of pollution along the line defined by $y = z = 0$ (along the floor and downwind from the smoke stack) is calculated for a system with no deposition and a system with deposition. The effective height of the smoke

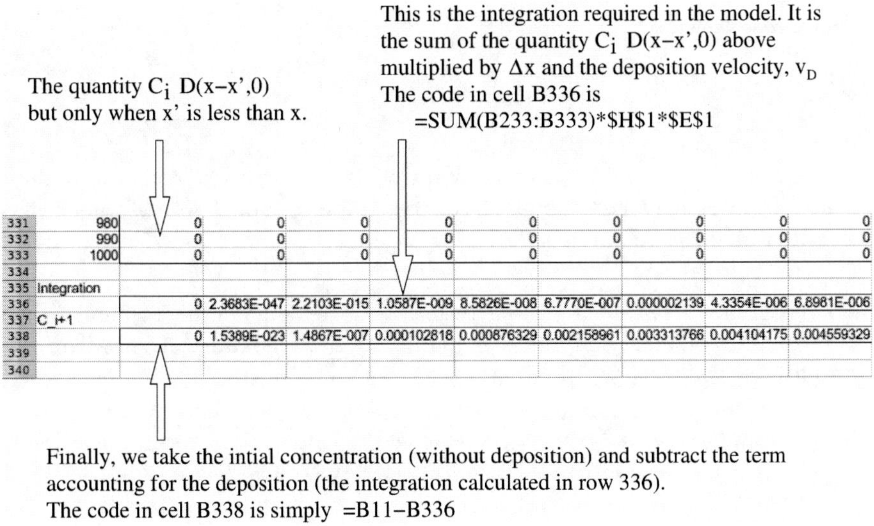

FIGURE 3.26
Spreadsheet implementation of Gaussian Plume Model with deposition of the pollutants (continued).

stack, $H = 20m$, the mean velocity field, $\bar{U} = 5m/s$, the source strength, $Q_o = 1$, and (in the system with deposition included) the velocity of deposition, $v_D = 0.1$. The different concentration profiles are contrasted in Figure 3.27. As expected, the concentration of ground level airborne pollution is greater in the system with no deposition (as less pollution is removed from the air by depositing on the floor). Furthermore, as you move further from the smoke stack the effect is more pronounced. As the concentration is deposited on the ground it can no longer propagate downwind. Also, the further you head downwind the more pollution that has been deposited on the ground. Therefore, the less pollution there is in the air.

The Gaussian plume model is a very powerful and popular model as, for the most part, the predictions of the model are relatively easy to calculate. However, there are obvious limitations with the technique. For example, the model does not take into consideration the time required for the pollution to travel downwind, and this can make it difficult to include time-dependent effects (such as chemical reactions). In addition, the model assumes a steady-state pollution source and constant wind speed. As one might expect the meteorological conditions to vary both spatially and with time, this limits the applicability of the model to a range, optimistically, of a few tens of kilometers from the source. That said, the Gaussian plume model is a popular model and has been successfully used to calculate the distribution of pollutants in a wide range of locations and conditions.

3.8 Exercises

1. The solar constant is known to vary over a 11-year solar cycle between values of 1365.5 to 1366.5 W/m^2. What effect does this have on the temperature of the planet? Use the global energy balance model to investigate such variation in solar constant.

2. One possible solution to global warming is geoengineering, where the climate of the planet is deliberately manipulated by human activity. To offset the effects of anthropogenic greenhouse gas emissions it has been suggested that we could paint a large area of the planet white. This would then reflect sunlight back into space, increasing the planet's albedo, and reducing global temperatures. The area of New York city (the largest city in the US) is $17,400km^2$. This is approximately 0.00034 % of the planet earth. Given that the current average albedo for the Earth is 0.3, calculate, using the global energy balance model, the effects of increasing 0.00034 % of the albedo to 1. How much would the planet be cooled by painting New York City white?

3. In the zonal energy balance model you can predict the planet's temperature as a function of CO_2. Generate a plot of the planet's temperature as a function of CO_2 concentrations.

4. According to solar physicists, the sun emitted a third less energy about 4 billion years ago and has been steadily brightening ever since. Yet for most of this time, Earth has been even warmer than today, a phenomenon sometimes called the faint sun paradox. Use the zonal energy balance model to determine the amount of CO_2 required to make the temperature of the planet 20 ^oC if the value of S is reduced by a third.

5. When describing the radiative model of a cloud a wide range of parameters was described. Investigate the range of parameters for the albedo, absorption, and emissivity of clouds in the radiative model and try to contrast the predicted differences between high and low altitude clouds.

6. In the radiative model for the cloud, we used the first two equations to generate an equation for ground temperature. However, we did not need to use the third equation which described the energy balance at the ground (as we had three equations and two unknowns). Obtain expressions for T_g using the first and third equations and the second and third equations. Do you get the same results as from the first and second equations?

7. Compare the numerical results from a model of an expanding puff of pol-

lution with the Gaussian analytical solution

$$C(r,t) = \frac{Q}{8\,(\pi D t)^{\frac{3}{2}}} \exp\left(-\frac{r^2}{4Dt}\right) \tag{3.40}$$

Are there any differences between the numerical prediction and the analytical solution?

8. From the gradient transport model which captures the pollutant distribution from a continuous point source in a uniform flow, obtain the location on the ground with the highest pollutant concentration as a function of time. Investigate how this location changes as you change the height of the smoke stack or the magnitude of the velocity. Produce a plot of location on the ground with the highest levels of pollution as a function of time for both different smoke stack heights and different wind velocities.

9. In the current implementation of the gradient transport model which captures the pollutant distribution from a continuous point source in a uniform flow the average velocity is constant everywhere and only in the x-direction. Implement a model where the velocity is no longer constant (but still in the x-direction) and have it linearly increase from zero at the floor to a maximum value at the top of the simulation. How does this affect your results from question 8?

10. In the Gaussian plume model the dispersion of the pollutants depends on the distance downwind, and the environmental conditions. In particular, the Paquill type of conditions (from stable to unstable) can be used to calculate the dispersion parameters as a function of distance. For an effective stack height of 1000 m and an average velocity of 5 m/s, calculate the location on the ground with the highest pollution concentration as a function of the stability conditions (Pasquill types A through F) and for urban and rural conditions.

11. For the deposition model calculate the reduction in pollution in the air at ground level downwind from the smoke stack for different values of deposition velocity, v_D. At what value of v_D does the solution become unstable? Can the stability of the system be extended by taking smaller discretization steps, Δx?

Additional Reading

Henderson-Sellers, A., & McGuffie, K. (1987). A climate modelling primer. Chichester: Wiley.

Arya, S. P. (1999). Air pollution meteorology and dispersion. Oxford University Press, Oxford.

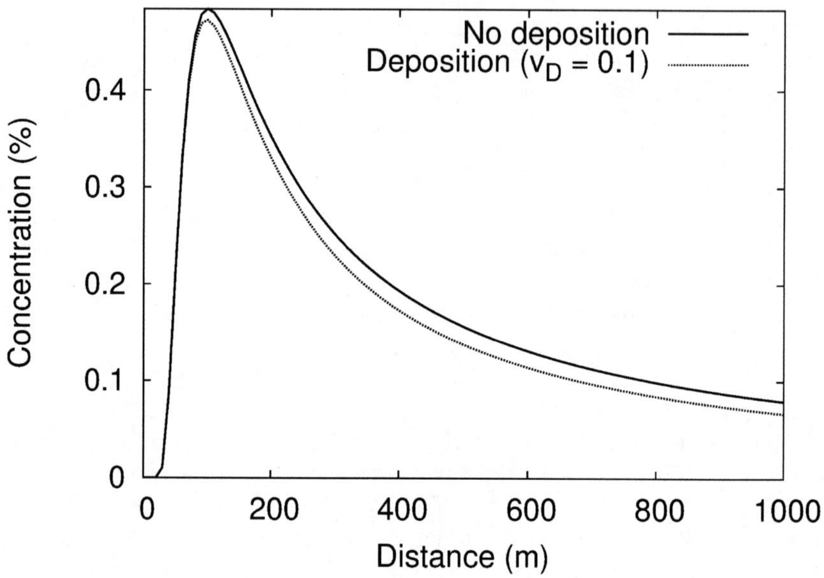

FIGURE 3.27
Pollution concentration as a function of distance along the ground. Comparison between a model incorporating the deposition of the pollution and one that doesn't. The system which doesn't account for deposition is represented by a solid line, while the system with deposition is represented by a dashed line.

4

Solar Cells

CONTENTS

Solar cells convert light into electricity. In most cases, a particle of light (called a photon) is absorbed by the solar cell material and causes the separation of charged particles (an electron and a 'hole'). The charged particles can then move to the electrodes, the electrons moving to the cathode and the holes to the anode, and these moving charges constitute the electric current. There are many different designs of solar cells as discussed in Chapter 1.2, but the basic principle is the same for all devices. Incoming photons of sunlight must be converted into free charges which can move around an external circuit. The important aspects of solar cell technologies are how much charge is produced by a solar cell if there is nothing attached across it but bare wire? How much voltage is there across the solar cell if there is nothing attached across the solar cell? And how much power can be extracted from the solar cell or, in other words, how efficient is the solar cell at converting light energy to electrical energy? These are the basic ways in which solar cells can be characterized, and computer simulations can be useful in elucidating the origins of these characteristics. In other words, the reason computer simulations are so popular in solar cell physics is that the processes inside the solar cell are difficult to observe directly. For example, it is difficult to directly observe the concentrations of electrons inside the solar cell in an experiment, but this is something which might be very easy to analyze in a computer model. Computer models are, therefore, an invaluable resource for predicting new designs and technologies, or shedding light on existing experimental systems.

There is a wide variety of simulation techniques for looking at solar cells. The most popular is arguably the drift diffusion method, followed by equivalent circuits and Monte Carlo methods. We will look at all three techniques here, along with the finite-difference time-domain (FDTD) method of modeling the propagation of light (which is becoming increasingly relevant to solar

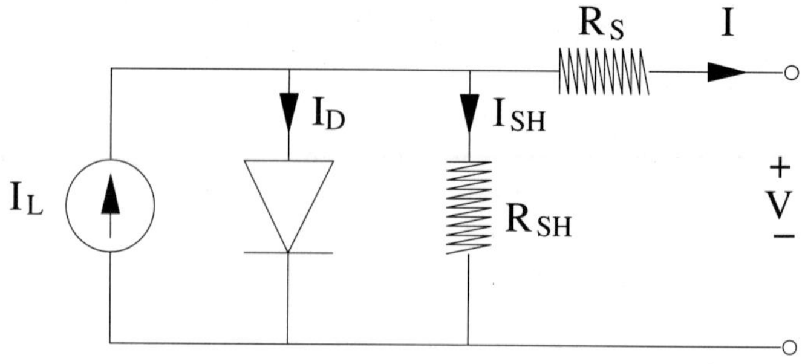

FIGURE 4.1
An equivalent circuit representation of a solar cell, where the solar cell is replaced by simpler electrical components.

cell physics as the structure or morphology of solar cell devices become more complex).

4.1 Equivalent Circuit of Solar Cells

The simplest model of solar cells involves conceptually replacing the solar cell with an equivalent circuit comprised of simpler circuit elements (diodes and resistors, whose behavior is well known). Essentially, while the equivalent circuit offers no insights into the reasons why a solar cell performs the way it does, it can help predict the voltage and current output from the solar cell under different conditions. In other words, a photovoltaic cell may be represented by an equivalent circuit, capable of analyzing and parameterizing the electrical characteristics of a solar cell at all points along the current-voltage curve. The parameters for this model can easily be obtained from data provided by module manufacturers, and allow a more versatile estimation of their performance.

The simplest equivalent circuit usually consists of a radiation-dependent current source, in parallel with a temperature-dependent diode and a resistance, and in series with a series resistance. More complicated circuits, including multiple diodes and resistors, have been proposed but offer very little towards the performance of the model. The equivalent circuit for a solar cell is shown in Figure 4.1. The current source, I_L, on the left-hand side of the circuit diagram represents the photogenerated electrical current in the solar cell. In particular, this current source represents the generation of charge carriers in the semiconductor region of the solar cell, as a result of incident solar radiation. The shunt diode, with a current of I_D, represents the recombination of these charge carriers inside the solar cell, before the charge carriers have a chance to leave the solar cell and be used in an external circuit or to power an external device. Because no solar cell is perfect, the shunt resistance, R_{SH}, is placed in parallel with the diode and accounts for the current loss, I_{SH}, through the solar cell due to mechanical defects and material dislocations. The series resistor accounts for electrical resistance that might arise throughout the solar cell device, primarily at the interface of the semiconductor and the metal contacts. The series resistance can also reflect the charge carrier mobility in the semiconductor, especially in organic solar cells where the charge carrier mobility is affected by space charges, traps, or other barriers. (In an organic semiconductor the charges generally hop from molecule to molecule, and the mobility is dependent on both the temperature and the local electric field.)

To capture the behavior of this equivalent circuit we consider a characteristic equation which describes the circuit. In particular, the current in the device, I, is the current arising from the photogenerated charge carriers, I_L, minus the shunt diode current, I_D, and the current through the shunt resistance, I_{SH}. The relationship between current and voltage in diode and resistors are well known, as so we can write down the current in the form

$$I = I_L - I_o \left\{ \exp\left[\frac{q\,(V + IR_S)}{nk_\beta T} \right] - 1 \right\} - \frac{V + IR_S}{R_{SH}} \tag{4.1}$$

where, in the first term, I_L is the current arising from the photogenerated charge carriers. In the second term, I_o is the reverse saturation current, n is the diode ideality factor, k_β is Boltzmanns constant, T is the cell temperature, and q is the charge of an electron. V is the voltage across the solar cell, I is the current through the solar cell, and R_S is the series resistance. In the final term, R_{SH} is the shunt resistance. Because solar cells can be made in different sizes, however, it is common to write the equation in terms of current density instead of current. Current density is simply the current divided by the area. The above characteristic equation becomes

$$J = J_L - J_o \left\{ \exp\left[\frac{q\,(V + JR_S)}{nk_\beta T} \right] - 1 \right\} - \frac{V + JR_S}{R_{SH}} \tag{4.2}$$

where J is the current density with units of amps per meter squared (A m^{-2}). Furthermore, to ensure the units match, the series and shunt resistance are now expressed as 'specific' resistance with units of Ω m^2. This equation is both an implicit (cannot be algebraically solved) and a nonlinear (the variables are inside the nonlinear exponential function) function of J.

In order to solve Equation 4.2 we can use a numerical method called the Bisection Method. In the Bisection Method we start by writing down our equation, such that the left-hand side is equal to zero.

$$0 = J_L - J_o \left\{ \exp\left[\frac{q\,(V + JR_S)}{nk_\beta T} \right] - 1 \right\} - \frac{V + JR_S}{R_{SH}} - J \tag{4.3}$$

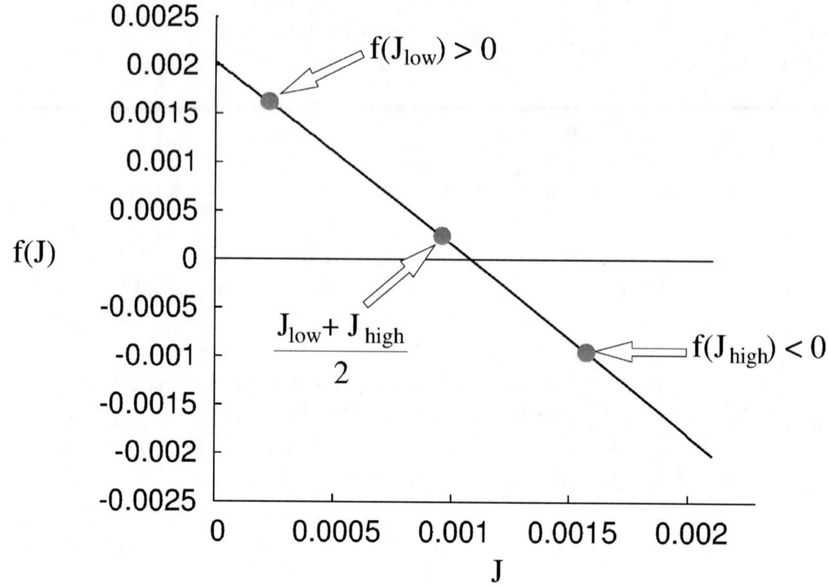

FIGURE 4.2
Depiction of the Bisection Method for solving a one-dimensional function.

The right-hand side of the equation is now a function, $A(J)$, which depends on J that we must solve to obtain the value of J that gives a value of $A(J)$ equal to zero. In other words, if we put in a value for J into the right-hand side of Equation 4.3 then we would be unlikely to 'guess' the correct value that would give us zero on the left-hand side of Equation 4.3. However, if I took a first guess at J that resulted in the left-hand side being a positive answer and a second guess of J that resulted in the left-hand side being a negative answer, then I would know the correct value of J must be in between these two guesses. This is the essence of the Bisection Method and is depicted in Figure 4.2.

We can choose a value for J which is too low, J_{low}, and a value for J which is too high, J_{high}, such that one of our guesses gives $A(J) > 0$ and the other $A(J) < 0$. In this particular system $A(J_{low}) > 0$ and $A(J_{high}) < 0$. Now, because we are looking for a value of J that will give us $A(J) = 0$ we know that the J we are looking for is somewhere in between J_{low} and J_{high}. So we can now pick a third guess, directly in the middle between J_{low} and J_{high}, such that $J_{new} = (J_{low} + J_{high})/2$. While this third guess is unlikely to be the correct answer, it will be closer to the correct answer than either J_{low} or J_{high}. In other words, we can use this third guess to replace either J_{low} or J_{high}, depending on whether or not $A(J_{new})$ is positive or negative. This essentially closes the bounds J_{low} and J_{high} such that they are closer to the true value. This procedure can be repeated until the difference between J_{low} and J_{high} is negligible and they both equal the correct value of J required to yield $A(J) = 0$.

The Bisection Method of the solar cell equivalent circuit can be easily implemented in a spreadsheet, as depicted in Figure 4.3. At the top of the spreadsheet is the list of constants required in the model. The constants, $q = 1.6 \times 10^{-19} C$ and $k_\beta = 1.38 \times 10^{-23}\ m^2kgs^{-2}K^{-1}$ are fixed for all solar cells. The constant voltage, V, is varied to obtain a complete current-voltage (I-V) curve. Below the voltage in the spreadsheet is the solution; the current density, J, copied from below such that the result of the Bisection Method is conveniently displayed next to the constants. The current describing the photogeneration of charge carriers, J_L, depends on the amount of light illuminating the solar cell. The other constants are dependent on the type of solar cell which is being simulated. Recall that the efficiency of solar cells is reduced by the shunt and series resistances. For an ideal cell, therefore, R_{SH} would be infinite and not provide a path for current to flow, beside through the external load, while R_S would be zero, resulting in no internal voltage drop inside the solar cell. The constants used here are from a recent paper on organic solar cells:

Parameter	Value
J_o	$1.54 \times 10^{-9}\ Acm^{-2}$
R_S	$4.05\ \Omega cm^2$
n	1.5
R_{SH}	$3.69 \times 10^{6}\ \Omega cm^2$

It should be noted, however, that organic solar cells might not be the best devices for this type of modeling. As the incident light is increased upon organic solar cells, the shunt resistance has been observed to decrease and the reverse saturation current density has been observed to decrease. In ideal inorganic solar cells, however, these parameters are constant except for the photogenerated current which is always taken to be a linear function of the light intensity.

In Figure 4.3, the spreadsheet implementation of the equivalent circuit model of the solar cell involves making two initial guesses for the current density, J. In particular, cell B9 contains a high value of J and cell D9 contains a low value of J, as initial guesses. The function which we are trying to set equal to zero, $A(J)$, as a function of J is calculated in cells C9 and E9, for the high and low values of J, respectively. For example, the code in cell C9 is

```
=$F$1-$F$2*(EXP(($H$1*($H$4+B9*$F$3))/($F$5*$H$2*$H$3)) - 1)

                    - (($H$4 + B9*$F$3)/($F$4)) - B9
```

where cell F1 contains the constant photogenerated current, the second term

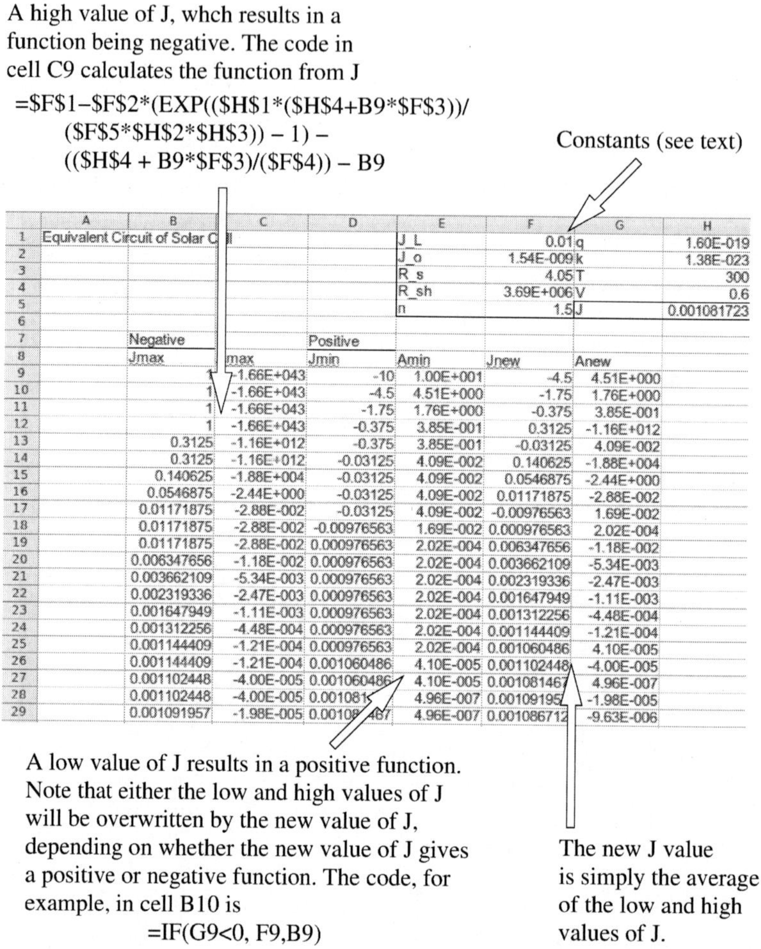

	A	B	C	D	E	F	G	H
1	Equivalent Circuit of Solar Cell				J_L	0.01	q	1.60E-019
2					J_o	1.54E-009	k	1.38E-023
3					R_s	4.05	T	300
4					R_sh	3.69E+006	V	0.6
5					n	1.5	J	0.001081723
6								
7		Negative		Positive				
8		Jmax	Amax	Jmin	Amin	Jnew	Anew	
9		1	-1.66E+043	-10	1.00E+001	-4.5	4.51E+000	
10		1	-1.66E+043	-4.5	4.51E+000	-1.75	1.76E+000	
11		1	-1.66E+043	-1.75	1.76E+000	-0.375	3.85E-001	
12		1	-1.66E+043	-0.375	3.85E-001	0.3125	-1.16E+012	
13		0.3125	-1.16E+012	-0.375	3.85E-001	-0.03125	4.09E-002	
14		0.3125	-1.16E+012	-0.03125	4.09E-002	0.140625	-1.88E+004	
15		0.140625	-1.88E+004	-0.03125	4.09E-002	0.0546875	-2.44E+000	
16		0.0546875	-2.44E+000	-0.03125	4.09E-002	0.01171875	-2.88E-002	
17		0.01171875	-2.88E-002	-0.03125	4.09E-002	-0.00976563	1.69E-002	
18		0.01171875	-2.88E-002	-0.00976563	1.69E-002	0.000976563	2.02E-004	
19		0.01171875	-2.88E-002	0.000976563	2.02E-004	0.006347656	-1.18E-002	
20		0.006347656	-1.18E-002	0.000976563	2.02E-004	0.003662109	-5.34E-003	
21		0.003662109	-5.34E-003	0.000976563	2.02E-004	0.002319336	-2.47E-003	
22		0.002319336	-2.47E-003	0.000976563	2.02E-004	0.001647949	-1.11E-003	
23		0.001647949	-1.11E-003	0.000976563	2.02E-004	0.001312256	-4.48E-004	
24		0.001312256	-4.48E-004	0.000976563	2.02E-004	0.001144409	-1.21E-004	
25		0.001144409	-1.21E-004	0.000976563	2.02E-004	0.001060486	4.10E-005	
26		0.001144409	-1.21E-004	0.001060486	4.10E-005	0.001102448	-4.00E-005	
27		0.001102448	-4.00E-005	0.001060486	4.10E-005	0.001081467	4.96E-007	
28		0.001102448	-4.00E-005	0.00108[illegible]	4.96E-007	0.00109195[illegible]	-1.98E-005	
29		0.001091957	-1.98E-005	0.0010[illegible]467	4.96E-007	0.001086712	-9.63E-006	

FIGURE 4.3
Spreadsheet implementation of the Bisection Method for obtaining the current density of an equivalent circuit of a solar cell.

is the diode current, and the third term is the current through the shunt resistance. Cell B9 is the value of J for which this function is being calculated.

Cell F9 calculates the average between the high and low values of J, to obtain a new guess for J. The code contained in cell F9 is simply

```
=(B9+D9)/2
```

Again, the function $A(J_{new})$ is calculated for this new guess of J_{new} in cell G9.

If the function $A(J_{new})$ is less than zero (if the value of J_{new} is too high) then we can replace the value of J_{high} with J_{new} to get a better estimate of our value for J. The code in cell B10, for example, is

```
=IF(G9<0, F9,B9)
```

which says that if cell G9 (containing $A(J_{new})$) is less than zero, then the value of J_{new} should be taken as the new J_{high} (cell F9), else we should stick with the old value of J_{high} (cell B9). Similarly, the code in cell D10 is

```
=IF(G9>0, F9, D9)
```

and the value of J_{low} is replaced with J_{new} in the event that $A(J_{new})$ is greater than zero. This process is repeated, as we go down the spreadsheet, with each row corresponding to an iteration step. The Bisection Method is iterated until the values of J_{low} and J_{high} are indistinguishable (about 100 iterations are found to be more than sufficient in the current system). The value of J which represents the solution to this system is copied at the top of the spreadsheet (in cell H5) so that the solution can be seen directly as a consequence of changing the parameters at the top of the spreadsheet. In particular, we can vary the value of the applied voltage and calculate the current density in the system.

Figure 4.4 depicts the current-voltage (I-V) curve for the solar cell modeled above. In particular, current density is plotted as a function of applied voltage. Where the applied voltage is zero, the current density is called the short circuit current density (J_{SC}). This is the current density that would be flowing through the solar cell if the cathode and anode electrodes were connected together with a conducting wire. Where the current density is zero, the voltage difference between the two electrodes is called the open circuit voltage (V_{OC}). This is the voltage from the solar cell if nothing is attached to the electrodes and the solar cell has no external load. A voltmeter, for example, would ideally have infinite resistance and the voltage measured across the solar cell would be its open circuit voltage. At some point (represented by the dashed lines in Figure 4.4) the voltage multiplied by the current density is a maximum. This represents the maximum power point and, in an operating solar cell system, an inverter will constantly adjust the load to seek out this particular point to ensure that the system yields the maximum power.

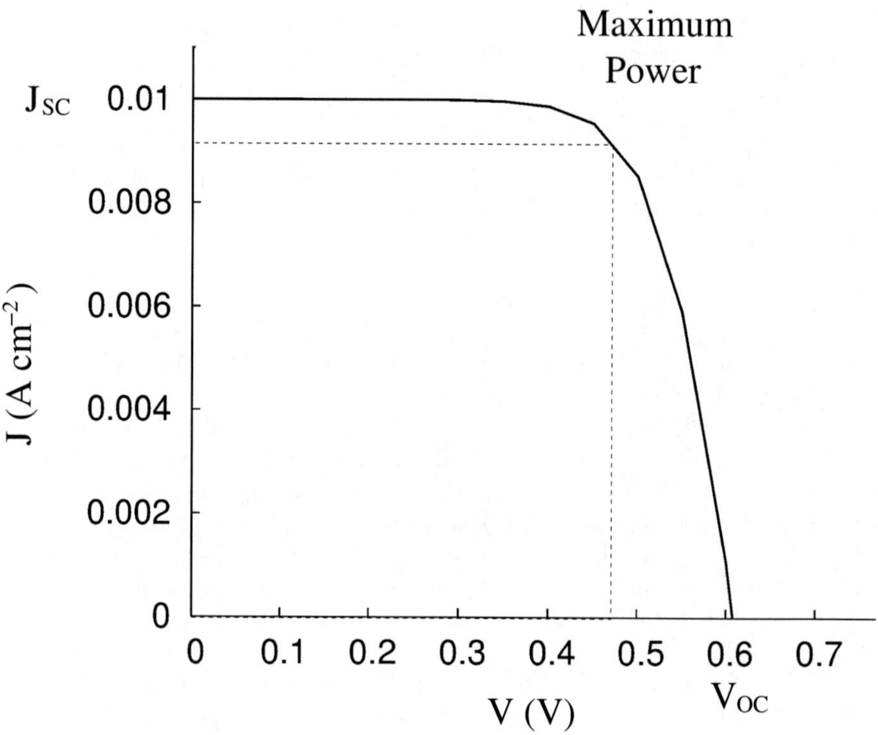

FIGURE 4.4
I-V curve for an organic solar cell calculated using an equivalent circuit model.

4.2 Drift-Diffusion Model of Photovoltaics

The most popular model that is used to capture the device physics of solar cells is the drift-diffusion model. The drift-diffusion model has been used extensively to capture the physics of inorganic devices, and has recently been extended to capture the physics of organic solar cells. The drift-diffusion model captures both the drift of charge carriers within the electric field of the device, moving the charge carriers towards the electrodes, and the diffusion of charge carriers from regions of high concentration to regions of low concentration. Typically, therefore, the basic drift-diffusion model consists of an equation containing two terms: the first term including the gradient in electric potential, and accounting for the drift of charges, and the second term including the gradient in charge carrier concentration, and accounting for the diffusion of charges. The equation is written for both the electrons and the holes (the two main charge carriers in semiconductor devices).

The electrons are subatomic particles, which can move to some extent in the semi-conductor materials, and is traditionally what we might think of as comprising the current in a conductor. The electron can be excited in the atom to an excited state; an excited state in which the electron is potentially mobile and can leave its atom behind and wander around the semiconducting material. In a semiconductor, however, we have a second charge carrier: the holes. The holes are defined as the lack of an electron at a position where an electron could exist in an atom or atomic lattice. In other words, as an electron is excited to a higher (potentially mobile) excited state, it leaves a hole in its old state, a hole that could be potentially filled with an electron, because the atom is now missing an electron. Holes can be considered as mobile because electrons can move from one atom that has its full company of electrons to an atom with a hole. However, upon filling this hole the electron has left behind a hole in its place. While the electron has moved in one direction, filling the hole, because it leaves a hole behind after it has moved it might appear as if the hole has moved in the opposite direction. In reality, they are two different holes and the electron has moved from one hole to the next, making it look like the empty hole moved. In this manner, a positive charge can move through the semiconductor in the form of one of these holes. A solar cell works by using the energy contained within a photon of light to excite an electron out of its hole, and transporting the electron to the cathode and the hole to the anode, such that the electric charge can be used in an external circuit.

The typical reduced dimensional form of the drift-diffusion equations are given as

$$\nabla \cdot (\epsilon \nabla \phi) = -q(p - n) \tag{4.4}$$

$$\frac{\partial n}{\partial t} = G - R - \frac{1}{q} \nabla \left[q n \mu_n \nabla \phi - k_\beta T \mu_n \nabla n \right] \tag{4.5}$$

$$\frac{\partial p}{\partial t} = G - R - \frac{1}{q}\nabla\left[-qp\mu_p\nabla\phi - k_\beta T\mu_p\nabla p\right] \tag{4.6}$$

where the top equation is Poisson's equation and solves the local electric potential, ϕ, in the solar cell. ϵ is the dielectric constant of the solar cell, which can vary between the different semiconducting materials comprising the solar cell. The electric field, which is the gradient of electric potential, inside the solar cell depends on the different work function of the electrodes and any externally applied voltage to the solar cell. However, the electric field can also be influenced (sometimes dominantly so) by the presence of charge carriers in the solar cell. The right hand side of the top equation, therefore, contains the contribution of the concentrations of the holes, p, and electrons, n. q is the elementary charge.

The second and third equations are the drift-diffusion equations for the electrons and holes, respectively. The left hand side of these equations is simply stating that the equations describe the local variation of concentrations with respect to time. The right-hand side of the equations contain a few terms which contribute to the local variation of concentrations with respect to time. The first, G, is the photogeneration of free charge in the semiconductor. In other words, this term describes the creation of the charge carriers as a consequence of the absorption of photons in the semiconductor. This is the same for both of the equations because the absorption of a photon excites an electron to a mobile state, which also leaves behind a mobile hole in its stead. The second term, R, is the recombination of the electrons and holes. As the the electrons and holes are moving around in the semiconductor, they have the potential to recombine. In other words, a free electron wandering around the semiconductor could occupy a mobile hole state, and this would result in the free electron and hole no longer existing. This could remove extractable energy from the system, and is something we would like to avoid. (This is by no means certain, as the recombination of charge can result in a photon being emitted from the semiconductor, which has the potential of being reabsorbed elsewhere in the semiconductor, and creating an entirely different free electron and hole.) The third term involves the gradients in electric field (electric field is $\nabla\phi$, where ϕ is the electric potential) and accounts for the drift of charge carriers in the solar cell device. μ_n and μ_p are mobilities of the electrons and holes, respectively, which characterizes how quickly the charge can be pulled through the semiconductor by the electric field. The last term contains the gradient in concentrations, ∇n or ∇p, and accounts for the diffusion of charge carrier concentrations from regions of high concentrations to regions of low concentrations.

In one-dimension these equations become

$$\frac{\partial}{\partial x}\left(\epsilon\frac{\partial}{\partial x}\phi\right) = -q(p-n) \tag{4.7}$$

$$\frac{\partial n}{\partial t} = G - R - \frac{1}{q}\frac{\partial}{\partial x}\left[qn\mu_n\frac{\partial}{\partial x}\phi - k_\beta T\mu_n\frac{\partial}{\partial x}n\right] \tag{4.8}$$

$$\frac{\partial p}{\partial t} = G - R - \frac{1}{q}\frac{\partial}{\partial x}\left[-qp\mu_p\frac{\partial}{\partial x}\phi - k_\beta T\mu_p\frac{\partial}{\partial x}p\right] \tag{4.9}$$

Discretizing the above equations starts with Scharfetter-Gummel method, which provides an optimum way to discretize the drift-diffusion equations. The Scharfetter-Gummel method considers the local current in charge carriers, which depends on the drift and diffusion of charge carriers, and locally replaces this with the analytical solution to the partial differential equations. These solutions can then be incorporated into the finite difference scheme used to solve these differential equations. For example, the Scharfetter-Gummel method when applied to the electron current yields

$$\begin{aligned} J_n &= n\mu_n\frac{\partial\phi}{\partial x} - \mu_n\frac{\partial n}{\partial x} \\ &= \frac{\mu_n}{\Delta x}\left[B(\phi_i - \phi_{i+1})n_i - B(\phi_{i+1} - \phi_i)n_{i+1}\right] \end{aligned} \tag{4.10}$$

where $B(x) = x/(e^x - 1)$ is the Bernoulli function. The benefit of this is that it makes the scheme much more stable in systems where the electric field locally changes. Using the Scharfetter-Gummel method, we can now write down the drift-diffusion equations as a set of discrete linear equations to be solved at each time step. The Poisson equation becomes

$$\epsilon_{i+\frac{1}{2}}\phi_{i+1}^{t+1} + \epsilon_{i-\frac{1}{2}}\phi_{i-1}^{t+1} - \left(\epsilon_{i+\frac{1}{2}} + \epsilon_{i-\frac{1}{2}}\right)\phi_i^{t+1} = p_i^t - n_i^t \tag{4.11}$$

where the superscripts represent time (either the current time, t, or the next time step, $t+1$). The subscripts represent the discrete locations in space (we solve for the electric potential at discrete locations, i's). Note that the dielectric constants are required at locations in between the discrete locations where the electric potential is solved. (In other words, the dielectric constant is required at the intermediate positions $i - \frac{1}{2}$ and $i + \frac{1}{2}$.) The purpose of the scheme is to obtain a set of linear equations where the unknown quantities are the electric potentials at different locations at the next time step. In this way, by solving the system of simultaneous equations we can obtain the electric potentials at time $t+1$ from information at t. The simulation progresses through the iterative solution of such simultaneous equations.

The equations for electron and hole concentrations are of the form

$$\begin{aligned} \frac{n_i^{t+1} - n_i^t}{\Delta t} &= G_i - R_i + \mu_{i+\frac{1}{2}}^n B\left(\phi_{i+1}^{t+1} - \phi_i^{t+1}\right) n_{i+1}^{t+1} \\ &+ \mu_{i-\frac{1}{2}}^n B\left(\phi_{i-1}^{t+1} - \phi_i^{t+1}\right) n_{i-1}^{t+1} \\ &- \left[\mu_{i+\frac{1}{2}}^n B\left(\phi_i^{t+1} - \phi_{i+1}^{t+1}\right) + \mu_{i-\frac{1}{2}}^n B\left(\phi_i^{t+1} - \phi_{i-1}^{t+1}\right)\right] n_i^{t+1} \end{aligned} \tag{4.12}$$

$$\begin{aligned} \frac{p_i^{t+1} - p_i^t}{\Delta t} &= G_i - R_i + \mu_{i+\frac{1}{2}}^p B\left(\phi_i^{t+1} - \phi_{i+1}^{t+1}\right) p_{i+1}^{t+1} \\ &+ \mu_{i-\frac{1}{2}}^p B\left(\phi_i^{t+1} - \phi_{i-1}^{t+1}\right) p_{i-1}^{t+1} \\ &- \left[\mu_{i+\frac{1}{2}}^p B\left(\phi_{i+1}^{t+1} - \phi_i^{t+1}\right) + \mu_{i-\frac{1}{2}}^p B\left(\phi_{i-1}^{t+1} - \phi_i^{t+1}\right)\right] p_i^{t+1} \end{aligned} \tag{4.13}$$

where it is assumed that the electric potential has just been solved at time $t+1$ and, therefore, the Bernoulli functions above are known quantities. Again, we have two sets of simultaneous equations which can be solved to obtain the charge concentrations at time $t+1$ from the electric potentials and the charge concentrations at time t.

In other words, the above linear set of equations can be solved iteratively, at each time step, to evolve the electric potential and charge carrier concentrations. At each time step, using information from the previous time step, we can update our values of the electric potential and charge carrier concentrations. Given the initial conditions (and appropriate boundary conditions) we can evolve this system of equations and capture how the solar cell device responds over time to a given input (for example, being initially exposed to sunlight). However, the system of equations above is computationally expensive to solve in a spreadsheet. Typically, the equations would be run over many iterations until the system reached a steady state (and neither the electric potential nor the charge carrier concentrations appreciably change over time). While this is a trivial system to solve in more powerful computer languages, the large number of iterations required to reach steady state makes this impractical to attempt in a spreadsheet.

We will not attempt to solve the above equations here. Instead, we will turn our attention to using the drift-diffusion model to capture the device performance of polymer solar cells. While polymer solar cells have many aspects to their operations which makes them more complicated than traditional inorganic solar cells, they are typically much thinner and the electric field inside a solar cell can be approximated by a constant. (This may also be true of other thin film solar cells.) Therefore, the drift diffusion equations simplify significantly and will allow us to capture the physics of polymer solar cells using spreadsheet models.

4.3 Drift-Diffusion Model: Polymer Solar Cells

Organic materials, such as polymers, have emerged as an attractive alternative to the more traditional inorganic semiconductor solar cells. One of the main advantages of using polymers is their high optical absorption which results in very small solar cell thicknesses, on the order of only 100 nm; this is a thousand times thinner than silicon-based solar cells and on the order of ten times thinner than inorganic thin film cells. The use of polymers can also significantly reduce the costs of producing the solar cells through solution processing and continuous deposition techniques. In other words, a smaller amount of these cheaper polymers need to be used and the devices can be manufactured using inexpensive low-temperature roll-to-roll printing processes. The cost of producing polymer solar cells, therefore, is significantly less than inorganic cells. While this is, perhaps, their main allure it is certainly not their only benefit. For example, polymer solar cells can be produced very easily on flexible substrates opening up the possibility of manufacturing photovoltaic fabrics.

The problem with polymer solar cells is that they are terribly inefficient. Polymer solar cells can convert solar energy into electrical energy with maximum efficiencies of around ten percent (although this is increasing rapidly as new materials are developed). These are the maximum efficiencies currently reported for laboratory devices (rather than mass-produced devices which typically have smaller efficiencies) and is less than inorganic solar cells (which might be between 20 and 30 percent). This means that while the cost of producing polymer solar cells is small, their low efficiencies limit their cost per Watt, a key measure of how economically viable they are.

Polymer solar cells are also a little different than inorganic solar cells in the way in which they work. In polymer solar cells the absorption of a photon does not result in a free electron and hole, as in inorganic devices, but rather the creation of a bound electron hole pair, called an exciton. In other words, the force of attraction between the negative electron and positive hole is greater in polymers than in inorganic semiconductors. This means that the electron stays close to the hole when it is excited to a more energetic state upon absorption of a photon. The excited state, called an exciton, consists of the electron bound to the hole and can wander around in the polymer semiconductor. In order to get electricity from these devices, however, the electron and hole must be separated (and separated before the electron combines with the hole and the absorbed energy from the photon is potentially lost).

To separate the electron from the hole, the exciton must then find an interface between two different materials with different electron donating and electron accepting properties. In other words, it is more energetically favorable for the electron to be in one material than the other. The act of separating the electron from the hole is called dissociation. The exciton is, therefore, dissociated into a free electron and hole, on either side of the interface between the two materials (with a free electron in one material and a hole in the other). These charges will then diffuse from regions of high concentration at the interface, and drift as a result of an electric field arising from the difference between the electrode work functions, until the charges reach their respective electrodes. Free charges which successfully reach the electrodes can be extracted and contribute to the electrical energy produced by the device. This process is depicted in Figure 4.5, where a photon is absorbed by the polymer material creating an exciton (bound electron and hole). The exciton

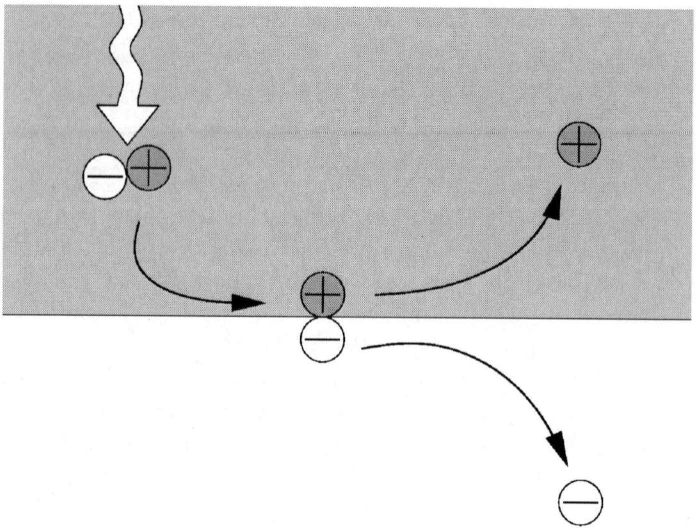

FIGURE 4.5
Illustration of how a photon enters the polymer material and is absorbed, resulting in a mobile excited state known as an exciton. The exciton then diffuses to the donor-acceptor interface inside the solar cell where the exciton is dissociated (separated) into a free electron and a free hole. The free charge carriers then move to the appropriate electrodes and the energy is extracted from the device.

diffuses to the interface between two different polymer materials which help dissociate the exciton into a free electron in one material and a hole in the other material.

We now turn our attention to solving the drift-diffusion equations for these polymer solar cells. While the way in which these devices operate can make the models a little more complicated, it is possible to make assumptions that allow the equations to be solved analytically.

Let's consider the above drift-diffusion equations for when the system is in steady state; in other words, when the system is not evolving over time and we can imagine that the term involving the derivative with respect to time can be set equal to zero. Furthermore, we can assume that the electric field is approximately constant. This results in the following equations:

$$-\frac{k_\beta T}{q}\mu_p\frac{\partial^2 p}{\partial x^2} + E\mu_p\frac{\partial p}{\partial x} + D - R = 0 \tag{4.14}$$

$$-\frac{k_\beta T}{q}\mu_n\frac{\partial^2 n}{\partial x^2} - E\mu_n\frac{\partial n}{\partial x} + D - R = 0 \tag{4.15}$$

$$-\frac{k_\beta T}{q}\mu_e\frac{\partial^2 e}{\partial x^2}+G-R_e-D=0 \tag{4.16}$$

where k_β is the Boltzmann constant, T is temperature, q is elementary charge, μ_p is the mobility of the holes, μ_n is the mobility of the electrons, and μ_e is the mobility of the excitons. p, n and e are the local concentrations of holes, electrons, and excitons, respectively. x is distance, E is the electric field, D is the dissociation of excitons, G is the photogeneration of excitons, R_e is the recombination of excitons, and R is the recombination of electrons with holes.

We could solve these equations numerically, and use a finite difference approximation to approximate the derivatives in distance. This would result in a set of linear equations which we would have to solve (an equation for each concentration and location in discrete space). However, we can also solve these equations analytically. The charge carriers can be created via the dissociation (separation of electron and hole) of excitons and this can be described by Onsagers theory for electrolyte dissociation and given by

$$D=q\frac{4}{\pi^{\frac{1}{4}}a_o^3}\int_0^\infty\frac{q<\mu>}{<\epsilon>}e^{-\frac{\Delta E}{k_\beta T}}\left[1+b+\frac{b^2}{3}+\cdots\right]a^2e^{\frac{-a^2}{a_o^2}}\,da \tag{4.17}$$

where $b=q^3E/(8\pi<\epsilon>k_\beta^2T^2)$. ΔE is the binding energy (the energy of interaction between the hole and electron that are bound within the exciton), and a_0 is a characteristic length of 1 nm. The recombination rate can be written as

$$R=\frac{q(\mu_n+\mu_p)pn}{\epsilon} \tag{4.18}$$

and is obviously proportional to the concentration of electrons and holes. The simplest expression for the photogeneration is the Beer-Lambert law.

$$G=G_oe^{-\alpha x} \tag{4.19}$$

where G_o is dependent on the amount of incident light, and α is the optical absorption coefficient of the polymer material. In reality, both the amount of incident light and the optical absorption coefficient would be dependent on the frequency of light. Furthermore, we might have to take into consideration more complicated optical effects as the light interacts with the different materials within the solar cell and reflects off the underlying substrate. We will consider this in more detail later, when we model the interactions of light with solar cell devices. The recombination, or decay, of the exciton (the electron fills the hole) is given by

$$R_e=\frac{e}{\tau_e} \tag{4.20}$$

which takes into consideration the average time that an exciton wanders around before it decays, $\tau_e=1\mu s$. In reality, even if the exciton decays it could result in the emission of a photon which might likely be reabsorbed by the surrounding polymer material. Therefore, we can ignore exciton decay which can be small in polymer solar cells anyway as the the different charge carriers are separated into different polymer phases. Furthermore, the equations can be further simplified if we assume a proportion of photogenerated excitons simply dissociate and consider the photogenerated excitons to be uniformly distributed throughout the solar cell. Therefore, with this assumption we do not need to solve for the spatial distribution of excitons. This results in the following equations

$$-\frac{k_\beta T}{q}\mu_p\frac{\partial^2 p}{\partial x^2}+E\mu_p\frac{\partial p}{\partial x}=PG \tag{4.21}$$

and

$$-\frac{k_\beta T}{q}\mu_n\frac{\partial^2 n}{\partial x^2}-E\mu_n\frac{\partial n}{\partial x}=PG \tag{4.22}$$

where G is a uniform photogeneration of excitons and P is the proportion which dissociate into free electrons and holes. These ordinary differential equations can be solved easily. If you are not familiar with ordinary differential equations there are number of symbolic mathematics software that can solve these systems. (For example, Maxima is an open source software that can easily solve these equations.) The solution to these two equations is

$$p(x)=B+C\exp\left(\frac{qEx}{k_\beta T}\right)+\frac{PGx}{\mu_p E} \tag{4.23}$$

and

$$n(x)=\acute{B}+\acute{C}\exp\left(\frac{-qEx}{k_\beta T}\right)-\frac{PGx}{\mu_n E} \tag{4.24}$$

where the constants (B, C, $\acute{B}$ and $\acute{C}$) can be obtained from the boundary conditions. The simplest boundary condition is to assume the charge carriers are zero at the electrodes. Under these boundary conditions it can be shown that $B=-C$ and

$$C=\frac{PGL}{\mu_p E\left[1-\exp\left(\frac{qEL}{k_\beta T}\right)\right]} \tag{4.25}$$

for the hole concentrations. For the electron concentration equation, $\acute{B}=-\acute{C}$ and

$$\acute{C}=\frac{-PGL}{\mu_n E\left[1-\exp\left(\frac{-qEL}{k_\beta T}\right)\right]} \tag{4.26}$$

where $x=0$ and $x=L$ are the boundaries of the solar cell. In other words, in the above equations for hole and electron concentrations, we put the equations equal to zero when $x=0$ and $x=L$. These four equations allow us determine the four constants, B, C, $\acute{B}$, and $\acute{C}$. These boundary conditions are overly simplistic, but we'll implement more complicated boundary conditions in a more sophisticated model in a moment.

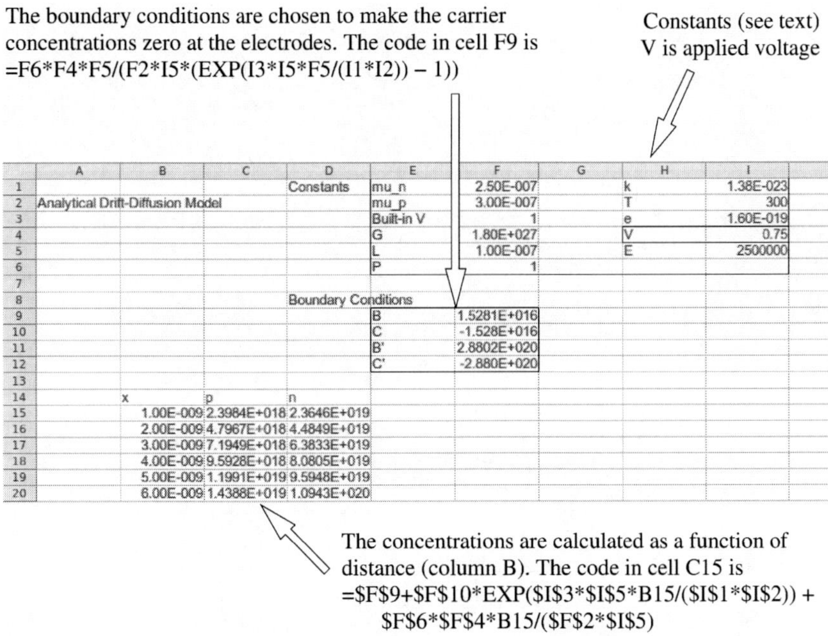

	A	B	C	D	E	F	G	H	I	J
1				Constants	mu_n	2.50E-007		k	1.38E-023	
2	Analytical Drift-Diffusion Model				mu_p	3.00E-007		T	300	
3					Built-in V	1		e	1.60E-019	
4					G	1.80E+027		V	0.75	
5					L	1.00E-007		E	2500000	
6					P	1				
7										
8				Boundary Conditions						
9					B	1.5281E+016				
10					C	-1.528E+016				
11					B'	2.8802E+020				
12					C'	-2.880E+020				
13										
14		x	p	n						
15		1.00E-009	2.3984E+018	2.3646E+019						
16		2.00E-009	4.7967E+018	4.4849E+019						
17		3.00E-009	7.1949E+018	6.3833E+019						
18		4.00E-009	9.5928E+018	8.0805E+019						
19		5.00E-009	1.1991E+019	9.5948E+019						
20		6.00E-009	1.4388E+019	1.0943E+020						

FIGURE 4.6
Spreadsheet implementation of an analytical drift-diffusion model for simulating a polymer solar cell.

The spreadsheet implementation for the above model is depicted in Figure 4.6. At the top of the spreadsheet are the constants. In particular, at the top of the spreadsheet we assign values for the mobility of the charge carriers, the built-in voltage (depending the work functions of the electrodes), the photogeneration of excitons, the proportion which dissociate, and the length of the solar cell (taken to be 100 nm). V is the applied voltage and E is the uniform electric field. In particular, the code in cell I5 is

```
=-(I4-F3)/F5
```

The first column in the spreadsheet, column B, contains the x values, increasing in increments of 1 nm up to the length of the solar cell, 100 nm. The next two columns, columns C and D, contain the calculation of the hole and electron concentrations, respectively. In particular, the code in cell C15 is

```
=$F$9+$F$10*EXP($I$3*$I$5*B15/($I$1*$I$2))

      +$F$6*$F$4*B15/($F$2*$I$5)
```

which is simply the analytical solution for the hole concentration from above.

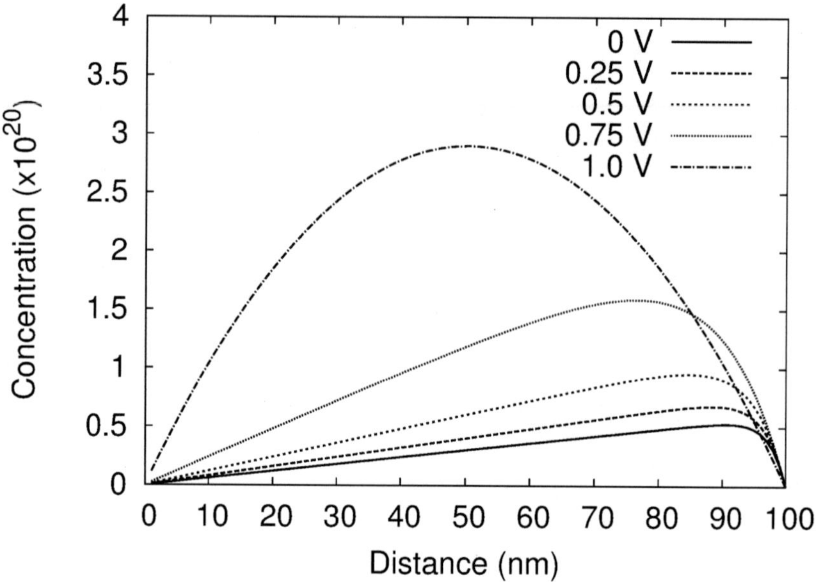

FIGURE 4.7
Concentration of electrons as a function of distance for different values of applied voltage.

The solution to the equations above is depicted in Figures 4.7 and 4.8. In particular, Figure 4.7 depicts the concentration of electrons as a function of distance for different values of applied voltage, while Figure 4.8 depicts the concentration of holes. For the case when the applied voltage is 1 V, the applied voltage opposes the built-in voltage and the overall electric field in the system is zero. This is the voltage you would have to apply to reduce the current from the solar cell to zero. In this case, the concentration of electrons and holes are largest, and symmetric, because there is no electric field in which the charge carriers will drift. As the applied voltage is reduced, the charges drift more towards the electrodes and the charge concentrations are reduced as charge makes its way out of the solar cell (and around an external circuit). With an applied voltage of zero, the internal electric field is entirely due to the built in voltage and this represents the case when the system would be operating in short circuit.

The above model is a wonderful example of how, through making approximations, we can simplify a model sufficiently to solve it analytically. The question, as always, is how accurate is the model given these assumptions? Let's consider an analytical model without as many assumptions.

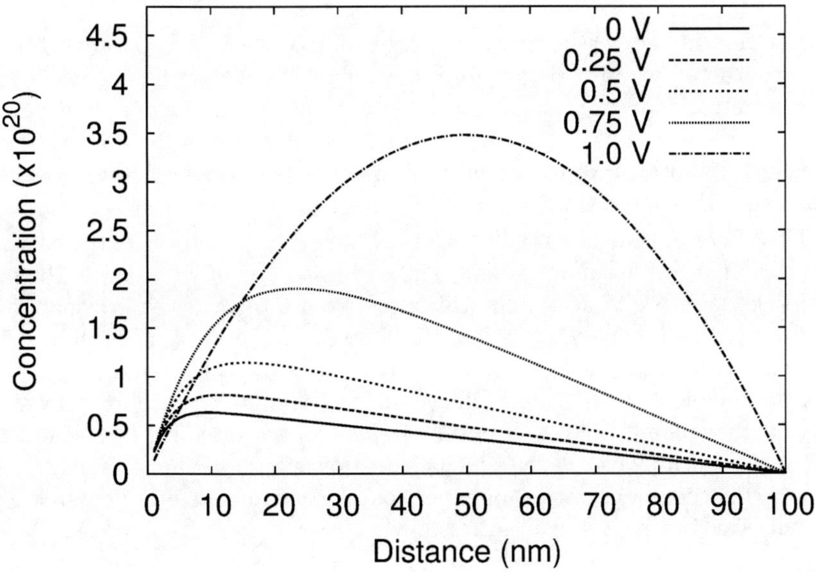

FIGURE 4.8
Concentration of holes as a function of distance for different values of applied voltage.

4.4 Drift-Diffusion Model: Nonuniform Exciton Dissociation

In polymer solar cells the excitons dissociate at the donor-acceptor interface. The donor-acceptor interface is not always uniformly distributed throughout the polymer solar cell (as was assumed in the previous model). For example consider the charge carrier concentrations, in Figure 4.9, which can be numerically obtained in a polymer solar cell consisting of a bilayer device. That is, a device which consists of a single layer of donor material and a single layer of accepter material simply sandwiched between the electrodes. The only donor acceptor interface, at which the excitons can dissociate into free electrons and holes, exists in the very center of the thin film device. The photogenerated excitons, therefore, only dissociate in the center of the device and this is why the concentration of excitons dips suddenly in the center of the polymer solar cell. The free electrons and holes which arise from the dissociation of the excitons are mainly within the acceptor and donor materials, respectively. This means that the charges will drift and diffuse towards the correct electrodes.

This is not optimum, as the light in polymer solar cells is usually absorbed over a distance of around 100 nm. Therefore, it is required to have polymer solar cells that are at least this thickness to ensure that the incident light is absorbed by the polymer solar cell. However, the excitons usually only diffuse over distances of 10 nm before the charges recombine and the exciton decays. Therefore, in a simple bilayer device as considered above, most of the absorbed light will not result in free electrons or holes in the device. This makes such bilayer devices inefficient. However, it demonstrates how the exciton concentrations within polymer solar cells can have quite complicated spatially varying exciton concentrations.

Let us consider an analytical model that has spatially varying exciton concentrations, but is still one-dimensional. This analytical model will still require assumptions, to enable the drift-diffusion equations to be solvable. Let us start with the basic drift-diffusion equations with uniform electric field and assuming that the recombination of electrons and holes can be neglected. These assumptions are required to simplify the drift-diffusion equations to solvable ordinary differential equations and separate the equations for the charge concentrations so that they can be solved separately. The equations we wish to solve are

$$-\frac{k_\beta T}{q}\mu_e\frac{\partial^2 e}{\partial x^2}+G(x)-\frac{e}{\tau_e}-De=0 \tag{4.27}$$

$$-\frac{k_\beta T}{q}\mu_p\frac{\partial^2 p}{\partial x^2}+E\mu_p\frac{\partial p}{\partial x}+De=0 \tag{4.28}$$

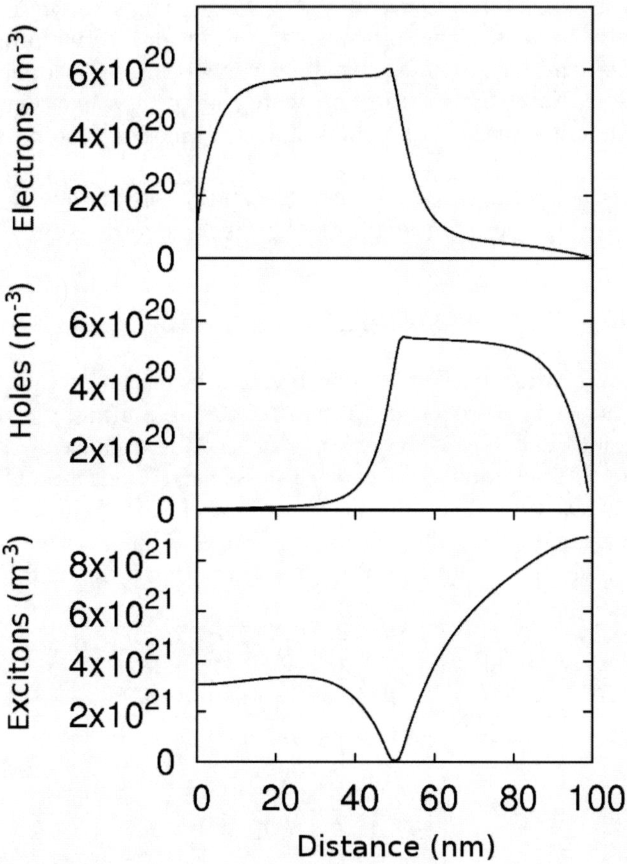

FIGURE 4.9
Concentration of electrons, holes, and excitons in a bilayer polymer photovoltaic device.

$$-\frac{k_\beta T}{q}\mu_n\frac{\partial^2 n}{\partial x^2} - E\mu_n\frac{\partial n}{\partial x} + De = 0 \tag{4.29}$$

where, because once the electric field is assumed to be constant, we can write the dissociation of the excitons as being a constant multiplied by the concentration of excitons; a constant which can be obtained from equation 4.17. Furthermore, the exciton decay is simply proportional to the exciton concentration. The drift diffusion equation for excitons, therefore, consists of a second derivative with respect to space (for diffusion), terms proportional to exciton concentration, e, and a spatially varying term, the photogeneration of excitons, $G(x)$. Note that treating the photogeneration of excitons as spatially varying is different than the previous analytical model. Also note that there is no drift term in the exciton equation, as excitons are charge neutral and do not drift in the electric field. The photogeneration of excitons, $G(x)$ can also be written to take into consideration the different intensity and absorption properties of light with different frequencies.

$$G(x) = \sum_i \Phi_i\alpha_i \exp(-\phi_i x) \tag{4.30}$$

where the summation is over different frequencies of light, Φ_i is the intensity of sunlight at a given frequency (usually obtainable in discrete form, hence the summation) and $-\phi_i$ is the polymer absorption coefficient (also depends on frequency). We can solve for the exciton concentrations first. Once we have solved for the exciton concentrations we can use this, in the dissociation term, when solving for the spatially varying electron and hole concentrations.

The general solution for the full exciton equation is of the form

$$\begin{aligned} e &= \sum_i \Phi_i\alpha_i \exp(-\alpha_i x)\frac{\tau_e q}{q(1+D\tau_e) - \tau_e k_\beta T\mu_e\alpha_i^2} \\ &+ K_1 \exp\left[x\left\{\frac{q(1+D\tau_e)}{\tau_e k_\beta T\mu_e}\right\}^{\frac{1}{2}}\right] \\ &+ K_2 \exp\left[-x\left\{\frac{q(1+D\tau_e)}{\tau_e k_\beta T\mu_e}\right\}^{\frac{1}{2}}\right] \end{aligned} \tag{4.31}$$

where the constants K_1 and K_2 are to be determined by the boundary conditions. In particular, if we put $e = 0$ when $x = 0$ and $x = L$, we can obtain the following two equations

$$K_1 = \sum_i \Phi_i\alpha_i \frac{\tau_e q}{\tau_e k_\beta T\mu_e\alpha_i^2 - q(1+D\tau_e)} - K_2 \tag{4.32}$$

and

$$K_2 = \frac{\sum_i \frac{\Phi_i\alpha_i\tau_e q}{q(1+D\tau_e)-\tau_e k_\beta T\mu_e\alpha_i^2}\left(\exp(-\alpha_i L) - \exp\left[L\left\{\frac{q(1+D\tau_e)}{\tau_e k_\beta T\mu_e}\right\}^{\frac{1}{2}}\right]\right)}{\exp\left[-L\left\{\frac{q(1+D\tau_e)}{\tau_e k_\beta T\mu_e}\right\}^{\frac{1}{2}}\right] - \exp\left[L\left\{\frac{q(1+D\tau_e)}{\tau_e k_\beta T\mu_e}\right\}^{\frac{1}{2}}\right]} \tag{4.33}$$

The above equations give a complete analytical solution for the exciton concentrations in a polymer solar cell. Note that the exciton dissociation only depends on the exciton concentrations, and any spatial distribution of donor-acceptor interfaces are not taken into consideration. In the ideal polymer solar cell, the donor-acceptor interface would be spread out over the thickness of the solar cell and the electrons and holes would have a direct path to the electrodes. Under these conditions, the above assumptions are reasonable.

The general solution for the hole drift-diffusion equation (incorporating the exciton solution directly into the dissociation term) is of the form

$$\begin{aligned} p &= \frac{D}{\mu_p}\sum_i \frac{q\exp(-\alpha_i x)}{(\tau_e k_\beta T\mu_e\alpha_i^2 - q(1+D\tau_e))(\alpha_i k_\beta T + Eq)} \\ &+ \frac{DK_1}{\mu_p}\left[\frac{\exp\left(x\left\{\frac{q(1+D\tau_e)}{\tau_e k_\beta T\mu_e}\right\}^{\frac{1}{2}}\right)}{\frac{1+D\tau_e}{\tau_e\mu_e} - \left\{\frac{q(1+D\tau_e)}{\tau_e k_\beta T\mu_e}\right\}^{\frac{1}{2}} E}\right] \\ &+ \frac{DK_2}{\mu_p}\left[\frac{\exp\left(-x\left\{\frac{q(1+D\tau_e)}{\tau_e k_\beta T\mu_e}\right\}^{\frac{1}{2}}\right)}{\frac{1+D\tau_e}{\tau_e\mu_e} + \left\{\frac{q(1+D\tau_e)}{\tau_e k_\beta T\mu_e}\right\}^{\frac{1}{2}} E}\right] \\ &+ K_3\exp\left(\frac{xEq}{k_\beta T}\right) + K_4 \end{aligned} \tag{4.34}$$

where the additional constants K_3 and K_4 can be obtained from the boundary conditions.

Rather than simply putting the concentrations to zero (which is not correct) we can capture the physics of the metal-semiconductor interfaces at the electrodes using the appropriate Schottky boundary conditions (see the paper by Scott and Malliaras for more details, Chem. Phys. Lett. 299 115, 1999). Rather than setting the concentrations to a particular value at the boundary, the Schottky boundary conditions set the flux of the charge carrier concentrations at the boundaries. In particular, the flux of holes can be written in the form

$$J_p = qE\mu_p p - k_\beta T\mu_p\frac{dp}{dx} \tag{4.35}$$

which we can set to the Schottky boundary condition fluxes. At $x = 0$ the boundary condition is

$$J_p = B\mu_p\left[P_0\exp\left(\frac{-\Phi^{AD}}{k_\beta T}\right)\exp(\sqrt{f}) - 0.25p\left(\frac{1}{\phi^2} - f\right)\right] \tag{4.36}$$

and at $x = L$ the boundary condition is

$$J_p = B\mu_p\left[-P_0\exp\left(-\frac{\Phi^{CD} + 0.25qEr_c}{k_\beta T}\right) - p\right] \tag{4.37}$$

where $B = 16pi^2\epsilon k_\beta^2 T^2/q^2$, P_0 is the density of chargeable sites in the donor, Φ^{AD} and Φ^{AD} is the difference between the anode and cathode, respectively, and the ionization potential of the donor material. $f = qEr_c/k_\beta T$ and $\phi = f^{-1} + f^{-1/2} - f^{-1}\sqrt{1+2\sqrt{f}}$. The distance r_c is given by $q^2/4\pi\epsilon k_\beta T$. Setting the flux at the boundary of the polymer solar cell equal to the theoretical Schottky boundary conditions allows us to solve for the two constants, K_3 and K_4. However, inserting the above solution for the hole concentrations into the above boundary conditions and solving at $x = 0$ and $x = L$ to find these constants can get a little messy. The final solution for these two constants is of the form

$$K_3 = -\frac{C_1 + C_4 + K_4 C_3}{C_2} \tag{4.38}$$

and

$$K_4 = \frac{C_6(C_1 + C_4) - C_5 C_2 - C_8 C_2}{C_2 C_7 - C_3 C_6} \tag{4.39}$$

where first C constant is given by

$$\begin{aligned} C_1 &= \frac{D}{\mu_p} \sum_i \frac{q^2 E\mu_p + k_\beta T\mu_p\alpha_i q + q\left(\frac{1}{4\phi^2} - \frac{f}{4}\right)}{(\tau_e k_\beta T\mu_e\alpha_i^2 - q(1 + D\tau_e))(\alpha_i k_\beta T + Eq)} \\ &+ \left[\frac{D}{\mu_p} \frac{\left(\frac{1+D\tau_e}{\tau_e\mu_e}\right)(K_1 + K_2) + \left(E\left\{\frac{q(1+D\tau_e)}{\tau_e k_\beta T\mu_e}\right\}^{\frac{1}{2}}\right)(K_1 - K_2)}{\left[\frac{1+D\tau_e}{\tau_e\mu_e}\right]^2 - \left[\frac{q(1+D\tau_e)}{\tau_e k_\beta T\mu_e}\right]}\right] \\ &\times \left(qE\mu_p + \left(\frac{1}{4\phi^2} - \frac{f}{4}\right)\right) \end{aligned} \tag{4.40}$$

the second constant by

$$C_2 = qE\mu_p - \frac{Eq}{k_\beta T} + \left(\frac{1}{4\phi^2} - \frac{f}{4}\right) \tag{4.41}$$

the third constant by

$$C_3 = qE\mu_p + \left(\frac{1}{4\phi^2} - \frac{f}{4}\right) \tag{4.42}$$

the fourth constant by

$$\begin{aligned} C_4 &= -\frac{Dq}{\mu_p k_\beta T} \frac{\left\{\frac{q(1+D\tau_e)}{\tau_e k_\beta T\mu_e}\right\}^{\frac{1}{2}}(K_1 - K_2) + \frac{Eq}{k_\beta T}(K_1 + K_2}{\frac{q(1+D\tau_e)}{\tau_e k_\beta T\mu_e} - \left(\frac{Eq}{k_\beta T}\right)^2} \\ &- B\mu_p\left(P_0 \exp\left(\frac{-\Phi^{AD}}{k_\beta T}\right)\exp(\sqrt{f})\right) \end{aligned} \tag{4.43}$$

The fifth constant is a doozy and given by

$$\begin{aligned}
C_5 &= \frac{D}{\mu_p}\sum_i \frac{(qE\mu_p - B\mu_p)q\exp(-\alpha_i L) + k_\beta T\mu_p\alpha_i q\exp(-\alpha_i L)}{(\tau_e k_\beta T\mu_e\alpha_i^2 - q(1+D\tau_e))(\alpha_i k_\beta T + Eq)} \\
&+ K_1\left[D(qE-B)\left(\frac{\exp\left(L\left\{\frac{q(1+D\tau_e)}{\tau_e k_\beta T\mu_e}\right\}^{\frac{1}{2}}\right)}{\frac{1+D\tau_e}{\tau_e\mu_e} - E\left\{\frac{q(1+D\tau_e)}{\tau_e k_\beta T\mu_e}\right\}^{\frac{1}{2}}}\right)\right. \\
&- \left. Dk_\beta T\left(\frac{\frac{q}{k_\beta T}\exp\left(L\left\{\frac{q(1+D\tau_e)}{\tau_e k_\beta T\mu_e}\right\}^{\frac{1}{2}}\right)}{\left\{\frac{q(1+D\tau_e)}{\tau_e k_\beta T\mu_e}\right\}^{\frac{1}{2}} - \frac{Eq}{k_\beta T}}\right)\right] \\
&+ K_2\left[D(qE-B)\left(\frac{\exp\left(-L\left\{\frac{q(1+D\tau_e)}{\tau_e k_\beta T\mu_e}\right\}^{\frac{1}{2}}\right)}{\frac{1+D\tau_e}{\tau_e\mu_e} + E\left\{\frac{q(1+D\tau_e)}{\tau_e k_\beta T\mu_e}\right\}^{\frac{1}{2}}}\right)\right. \\
&+ \left. Dk_\beta T\left(\frac{\frac{q}{k_\beta T}\exp\left(-L\left\{\frac{q(1+D\tau_e)}{\tau_e k_\beta T\mu_e}\right\}^{\frac{1}{2}}\right)}{\left\{\frac{q(1+D\tau_e)}{\tau_e k_\beta T\mu_e}\right\}^{\frac{1}{2}} - \frac{Eq}{k_\beta T}}\right)\right] \qquad (4.44)
\end{aligned}$$

The sixth constant is given by

$$C_6 = (qE\mu_p - B\mu_p - \frac{Eq}{k_\beta T})\exp\left(\frac{LEq}{k_\beta T}\right) \qquad (4.45)$$

The seventh constant is given by

$$C_7 = qE\mu_p - B\mu_p \qquad (4.46)$$

And last, but not least, the eighth constant is given by

$$C_8 = B\mu_p P_0\exp\left(-\frac{\Phi^{CD} + 0.25qEr_c}{k_\beta T}\right) \qquad (4.47)$$

The above set of equations allows us to obtain an analytical solution for the hole concentrations in the polymer solar cell. The electron concentrations can be obtained in a similar way. However, solving this analytically has gotten very complicated (and this still assumes electric field is constant). Solving this numerically, however, is very easy, if a little beyond the capabilities of a spreadsheet model.

4.5 Monte Carlo Model of Photovoltaics

Monte Carlo models are stochastic in nature. Typically, a Monte Carlo (abbreviated MC) model will randomly sample the phase space of a system (for example, by allowing the constituents in a system to randomly relocate). As the phase space is randomly sampled, a new configuration is either accepted or rejected based on a stochastic scheme (called the Metropolis algorithm) which takes into consideration the energy difference between the two states. In other words, a component of a system might be relocated to a new position and the difference in energy between the old and new configurations is calculated. If the energy of the new configuration is less than the old configuration then the move is accepted. However, if the energy of the new configuration is more than the old configuration then the move is subject to a random criteria for acceptance, with the move being more likely to be rejected for larger energy differences. In this way the energy of the system is minimized over time; such a scheme will, therefore, optimize the system and converge toward equilibrium. Monte Carlo models of photovoltaics are typically applied to organic solar cells and are dynamic models. A dynamic Monte Carlo model is similar to a more traditional Monte Carlo model, but the movement of constituents is limited to local moves.

Dynamic Monte Carlo models of polymer solar cells progress through successive rate-dependent events in the system, such as charge hopping or exciton dissociation. The different events which could occur within the simulation are depicted in Figure 4.10. In particular, excitons can be photogenerated, excitons can diffuse, excitons can decay, excitons can dissociate, charges can move, charges can recombine and charges can be extracted from the device.

In other words, individual electrons, holes, and excitons are updated on a regular lattice depending upon the rates at which different events are likely to occur. We assign these rates within the model and evolve the system by successively updating the electrons, holes, and excitons in the system. This technique has recently emerged as a promising, and quite popular, method for capturing the physics of polymer solar cells. The dynamical Monte Carlo model of solar cells is a microscopic approach and it is important to maintain a proper temporal sequence of events. At each time step, therefore, the rates at which different events are likely to occur have to be identified so that the most likely events are preferentially selected.

An exciton can hop from its current site i to a neighboring site j with a rate

$$w_{ij}^{e} = w_o \left(\frac{R_0}{R_{ij}} \right) f(E_i - E_j) \tag{4.48}$$

where R_{ij} is the distance between sites i and j, R_0 is the exciton localization radius, and w_e is the rate at which hopping is attempted. The function $f(E_i -$

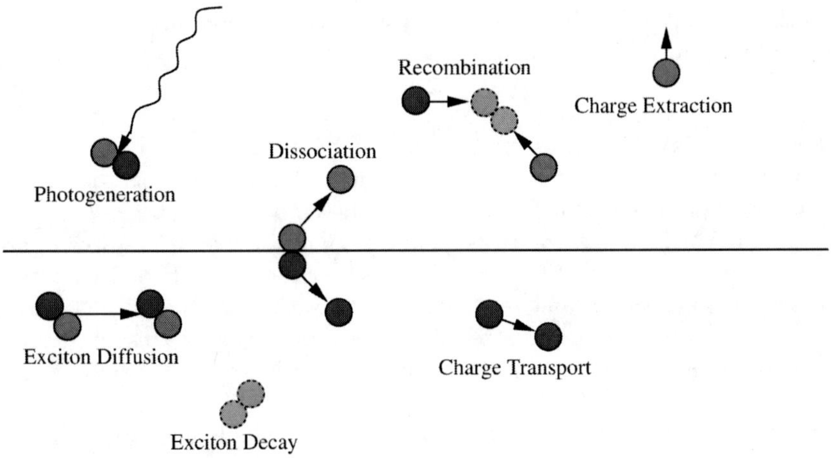

FIGURE 4.10
A schematic showing the different possible events which can occur within a dynamic Monte Carlo (MC) model of polymer solar cells.

E_j) can be given by

$$f(E_i - E_j) = \begin{cases} \exp\left(-\frac{E_j - E_i}{kT}\right) & E_j > E_i \\ a & E_j < E_i \end{cases} \tag{4.49}$$

where E_i and E_j are the energies of occupation of sites i and j, respectively. An exciton may also recombine at a rate of w_{cr}, or dissociate at a DA interface with a rate of w_{diss}. Charge carriers may also hop with a rate, in the case of an electron, of

$$w_{ij}^n = w_n \exp(-2\gamma R_{ij}) f(E_i - E_j) \tag{4.50}$$

where w_n is the rate at which hopping is attempted for an electron. γ is a localization constant which can be related to the charge mobility. Similarly for a hole, the rate of hopping from site i to j is

$$w_{ij}^p = w_p \exp(-2\gamma R_{ij}) f(E_i - E_j) \tag{4.51}$$

where w_p is the rate at which hopping is attempted for a hole. Oppositely charged particles on adjacent sites may typically recombine at a rate of w_{cr} (although this effect can be assumed to be small for polymer solar cells). The charge can also leave the system if the charge is next to the appropriate electrode, at a rate of w_{ce}. This extraction of charge is the current that is produced by the photovoltaic device.

The dynamical MC method for polymer solar cells is typically evolved in time using the first reaction method (FRM). The first reaction method evolves by calculating a waiting time for each event, from the rates described above.

In other words, for each rate in the system a corresponding waiting time is calculated from

$$\tau = -\frac{1}{w} \ln\left(\epsilon[0:1]\right) \tag{4.52}$$

where $\epsilon[0:1]$ is a random number uniformly distributed between 0 and 1. The event with the smallest τ for a given particle is then chosen as the event for that particle which is added to a list of events. Only one event for each particle is added to the list. The events list contains all of these events, stored in order of increasing τ. At each time step the event at the start of the list, with the smallest τ, is executed and removed from the list. An amount of time τ is, therefore, assumed to have passed during the execution of this event and all other events in the list have their waiting time reduced by this value. All new events are slotted into the list according to their waiting times, maintaining the correct temporal sequence. This algorithm is continually evolved until steady state is obtained. In particular, when there is a balance between charges being created from the exciton dissociation events and charge being removed from the system (either recombining with other charge or being extracted via the electrodes) then the system is considered to be in a steady state.

The first reaction method is not the only method to update the events in the dynamic Monte Carlo method, however, and here we consider a different method. The rates at which events are estimated to occur can be placed in an array, or series. The total rates in the system is defined as

$$W = \sum_{i}^{N} w_i \tag{4.53}$$

where w_i is the rate of the ith event and N is the total number of events. Rather than assigning a time to each event, we can assume that one of the events will occur during a time step and that this time step is given by

$$\Delta t = \frac{1}{\sum_i^N w_i} \tag{4.54}$$

An event must be chosen to occur in this time interval, taking into consideration the rates associated with the different events. The event chosen is the jth event in the sequence for which

$$\epsilon[0:1] > \frac{\sum_i^j w_i}{\sum_i^N w_i} \tag{4.55}$$

In other words, we take the total of all the rates and use this to normalize the summation $\sum_i^j w_i$. Therefore, the summation $\sum_i^j w_i / \sum_i^N w_i$ goes from zero to one as j goes from zero to N. However, the random number $\epsilon[0:1]$ is used to select which event possesses the rate which, as the rates are summed, takes the tally over the random number. In this manner, the likelihood of an event being selected is proportional to its rate. An alternative way of visualizing

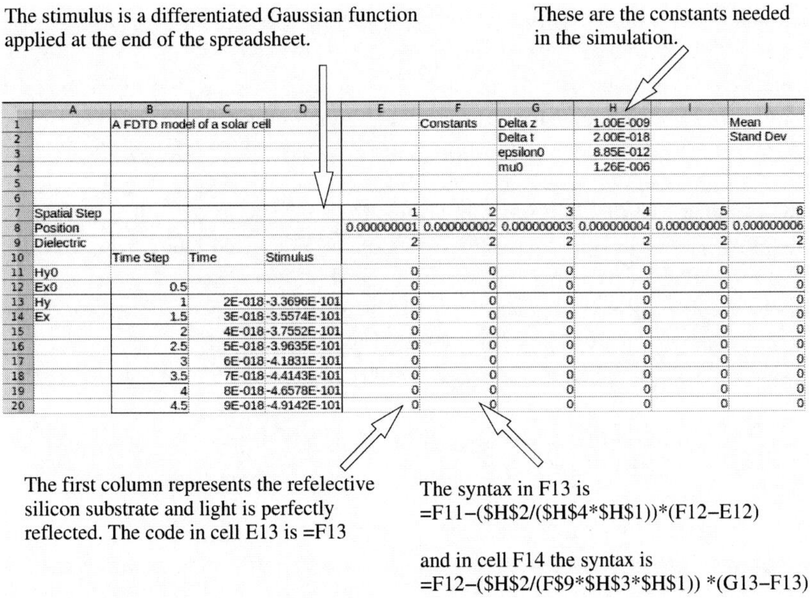

	A	B	C	D	E	F	G	H	I	J
1		A FDTD model of a solar cell				Constants	Delta z	1.00E-009		Mean
2							Delta t	2.00E-018		Stand Dev
3							epsilon0	8.85E-012		
4							mu0	1.26E-006		
5										
6										
7	Spatial Step				1	2	3	4	5	6
8	Position				0.000000001	0.000000002	0.000000003	0.000000004	0.000000005	0.000000006
9	Dielectric				2	2	2	2	2	2
10		Time Step	Time	Stimulus						
11	Hy0				0	0	0	0	0	0
12	Ex0	0.5			0	0	0	0	0	0
13	Hy	1	2E-018	-3.3696E-101	0	0	0	0	0	0
14	Ex	1.5	3E-018	-3.5574E-101	0	0	0	0	0	0
15		2	4E-018	-3.7552E-101	0	0	0	0	0	0
16		2.5	5E-018	-3.9635E-101	0	0	0	0	0	0
17		3	6E-018	-4.1831E-101	0	0	0	0	0	0
18		3.5	7E-018	-4.4143E-101	0	0	0	0	0	0
19		4	8E-018	-4.6578E-101	0	0	0	0	0	0
20		4.5	9E-018	-4.9142E-101	0	0	0	0	0	0

FIGURE 4.11
A plan of how a dynamic Monte Carlo (MC) model of a polymer solar cell might be implemented in a spreadsheet.

this method is to imagine stacking the rates on top of each other, as if the rates were boxes and the height of the boxes depended on their rates. Next a height is selected at random and the box at this height is chosen. As you look at the stack of boxes, the boxes with the greater heights cover more of the total height of the stack of boxes, and are therefore more likely to be selected. As new events are added to the system, their rates are simply added to the series and the simulation progresses while maintaining the correct temporal sequence of events.

Figure 4.11 depicts a possible layout for a spreadsheet implementation of the above dynamic Monte Carlo model. The constants required in the simulation are stored at the top of the spreadsheet (labeled number 1). Below the constants is a list of the excitons, electrons, and holes which store the positions of the particles and calculates the different rates associated with these particles (labeled number 2). It is from these rates that an event will be selected. In particular, the rates are listed (area labeled number 3) and an event is randomly selected, where the probability of the event being selected is proportional to the rate at which the event is designated to occur. An integral part of determining these rates is the occupancy of neighboring sites. In other words, when calculating the rate at which a particle might hop to a given neighboring location, we must check that the location is unoccupied. If the location is occupied, then the rate at which particle might hop into this location is put to zero to prevent overlap in particle positions. A list of sites along with their occupancy is therefore maintained (area labeled number 4). Once an event is selected, the updated positions are calculated (area labeled number 5). If a particle does not exist, then its position in the x-direction is set equal to -1. An iteration consists of copying the updated positions and pasting them into a new area (labeled number 6), sorting the particles in order of their positions (such that the particles that do not exist, with a position of -1, are sorted to the bottom), and copying the new list of positions to the top of the simulation (area labeled number 2). The rates are, of course, recalculated given the updated positions, and the simulation progresses an iteration with the next iteration's positions being calculated.

Now let's turn our attention to the actual spreadsheet implementation of the Monte Carlo model of a polymer solar cell, and put the above plane into operation. The spreadsheet used to implement this model is shown in Figures 4.12 to 4.19. At the very top of the spreadsheet, shown in Figure 4.12, the constants used in the model are entered. In particular, the size of the system (here 20×20 system size is considered), the electric field (which is assumed to be constant), and the various rate constants are required. In rows 10 through 110 the positions and rates for the excitons, electrons, and holes are stored. Notice that in the current model the number of these particles is capped at 100 (although this can easily be changed by extending this range). In column A the number of the exciton is stored (a simple index to keep track of the particles). In columns B and C the x and y positions of these excitons are stored. To incorporate the fact that not all 100 particles will be used in the system (the system can be started with no particles, and excitons photogenerated over time) particles that don't exist are given an x-position of -1. This is simply a flag to reveal that this particle is not used, or does not exist. In column D is an index to the site (which is calculated from the x- and y-positions) of the form $I = x + y \times NX$, where x and y are the x- and y- cordinates in columns B and C, and NX is the number of sites in the x-direction (here, it's 20 because the system size is 20×20). In other words, the code in cell D11 is

```
=IF(B11>-1, B11+$C$3*C11, -1)
```

If the x-position is equal to -1 then the particle does not exist and so the value in cell D11 should be -1. However, if the x-position is greater than -1 then calculate the index I using the x- and y-coordinates and the system size in the x-direction (stored in cell C3). Next, in columns E through J are the rates at which different events can occur (a rate of zero indicates that the event will never occur). Let's look at column E, which is the rate at which the excitons might try to move up. The code in cell E11 is

```
=IF(B11=-1, 0, IF(C11>$C$4-2, 0, IF((D11 +$C$3 < 0)
OR(D11 +$C$3 > 399), 0, IF((INDEX( $AL$11:$AL$410,
(D11 +$C$3+1)) = 0), $F$4, 0) ) ))
```

The first part states that if the exciton does not exist, then the rate at which it moves should be zero (because it does not exist). If it does exist then we move on to the next part of the code

```
IF(C11>$C$4-2, 0, IF((D11 +$C$3 < 0)OR
(D11 +$C$3 > 399), 0, IF((INDEX( $AL$11:$AL$410,
(D11 +$C$3+1)) = 0), $F$4, 0) ) )
```

This part of the code simply states that if the exciton is at the edge of the simulation, `C11>$C$4-2` asks if the y-position is greater than the last grid point minus 1. It is worth noting that if the system size is 20 then the last point is 19 (range from 0 to 19) and if the y-position was greater then 18 then it would not be possible for the exciton to move up any further. The remaining part of the code is

```
IF((D11 +$C$3 < 0)OR(D11 +$C$3 > 399), 0,
IF((INDEX( $AL$11:$AL$410, (D11 +$C$3+1)) = 0), $F$4, 0) )
```

The next step is to add the number of x grid points to the index (as you move up 1 space in the y-direction the index increase by 20 in the current system size) and check that the new index is within the bounds. If the new index would be less than zero or greater than 399 then the move would not be valid. The previous part of the code, however, makes sure that the exciton does not leave the system's boundaries, and so this is a redundant check. The last part of the code is

```
IF((INDEX( $AL$11:$AL$410, (D11 +$C$3+1)) = 0), $F$4, 0)
```

and this is perhaps the most complicated section. If the site that the exciton is potentially moving up into is already occupied by a particle then the rate is zero, and the move will never occur. This is achieved by indexing the sites in cells `$AL$11:$AL$410` (this will be explained in more detail when we discuss Figure 4.15). In other words, within the spreadsheet is an index of the sites which is equal to zero if the site is empty. If the cell is empty, `INDEX( $AL$11:$AL$410, (D11 +$C$3+1)) = 0)`, then make the rate equal to the rate of an exciton hopping from one site to the next. Note that the index starts off with one being the first cell, whereas our index runs from 0 to 399; therefore we look up the cell which is the index of the site the exciton is moving into plus one. The code in columns F, G and H is very similar but modified to account for exciton movement in the down, left and right directions, respectively.

Column I contains the rates at which the excitons will decay (the coulombically bound electron and hole recombine). The code in cell I11 is simply

```
=IF($B11=-1,0,$F$7)
```

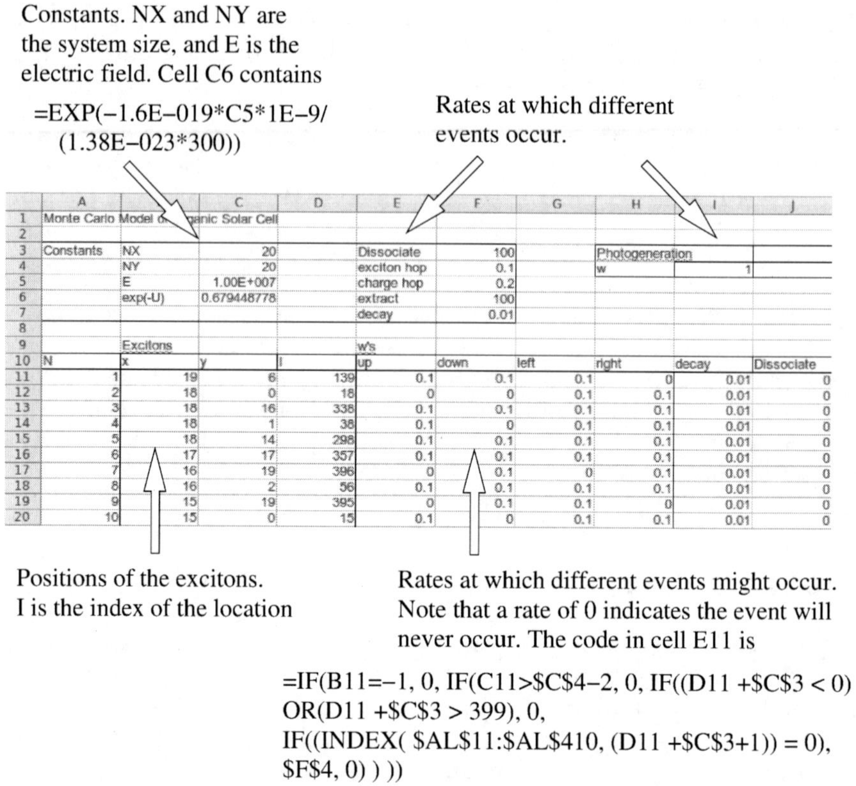

	A	B	C	D	E	F	G	H	I	J
1	Monte Carlo Model	o	ganic Solar Cell							
2										
3	Constants	NX	20		Dissociate	100		Photogeneration		
4		NY	20		exciton hop	0.1		w	1	
5		E	1.00E+007		charge hop	0.2				
6		exp(-U)	0.679448778		extract	100				
7					decay	0.01				
8										
9		Excitons			w's					
10	N	x	y	I	up	down	left	right	decay	Dissociate
11	1	19	6	139	0.1	0.1	0.1	0	0.01	0
12	2	18	0	18	0	0	0.1	0.1	0.01	0
13	3	18	16	338	0.1	0.1	0.1	0.1	0.01	0
14	4	18	1	38	0.1	0	0.1	0.1	0.01	0
15	5	18	14	298	0.1	0.1	0.1	0.1	0.01	0
16	6	17	17	357	0.1	0.1	0.1	0.1	0.01	0
17	7	16	19	396	0	0.1	0	0.1	0.01	0
18	8	16	2	56	0.1	0.1	0.1	0.1	0.01	0
19	9	15	19	395	0	0.1	0.1	0	0.01	0
20	10	15	0	15	0.1	0	0.1	0.1	0.01	0

FIGURE 4.12
Spreadsheet implementation of a dynamic Monte Carlo (MC) model of a polymer solar cell. Constants, exciton positions, and rates.

If the particle does not exist then the rate of decay is zero, whereas if it does exist then it is assigned a value (from cell F7). Finally, column J contains the rate of dissociation and the code in cell J11 is

```
=IF($B11=-1,0,IF((($C11=$C$4/2) OR ($C11=$C$4/2-1)),$F$3, 0))
```

If the exciton does not exist, and the x-position is equal to -1, then the rate of dissociation is zero. If it does exist, then if the y-position is either side of the donor-acceptor interface, the rate of exciton dissociation is given by the constant in cell F3. In the current model, the donor-acceptor interface is through the center of the system (y-positions equal to the system size in the y-direction divided by two). The rate of dissociation for excitons at the donor-acceptor interface is large to ensure this event is preferentially selected.

Figure 4.13 shows the next section along the spreadsheet to the right of what is shown in Figure 4.12. At the top, in cells M3 to M5, are a tally of the number of excitons, the number of electrons, and the number of holes currently present in the simulation. In particular, column AM contains a record of which sites contain an exciton (with cells containing a value of 1 if the site is occupied or 0 if it is unoccupied by an exciton). We'll come back to how this column is calculated in a moment (when we discuss Figure 4.16 which depicts this section of the spreadsheet). The code in cell M3 is simply

```
=SUM(AM11:AM410)
```

which sum over the sites that contain a 1 if an exciton exists, to obtain the total number of excitons present. Cells M4 and M5 calculate the number of electrons and holes in a similar manner. Also shown at the top of Figure 4.13 are two buttons. The first button runs a macro that constitutes an iteration in the simulation, while the second button runs the same macro 100 times and performs 100 iterations. Recall that an iteration, which consists of running a macro upon pushing the bottom, in the spreadsheet copies position from the bottom of the spreadsheet up to the top of the spreadsheet (see Figure 4.11 for the plan of how to capture this model within the spreadsheet). The position and rates for the electrons are also shown in Figure 4.13. This is similar to how the positions and the rates associated with the excitons are stored and calculated, respectively. The code in cell N11, which calculates the rate at which an electron might hop upwards, is

```
=IF(K11=-1,0, IF(L11>$C$4-2, 0,  IF((M11 +$C$3 < 0)OR
  (M11 +$C$3 > 399), 0, IF((INDEX( $AL$11:$AL$410,
            (M11 + $C$3+1)) = 0),
  IF(L11<$C$4/2, $C$6*$F$5, $C$6*0.1*$F$5), 0))))
```

Let's look at this code, layer by layer. The first part simply checks if the particle exists (if the x-position in cell K11 is equal to -1 then it does not). If it does exist then we are left with the code

```
IF(L11>$C$4-2, 0,  IF((M11 +$C$3 < 0)OR
(M11 +$C$3 > 399), 0, IF((INDEX( $AL$11:$AL$410,
          (M11 + $C$3+1)) = 0),
IF(L11<$C$4/2, $C$6*$F$5, $C$6*0.1*$F$5), 0))
```

If the y-position is too large, then it is not physically possible for the electron to move up (because no such site exists) and the rate is set to zero. If the y-position exists then we check that the new location exists (again a slightly redundant check). We are, therefore, left with the remaining code

```
IF((INDEX( $AL$11:$AL$410,
```

```
(M11 + $C$3+1)) = 0),

IF(L11<$C$4/2, $C$6*$F$5, $C$6*0.1*$F$5), 0)
```

Now we check to make sure that the site the electron is looking to potentially move into is unoccupied. This is why we keep a record of which sites are occupied in column AL. Finally, the rate of hopping depends on the domain in which the electron finds itself (although it should be in the acceptor where it has the higher mobility) and also the rate of hopping will depend on the electric field. Here we assume a constant electric field. In other words, the rate of hopping is given by the remaining code from cell N11

```
IF(L11<$C$4/2, $C$6*$F$5, $C$6*0.1*$F$5)
```

If the y-position is less than the midway point (system size divided by two, which is where the donor-acceptor interface is assumed to be) then the electron is in the acceptor material, else we multiply the rate by 0.1 to decrease the mobility. The mobility is given by `$C$6*$F$5` which is the rate at which charge hop (in cell F5) multiplied by a factor which accounts for the energy in moving against the electric field (in cell C6).

Figure 4.14 depicts the part of the spreadsheet which calculates the rates from the events associated with the holes. The rates are calculated for the holes in a manner very similar to that of the electrons. To the right of Figure 4.14 is the list of all events. This list identifies the event with three numbers. The first number is the number of the particle (a number between 1 and 100) for the potential 100 excitons, electrons or holes. Note that a number of 0 represents the photogeneration of an exciton, an event not associated with any of the particles. The second number, in column AC, indicates the type of particle for which the event corresponds. Again the number 0 is reserved for photogeneration, 1 is for an event associated with an exciton, 2 is for an event associated with an electron, and 3 is for an event associated with a hole. The third number indicates the type of event. So for all of the different particles, the numbers 1 through 4 indicate hopping in the four lattice directions (up, down, left, and right). This is expanded upon further in Figure 4.15.

In column AE is a list of all the rates corresponding to the three numbers in rows AB, AC, and AD. These are simply linked to the rates calculated for the excitons, electrons, and holes (depicted in Figures 4.12, 4.13, and 4.14). The cell AE8 sums over all these rates to get the total rate of something happening in the system

```
=SUM(AE11:AE1611)
```

Cell AF8 is simply a random number (`=RAND()`) which we can use to determine the event to be chosen. In particular, the column AF contains the cumulative sum of the rates from column AE, normalized to the total rate. For example, the code in cell AF11 is

```
=SUM($AE$11:AE11)/$AE$8
```

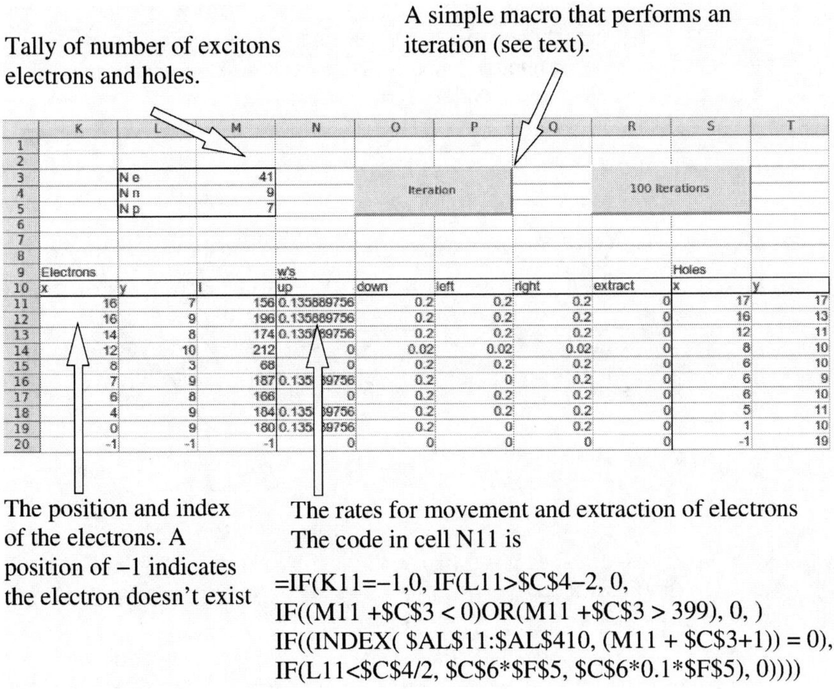

	K	L	M	N	O	P	Q	R	S	T
1										
2										
3		Ne	41							
4		Nn	9							
5		Np	7							
6										
7										
8										
9	Electrons			w's					Holes	
10	x	y	l	up	down	left	right	extract	x	y
11	16	7	156	0.135889756	0.2	0.2	0.2	0	17	17
12	16	9	196	0.135889756	0.2	0.2	0.2	0	16	13
13	14	8	174	0.135[illegible]89756	0.2	0.2	0.2	0	12	11
14	12	10	212	0	0.02	0.02	0.02	0	8	10
15	8	3	68	0	0.2	0.2	0.2	0	6	10
16	7	9	187	0.135[illegible]9756	0.2	0	0.2	0	6	9
17	6	8	166	0	0.2	0.2	0.2	0	6	10
18	4	9	184	0.135[illegible]9756	0.2	0.2	0.2	0	5	11
19	0	9	180	0.135[illegible]9756	0.2	0	0.2	0	1	10
20	-1	-1	-1	0	0	0	0	0	-1	19

FIGURE 4.13
Spreadsheet implementation of a dynamic Monte Carlo (MC) model of a polymer solar cell. Buttons, electron positions and rates, and hole positions.

N is the number of the particle, n1 is the type of particle (0 for photogeneration, 1 for exciton, 2 for electron and 3 for hole), and n2 is the action.

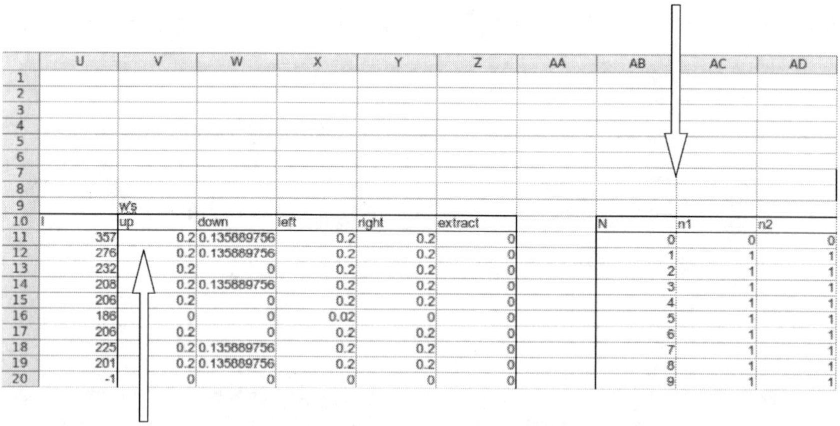

The rates for various potential events. The code in cell V11 is
=IF(S11=–1, 0, IF(T11 > $C$4–2, 0, IF((U11 + $C$3 < 0)OR(U11 + $C$3 > 399)
, 0, IF((INDEX($AL$11:$AL$410, (U11 + $C$3+1)) = 0),
IF(T11>$C$4/2–1, $F$5, 0.1*$F$5), 0))))

FIGURE 4.14
Spreadsheet implementation of a dynamic Monte Carlo (MC) model of a polymer solar cell. Rates of holes.

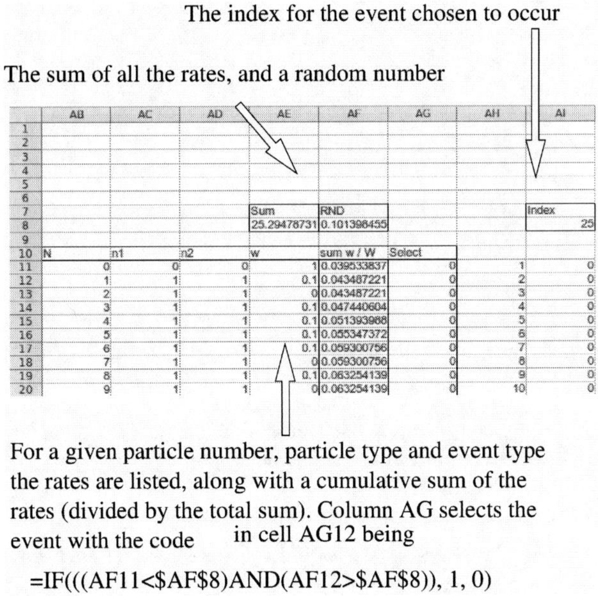

FIGURE 4.15
Spreadsheet implementation of a dynamic Monte Carlo (MC) model of a polymer solar cell. Rates are listed for all events in the system, and based on these rates an event is selected.

In other words, the sum of the rates from the first rate through to the current rate (as we go down the list in column AE) divided by the total sum. The values in this column, therefore, increase from 0 to 1 as we go down the column and the increments occur when there are events with finite rates (with the increment proportional to the rate). In column AG one of the events is selected based on the random number in cell AF8. The code in cell AG12 is

```
=IF(((AF11<$AF$8)AND(AF12>$AF$8)), 1, 0)
```

If the simulation of an event in the cumulative list in column AF11 is such that it steps over the random number, then that is the event chosen to occur. Because these summations are proportional to the rates at which events are expected to occur, the events with higher rates are proportionately more likely to be selected.

An important aspect in determining the rates is to ensure single site occupancy, and prohibit two particles from occupying the same site. Therefore, a lookup table of site occupancy is required. Figure 4.16 shows the section of the spreadsheet where this is stored. Column AK is a list of the sites, with an index ranging from 0 to 399, corresponding to the 20×20 system size. Column AL checks to see if the site index in column AK is occupied and returns zero

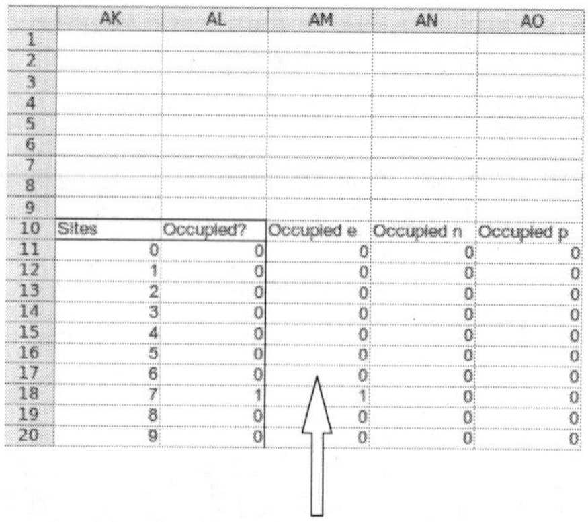

	AK	AL	AM	AN	AO
1					
2					
3					
4					
5					
6					
7					
8					
9					
10	Sites	Occupied?	Occupied e	Occupied n	Occupied p
11	0	0	0	0	0
12	1	0	0	0	0
13	2	0	0	0	0
14	3	0	0	0	0
15	4	0	0	0	0
16	5	0	0	0	0
17	6	0	0	0	0
18	7	1	1	0	0
19	8	0	0	0	0
20	9	0	0	0	0

FIGURE 4.16
Spreadsheet implementation of a dynamic Monte Carlo (MC) model of a polymer solar cell. Check if site is occupied.

if it is unoccupied, 1 if occupied by an exciton, 2 if occupied by an electron, and 4 if occupied by a hole. The code in cell AL11 is

```
=IF(ISNA(MATCH(AK11,$D$11:$D$110,0)),0,1) +

IF(ISNA(MATCH(AK11,$M$11:$M$110,0)),0,2)  +

 IF(ISNA(MATCH(AK11,$U$11:$U$110,0)),0,4)
```

The first part compares the index in cell AK11 with the indices of the exciton particles (in column D from D11 to D110). In particular, the function `MATCH` searches for a value in an array and returns the relative position of that item. However, if the function `MATCH` does not find the value then it returns `#N/A`. The function `ISNA` will return `TRUE` if the function `MATCH` returns `#N/A`, otherwise, it will return `FALSE`. In other words, the code checks if nothing was found and returns a zero if nothing was found, or a 1 if something was found. Similarly, the second and third parts of the code check to see if electrons or holes are present, and return either a 2 or a 4, respectively. Therefore, we can use column AL to check site occupancy, and prevent particle overlap, and columns AM, AN, and AO are used to tally the number of excitons, electrons, and holes in the system.

The number of the particle, the type of particle, and the type of event selected are stored as three numbers in cells N118, O118, and P118 (see Figure 4.18). In Figures 4.17 and 4.18 we show how these values are used to update the positions of excitons, electrons, and holes. The numbers associated with the particles are in column A (with values running from 1 to 100 corresponding to the potentially 100 excitons, electrons, or holes in the system). Columns B and C (between rows 118 and 217) calculate the new x- and y-positions of the excitons (assuming there is any change). In particular, the code in cell B118 is

```
=IF(($N$118=A118)AND($O$118=1)AND($P$118>4), -1,
IF(($N$118=A118)AND($O$118=1), (IF($P$118=3, B11-1,
(IF($P$118=4, B11+1, B11)))), B11 ))
```

The first part of this code check to see if the number of the particle and the type of the particle are the same (`($N$118=A118)AND($O$118=1)`). If they are the same and the type of event is greater than 4 (which can only mean that the event is either exciton decay or exciton dissociation) then the exciton no longer exists and the x-position is set equal to -1 (which indicates that the exciton no longer exists). Else if the particle number and the number indicating the particle type both match up with that of the selected event, but the event number is either 3 or 4, then the position of the particle should be moved either to the left (`B11-1`) or the right (`B11+1`). If none of the above conditions apply, then the position stays the same and the value of B11 is kept. The code for the electrons and holes are similar, and the updated positions represent the changes due to the performed event.

Of course, sometimes particles are created, and this is done in the last cell of the array. For example, in cell B217 is the code

```
=IF($N$118=0, INT(RAND()*$C$3), -1)
```

which states that if an exciton is to be photogenerated (indicated by the particle number being zero) then create a new particle with a random x-position. Likewise, the y-position is similarly created. Electrons and holes are dissociated from excitons which have reached the donor-acceptor interface and in this event we create both a new electron and a new hole. The code in cell F217 is

```
=IF(($O$118 = 1)AND($P$118 = 6),
INDEX($B$11:$B$110, $N$118), -1)
```

and in cell G217

```
=IF(($O$118 = 1)AND($P$118 = 6), $C$4/2-1, -1)
```

Depending on the values of N, n1 and n2 (in cells N118, O118 and P118) update positions. The code in cell F118 is

```
=IF(($N$118=A118)AND($O$118=2)AND($P$118>4), –1,
IF(($N$118=A118)AND($O$118=2), (IF($P$118=3, K11–1,
(IF($P$118=4, K11+1, K11)))), K11 ))
```

	A	B	C	D	E	F	G	H
113								
114	Updated Positions							
115								
116	Excitons					Electrons		
117	N	x	y	l		x	y	l
118	1	19	6	139		16	7	156
119	2	18	0	18		16	9	196
120	3	18	16	338		14	8	174
121	4	18	1	38		12	10	212
122	5	18	14	298		8	3	68
123	6	17	17	357		7	9	187
124	7	16	19	396		6	8	166
125	8	16	2	56		4	9	184
126	9	15	19	395		0	9	180
127	10	15	0	15		-1	-1	-1
128	11	13	19	393		-1	-1	-1
129	12	12	6	132		-1	-1	-1
130	13	11	5	111		-1	-1	-1
131	14	11	12	251		-1	-1	-1
132	15	11	15	311		-1	-1	-1

The code in cell B118 is

```
=IF(($N$118=A118)AND($O$118=1)AND($P$118>4), –1,
IF(($N$118=A118)AND($O$118=1), (IF($P$118=3, B11–1,
(IF($P$118=4, B11+1, B11)))), B11 ))
```

FIGURE 4.17
Spreadsheet implementation of a dynamic Monte Carlo (MC) model of a polymer solar cell. Exciton and electron positions are updated.

If an exciton (indicated by cell `$O$118 = 1`) is dissociated (indicated by cell `$P$118 = 6`) then create a new electron whose x-coordinate is the same as that of the exciton and whose y-position is at the donor-acceptor interface. The position of the dissociating exciton is obtained using the function `INDEX` and the donor-acceptor (DA) interface is assumed to run through the center of the simulation.

Finally, the iteration in the simulation consists of copying the updated position of the excitons, electrons, and holes to a new space (rows 225 to 324) in the spreadsheet. Next the excitons, electrons, and holes are sorted in order of decreasing x-position such that particles that do not exist (with x-positions of -1) will be moved to the bottom of the pile and newly created particles will move up. Then the updated (and sorted) positions are copied to the original positions at the top of the simulation. Now the calculation of rates and the selection of an event which is to occur next will all be based on the new particle positions. In this manner, the simulation has evolved an iteration.

The above copying, sorting, and pasting is done only once and recorded in a macro. In other words, you can start recording your macro, then perform the copying, sorting, and pasting once, before stopping the macro from recording any more. Now the macro contains these commands, and running the macro will perform all of the necessary steps to iterate the simulation.

An example of the simulation at work is depicted in Figure 4.20. A single

The cells N118, O118 and P118 are indexed to pick out the randomly selected event. The code in N118 is

=INDEX(AB11:AB1513,$AI$8)

	J	K	L	M	N	O	P
113							
114							
115							
116	Holes				Selected		
117	x	y	i		N	N1 (type)	N2 (action)
118	17	17	357		24	1	1
119	16	13	276				
120	12	11	232				
121	8	10	208				
122	6	10	206				
123	6	9	186				
124	6	10	206				
125	5	11	225				
126	1	10	201				
127	-1	19	-1				
128	-1	-1	-1				
129	-1	-1	-1				
130	-1	-1	-1				
131	-1	-1	-1				
132	-1	-1	-1				
133	-1	-1	-1				

The holes can also be moved (or created and extracted) and the code in cell J118 is

=IF(($N$118=A118)AND($O$118=3)AND($P$118>4), −1,
IF(($N$118=A118)AND($O$118=3), (IF($P$118=3, S11−1,
(IF($P$118=4, S11+1, S11)))), S11))

FIGURE 4.18
Spreadsheet implementation of a dynamic Monte Carlo (MC) model of a polymer solar cell. Hole positions are updated. The randomly selected event is shown.

	A	B	C	D	E	F	G	H	I	J	K	L
220	Copy positions as values to below, sort and move back to the top.											
221												
222												
223	Excitons					Electrons				Holes		
224	N	x	y	i		x	y	i		x	y	i
225	1	19	6	139		16	7	156		17	17	357
226	2	18	0	18		16	9	196		16	13	276
227	3	18	16	338		14	8	174		12	11	232
228	4	18	1	38		12	10	212		8	10	208
229	5	18	14	298		8	3	68		6	10	206
230	6	17	17	357		7	9	187		6	9	186
231	7	16	19	396		6	8	166		6	10	206
232	8	16	2	56		4	9	184		5	11	225
233	9	15	19	395		0	9	180		1	10	201
234	10	15	0	15		-1	-1	-1		-1	19	-1
235	11	13	19	393		-1	-1	-1		-1	-1	-1
236	12	12	6	132		-1	-1	-1		-1	-1	-1
237	13	11	5	111		-1	-1	-1		-1	-1	-1
238	14	11	12	251		-1	-1	-1		-1	-1	-1
239	15	11	15	311		-1	-1	-1		-1	-1	-1
240	16	11	14	291		-1	-1	-1		-1	-1	-1

The final step is to copy the updated positions from above to an area below. Then the positions are sorted in descending order. This puts the particles with an x position of −1 at the bottom, as these particles no longer exist. Then these updated positions are copied to the very top so that the process can repeat again. This constitutes an iteration in the model and the button at the top runs a macro which performs these operations.

FIGURE 4.19
Spreadsheet implementation of a dynamic Monte Carlo (MC) model of a polymer solar cell. Updated exciton, electron and hole locations are sorted before being copied and pasted to the top of the spreadsheet (using a macro).

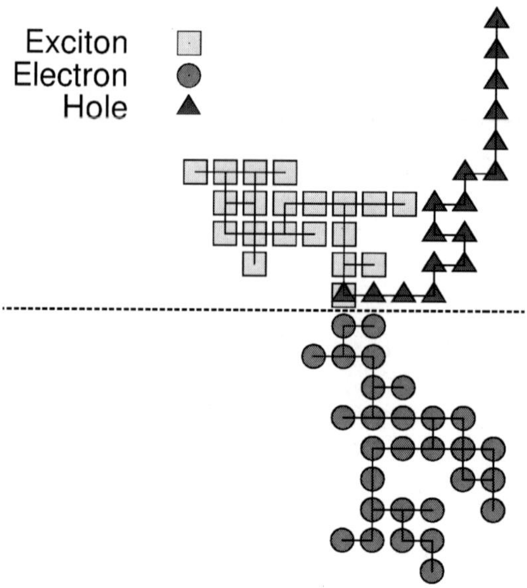

FIGURE 4.20
A simulation in which a single electron diffuses to the Donor-Acceptor (DA) interface, dissociates into free charge carriers, and the free charge carriers drift to the appropriate electrodes.

exciton is placed in the system and, in the current simulation, the rate of photogeneration is put to zero so that we can just focus on the behavior of one particle. The exciton diffuses around randomly (the squares) until it randomly finds the interface. Note that if there were many excitons in the system then the concentration near the donor-acceptor (DA) interface would be reduced as they dissociate into free electrons and holes. This would make it easier for the excitons to diffuse towards the interface. However, for now we just focus on one particle which randomly moves around until it either decays or finds the donor-acceptor (DA) interface and dissociates. Upon dissociation, an electron (the circles) and a hole (the triangles) are created which preferentially move slightly more towards the cathode and anode, respectively. The path of the particles is quite random, with the particles hopping over the same sites multiple times. Therefore, larger system sizes can get computationally more expensive as the particles can spend a long time revisiting similar territory without doing anything particularly purposeful.

Eventually, however, the electron and hole reach the electrodes and electricity is extracted from the device. Typically a simulation can be quite large and consist of many charges moving around. However, it is customary to think of the excitons and charges in such a simulation as representatives of the many more particles that might exist in a real system. Rather than the simulation capturing every single particle in the device the simulation captures a representative sample, which keeps the model computationally tractable.

4.6 Finite-Difference Time-Domain of Solar Cells Optics

Solar cells capture incoming photons of light and, ideally, convert these photons into free charge that can be extracted as electricity from the device. There are many steps in the process that influence the efficiency of the device, but the first step is how sunlight interacts with the solar cell. Initially, on reaching the top surface of the solar cell some of the sunlight will enter the solar cell, but some might be reflected. Any sunlight that is reflected at this top surface is effectively lost from the solar cell and reduces the efficiency of the device. For this reason antireflective coatings are often added to the top surface of the solar cell. Furthermore, the light will traverse the solar cell, reflect from the metallic substrate, traverse back through the solar cell and, whatever light is not absorbed, will leave the solar cell. Recall, that any light that is not absorbed by the solar cell is lost, and this will reduce the efficiency. An interesting new technique to increase absorption in the solar cell is to effectively trap light inside the solar cell structure. This is achieved using photonic crystals, regular repeating structural units with differing dielectric constants. This can cause light to be redirected along the plane of the solar cell or bounce the light around within its internal structure. The optical properties of solar cells is therefore a complicated area of active research. To introduce this important topic, let's consider the Finite-Difference Time Domain Method.

The FDTD method involves the discretization of Maxwell's equations of electromagnetism. The differential form of Maxwell's equations are

$$\nabla \times \vec{H} = \epsilon \frac{\partial \vec{E}}{\partial t} + \sigma \vec{E} \tag{4.56}$$

and

$$\nabla \times \vec{E} = -\mu \frac{\partial \vec{H}}{\partial t} + \sigma_* \vec{H} \tag{4.57}$$

where $\vec{E}$ is the electric field, $\vec{H}$ is the magnetic field, ϵ and μ are the permittivity and permeability, respectively, and σ and σ_* are the respective electric and magnetic conductivities. Simplifying this for one-dimensions, we can write down the following equations

$$\frac{\partial H_y}{\partial z} = \epsilon \frac{\partial E_x}{\partial t} \tag{4.58}$$

and

$$\frac{\partial E_x}{\partial z} = -\mu \frac{\partial H_y}{\partial t} \tag{4.59}$$

where the solution is now only in the z-direction, and the electric and magnetic fields are both perpendicular to each other and this direction of propagation. Note that the electric and magnetic conductivities are removed which means that the light will traverse the material without loss (we'll discuss including loss later).

We can now turn our attention to solving this system of equations with the finite-difference time-domian (FDTD) method, which is a flexible, numerical means of analyzing interactions between electromagnetic waves and complex structures. The technique involves approximating the integration of Maxwell's equations in real space by the use of finite differences. Specifically, the FDTD is a time marching algorithm used to solve the above equations at each point on a grid. In particular, the simulation region is usually considered to be divided into "Yee cells", where the electric fields are given at the edges of the Yee cell and the magnetic fields are given at the faces of the Yee cell. However, in our one-dimensional model the electric field is solved at discrete spatial steps and the magnetic field is solved at spatial steps which are in between the spatial steps of the electric field (essentially, adding or subtracting $\frac{1}{2}$ from the spatial discretization for the electric field). The FDTD method is then used to propagate the electric and magnetic fields in time by integrating the discretized equations through a time-stepping (leapfrog) methodology where the magnetic fields are calculated half a time step later than the electric fields. The updated electromagnetic field components are, therefore, only dependent on the values at the previous time step. This is easier to understand if we look at the finite-difference approximation of Maxwell's equations, given by the FDTD method

$$\epsilon \frac{E_z^{t+\frac{1}{2}} - E_z^{t-\frac{1}{2}}}{\Delta t} = \frac{H^t_{z+\frac{1}{2}} - H^t_{z-\frac{1}{2}}}{\Delta z} \tag{4.60}$$

and

$$\mu \frac{H^{t+1}_{z+\frac{1}{2}} - H^t_{z+\frac{1}{2}}}{\Delta t} = -\frac{E^{t+\frac{1}{2}}_{z+1} - E^{t+\frac{1}{2}}_z}{\Delta z} \tag{4.61}$$

where the subscripts on the electromagnetic fields represent the spatial coordinate and the superscripts represent time. Note that the electric fields are staggered from the magnetic fields by half a discrete step size in both space and time. To put this into a spreadsheet, we rearrange for the updated quantities

$$E_z^{t+\frac{1}{2}} = E_z^{t-\frac{1}{2}} + \frac{\Delta t}{\epsilon \Delta z}\left(H^t_{z+\frac{1}{2}} - H^t_{z-\frac{1}{2}}\right) \tag{4.62}$$

and

$$H^{t+1}_{z+\frac{1}{2}} = H^t_{z+\frac{1}{2}} - \frac{\Delta t}{\mu \Delta z}\left(E^{t+\frac{1}{2}}_{z+1} - E^{t+\frac{1}{2}}_z\right) \tag{4.63}$$

where now we have the electric field at $t + \frac{1}{2}$ and the magnetic field at time $t + 1$ in terms of the electromagnetic fields at earlier times. The field is set to zero at the initial time step, and at the next time step, a source is turned on

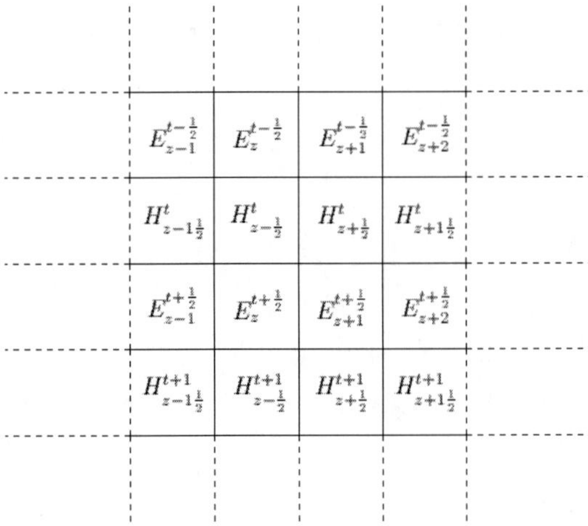

FIGURE 4.21
The storage of electric and magnetic fields in the cells of a spreadsheet model of light propagation.

to generate an optical signal. From this initial condition, the wave fields at all points on the grid at any later time can be calculated through a simple iterative scheme. The implementation of this scheme on a spreadsheet is depicted in Figure 4.21. A section of the spreadsheet grid is represented along with the electromagnetic fields inside the cells in the spreadsheet. Note that as you go down the rows represent different times and alternate between electric fields and magnetic fields. Recall, that the FDTD is a leap-frog algorithm where the electric and magnetic fields are staggered in time by half a time step. Furthermore, in a one-dimensional implementation the electromagnetic fields are also spatially staggered by half of a spatial step. To do this in a spreadsheet, the $z - \frac{1}{2}$ column for the magnetic fields are aligned with the z column for the electric fields.

Let's look at how a spreadsheet for this model might look. Figure 4.22 shows a snapshot of a spreadsheet implementation of the FDTD. There are constants in the top right hand side of the snapshot. In particular, the permittivity and permeability of free space are given, as well as the spatial and temporal time steps. The spatial resolution of the grid is 500 sites, which given the spatial discretization of 1 nm corresponds to a simulation domain of 500 nm. On the far left of the spreadsheet we have a column with the time step (notice it increases by $\frac{1}{2}$ each row as the electric and magnetic fields are updated using the leap-frog approach). Moving horizontally across the simulated domain in the spreadsheet, therefore, shows cells with increasing position and moving down the spreadsheet shows cells at increasingly later times. Also on the left of the spreadsheet we have the actual time (in seconds), and the value for the stimulus that we apply to initiate the incident light. The light that is

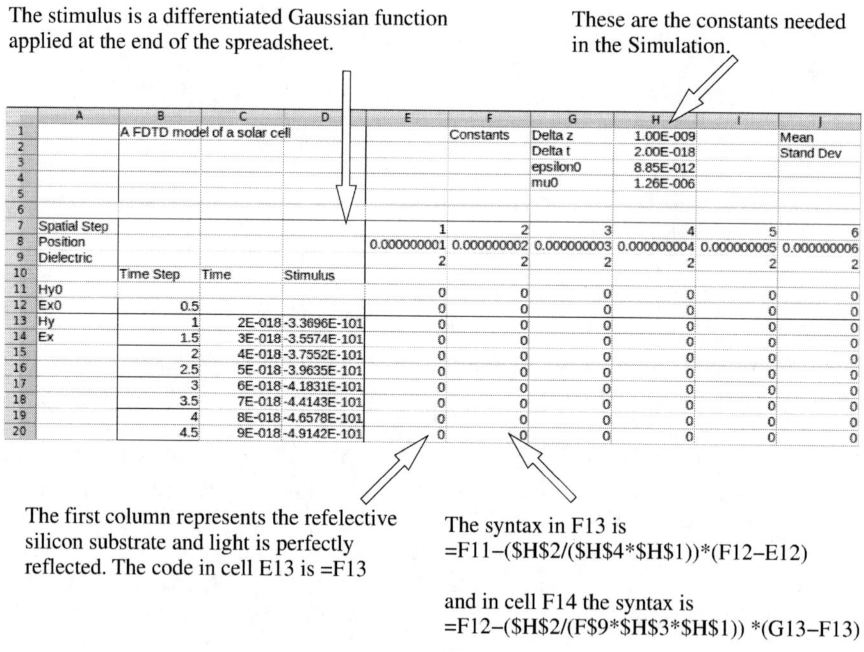

	A	B	C	D	E	F	G	H	I	J
1		A FDTD model of a solar cell				Constants	Delta z	1.00E-009		Mean
2							Delta t	2.00E-018		Stand Dev
3							epsilon0	8.85E-012		
4							mu0	1.26E-006		
5										
6										
7	Spatial Step				1	2	3	4	5	6
8	Position				0.000000001	0.000000002	0.000000003	0.000000004	0.000000005	0.000000006
9	Dielectric				2	2	2	2	2	2
10		Time Step	Time	Stimulus						
11	Hy0				0	0	0	0	0	0
12	Ex0	0.5			0	0	0	0	0	0
13	Hy	1	2E-018	-3.3696E-101	0	0	0	0	0	0
14	Ex	1.5	3E-018	-3.5574E-101	0	0	0	0	0	0
15		2	4E-018	-3.7552E-101	0	0	0	0	0	0
16		2.5	5E-018	-3.9635E-101	0	0	0	0	0	0
17		3	6E-018	-4.1831E-101	0	0	0	0	0	0
18		3.5	7E-018	-4.4143E-101	0	0	0	0	0	0
19		4	8E-018	-4.6578E-101	0	0	0	0	0	0
20		4.5	9E-018	-4.9142E-101	0	0	0	0	0	0

FIGURE 4.22
Spreadsheet implementation of the finite-difference time-domain (FDTD) model.

transmitted towards the solar cell should encompass as many frequencies as possible. For this reason it is common to use a differentiated Gaussian as the input stimulus and the functional form of the electromagnetic radiation incident on the simulated solar cell. The differentiated Gaussian is simply obtained by taking the derivative of a Gaussian function. If we take the fast Fourier transform (FFT) of the differentiated Gaussian then the resultant frequency spectrum encompasses a wide range of frequencies that may be emitted by the light source. (The FFT converts the differentiated Gaussian signal from time-domain where the signal changes as a function of time, to a spectrum showing which frequencies are present in the signal.) This is given in column D, and the formula in the cell D13 is

```
=1E-030*((C13-$K$1)/SQRT(2*PI())) * $K$2 * $K$2 * $K$2

    * EXP(-(((C13-$K$1)*(C13-$K$1))/(2*$K$2*$K$2)))
```

where cell K1 contains the mean for the differentiated Gaussian (5.6×10^{-16}) and cell K2 contains the standard deviation for the differentiated Gaussian (1.0×10^{-16}). While the stimulus is calculated here it is not used here. In particular, the solar cell is considered to be on the left hand side of the simulation domain and the differentiated Gaussian is sent in from the far right hand side.

The code in cell F13 updates the magnetic field using the finite difference equations mentioned earlier

```
=F11-($H$2/($H$4*$H$1))*(F12-E12)
```

Similarly, cell F14 updates the electric field

```
=F12-($H$2/(F$9*$H$3*$H$1)) *(G13-F13)
```

The first cells, in column E, reflect light back into the domain by setting the gradients in electric and magnetic field to zero. To do this the formula in column E simply puts the values equal to those in column F (the slope with respect to position, therefore, is zero). This mimics the metallic or reflective substrates often used at the back of solar cells. This is a good property for solar cells to have as, instead of light having traversed the solar cell leaving, the light is reflected back into the solar cell which makes it more likely to be absorbed and generate electricity.

Along the top of the spreadsheet runs the spatial grid for the electric field (recall the magnetic field is offset from the electric field by half a space step). Below the row containing position (in meters) is the cell containing the relative permittivity. This is multiplied by the permittivity of free space to calculate the permittivity in the solar cell. In air the relative permittivity is equal to 1, but is higher inside the materials used to construct solar cell devices.

The far right hand side of the simulation is depicted in Figure 4.23. Along the top of the spreadsheet is the dimensionless position (in terms of columns),

As before, the top line is the coordinate z, the second line is the actual position and the third line is the dielectric constant.

	SB	SC	SD	SE	SF	SG	SH	SI	SJ	SK
1										
2										
3										
4										
5										
6										
7	492	493	494	495	496	497	498	499	500	
8	0.000000492	0.000000493	0.000000494	0.000000495	0.000000496	0.000000497	0.000000498	0.000000499	0.0000005	
9	1	1	1	1	1	1	1	1	1	
10										
11	0	0	0	0	0	0	0	0	0	
12	0	0	0	0	0	0	0	0	0	
13	0	0	0	0	0	0	0	0	0	
14	0	0	0	0	0	0	0	0	-3.5574E-101	
15	0	0	0	0	0	0	0	0	5.6466E-104	
16	0	0	0	0	0	0	0	-1.2761E-101	-6.2448E-101	
17	0	0	0	0	0	0	0	2.0255E-104	[illegible]3534E-103	
18	0	0	0	0	0	0	-4.5774E-102	-3.8767E-101	-[illegible]007E-101	
19	0	0	0	0	0	0	7.2657E-105	7.4525E-104	1[illegible]9445E-103	
20	0	0	0	0	0	-1.6420E-102	-1.9777E-101	-6.5868E-101	-[illegible]1207E-101	

In the final cell the electric field is modified to add on the stimulus calculated earlier. The code in cell SJ14 is =SJ12–($H$2/(SJ$9*$H$3*$H$1)) *(SK13–SJ13) + D14
Note that the stimulus is only added to the electric field.

FIGURE 4.23
Spreadsheet implementation of the finite-difference time-domain (FDTD) model (continued).

the actual position in meters and the relative permittivity which is set to 1 outside the solar cell. However, in the simulation domain the stimulus (calculated previously in column D) is now added to the electric field along the cells at the very edge of the simulation. In particular, the differentiated Gaussian signal is initiated on the right hand side of the spreadsheet simulation and this signal propagates to the left. The top two rows of the simulation domain which solves the FDTD (rows 11 and 12) can be copied down to populate the entire spreadsheet, and calculate the propagation of the signal for subsequent times. However, for large time scales this may be too large of a spreadsheet to run efficiently (depending on the spreadsheet chosen and the computer it is running on). An alternative to copying the data downwards would be to use a macro, as discussed earlier, to simulate a smaller system size (say 100 iterations), and then copy and paste the outputted electromagnetic fields after 100 iterations into the initial conditions to simulate the next 100 iterations. Automating this process using a macro could be a faster and more efficient way to simulate larger timescales using this spreadsheet method (using a spreadsheet's plotting capabilities you could also set up an animation of the simulation as it runs). However, as always spreadsheets are never the most efficient way to implement large simulations and if system size becomes an issue then a real programming language should be adopted instead.

In the simple system considered here the differentiated Gaussian pulse is

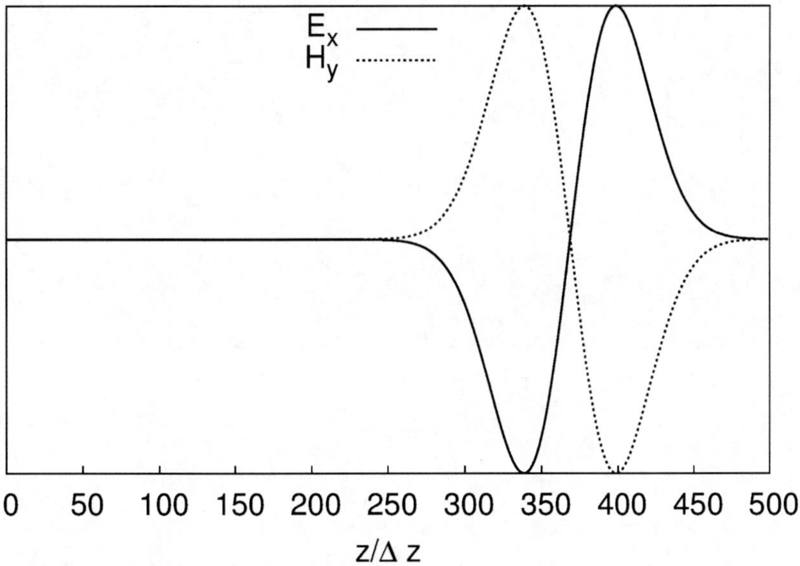

FIGURE 4.24
The electric and magnetic fields for a differentiated Gaussian pulse in a FDTD model.

initiated on the right and propagates to the left into the simulated domain. The solar cell is positioned from $z = 0$ to $z = 100nm$. The propagating signal is shown in the simulated domain, prior to interacting with the solar cell, in Figure 4.24.

Both the magnetic and electric fields are shown. Recall that electromagnetic waves propagate because a change in the electric field causes a change in the magnetic field, and vice versa, a change in the magnetic field causes a change in the electric field. The magnitude of the signal is not depicted as the size of the stimulus is somewhat arbitrary. The functional form of the differentiated Gaussian can be clearly be seen as the signal propagates.

Looking at this propagation a little closer, Figure 4.25 depicts the electric field at times of 1 fs, 1.6 fs, and 2.2 fs. After 1 fs the differentiated Gaussian pulse has been created and is propagating to the left. After 1.6 fs the pulse is close to the top surface of the spreadsheet (100 nm), and has clearly propagated across the system. After 2.2 fs the pulse has interacted with the solar cell structure. Notice that now there appears to be two separate pulses in the system. One signal is traveling to the left inside the solar cell, but the second is smaller and moving away from the solar cell. It is hard to infer these dynamics in a snapshot of the simulation, but the pulse to the left of 100 nm is traveling to the left and pulse to the right of 100 nm is traveling to the

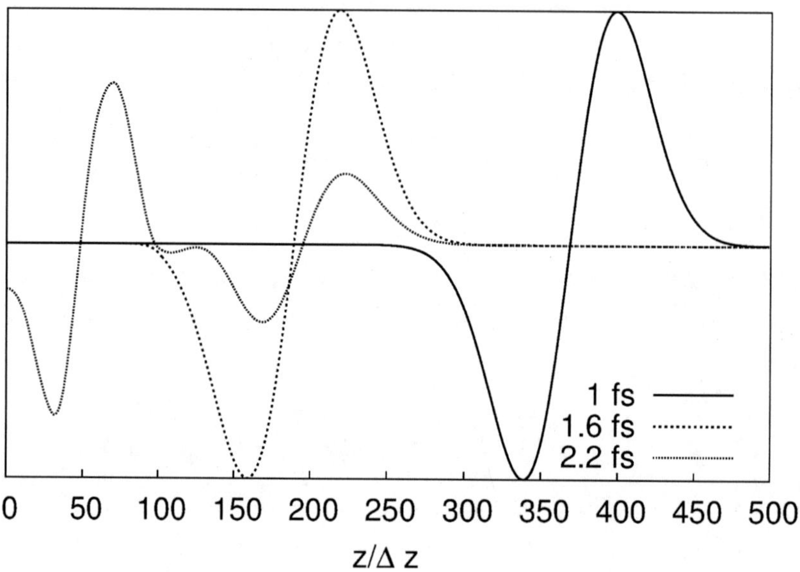

FIGURE 4.25
Interaction of an electromagnetic pulse (traveling from right to left) with a solar cell surface at $z/\Delta z = 100$ nm. After 2.2 fs some of the light has entered the solar cell, but some of the light has been reflected.

right (having reflected back from the upper surface of the solar cell). In other words, some of the light never even made it into the solar cell!

4.7 Exercises

1. Using the equivalent circuit model of a solar cell, test the sensitivity of the model to the different input parameters for the organic solar cell constants provided. For example, by what percentage might the maximum power, open circuit voltage or short circuit current vary if one of the parameters was varied by, say, 1%, 2%, 3%, etc... ?

2. Implement the simple drift diffusion model for a polymer solar cell, where the photogenerated excitons are uniformly distributed throughout the cell. a) How does varying the mobility of the holes and electrons affect the charge distributions inside the cell? Perform a systematic investigation of the role mobility plays in these devices. b) Extend this simple drift diffusion model to output the current density from the device. The electron current density, in the drift diffusion model, is defined as

$$J_n = n\mu_n E - \mu_n \frac{dn}{dx} \tag{4.64}$$

where n is the concentration of electrons, E is electric field, μ_n is the electron mobility, and x is distance.

3. Show that the electric field in a thin polymer solar cell can be approximated as a constant. Take the charge carrier profile from the simple drift diffusion spreadsheet model (with uniform exciton distribution) and solve for the electrical potential using

$$\epsilon_{i+\frac{1}{2}}\phi_{i+1}^{t+1} + \epsilon_{i-\frac{1}{2}}\phi_{i-1}^{t+1} - \left(\epsilon_{i+\frac{1}{2}} + \epsilon_{i-\frac{1}{2}}\right)\phi_i^{t+1} = (p_i^t - n_i^t)q \tag{4.65}$$

In particular, use either the relaxation method or the conjugate gradient method. Is the resulting electric potential a linear function with respect to distance?

4. The hole concentration in a polymer solar cell is given for a model with non-uniform exciton concentration. Use a spreadsheet model to solve for the hole concentration in this system. Start by calculating the required constants and then incorporate these into the derived solution. How might you organize the spreadsheet? What free parameters must you account for?

5. The hole concentration in a polymer solar cell is given for a model with

non-uniform exciton concentration. Perform a similar analysis for the electron concentration and obtain similar expressions for the electron concentrations in a polymer solar cell. Use the appropriate Schottky boundary conditions, which can be found in

Scott, J. C., & Malliaras, G. G. (1999). Charge injection and recombination at the metalorganic interface. Chemical physics letters, 299(2), 115-119.

This is a challenging problem which will require a lot of math!

6. In the two-dimensional Monte Carlo model of a polymer solar cell the movement of excitons, electrons, and holes are captured. Run the simulation over adequately long time frames and obtain a density or concentration plot of excitons, electrons, and holes within this device. This can be collapsed into one-dimension, as we are only interested in the direction through the device (perpendicular to both the surfaces and the DA interface). How does this compare to the concentrations obtained from the drift diffusion model (see Figure 4.9)?

7. The amount of light reflected back from the surface should be dependent on the contrast in relative permittivity between the solar cell and the outside air (relative permittivity of 1). In order to investigate this reflection, calculate the amount of energy reflected back from the surface. The energy stored in the system can be calculated directly from the fields

$$u = \frac{1}{2}\left(\epsilon E^2 + \mu H^2\right) \tag{4.66}$$

where u is the energy per volume. Calculate this quantity for the pulse going in and the pulse reflected out to find the percentage of energy reflected back from the surface.

Try changing the relative permittivity of the solar cell (assume it's uniform for now) to see if the amount of reflected energy increases with the relative permittivity of the solar cell.

While many anti-reflective coatings will use geometry, in the form of surface texturing, in one-dimensional systems they can consist of a thin layer with a specially chosen thickness so that interference effects in the coating cause the wave reflected from the anti-reflection coating top surface to be out of phase with the wave reflected from the semiconductor surface. These destructively interfering reflected waves result in zero net reflected energy, but only for a specific wavelength. Try adding an anti-reflective coating with a relative permittivity in between that of the air and the solar cell and a thickness equal to

$$d = \frac{\lambda}{4\sqrt{\epsilon_r}} \tag{4.67}$$

where d is thickness, λ is the wavelength you're trying to eliminate, and ϵ_r is the relative permittivity of your anti-reflective coating. From an energy point of view which wavelength is it best to eliminate?

8. While looking at the light both incident and reflected from the solar cell, try taking these pulses of light as a function of time and calculating the frequencies contained within the pulses using the a fast Fourier transform (FFT). Most spreadsheets will have a tool or extension for calculating the FFT (or you could use an external package such as Octave).

9. Incorporate a photonic crystal structure into your solar cell to see if light absorption could be enhanced by trapping light within the structure. Note that you are limited to a one-dimensional structure because of the one-dimensional nature of the current model. This would consist of alternating stripes of different regions with different relative permittivities. Calculate the FFT for the differentiated Gaussian entering the solar cell and the signal as a function of time which is reflected out afterwards. A potential problem, is that some light will bounce back and forth inside the photonic crystal and it may take a long time for the reflected light to gradually come out of the solar cell. However, once the light reaches the boundary at $z = 500nm$ it will be reflected back in towards you simulation domain, and muddle up your results.

10. The boundary conditions can be modified to absorb rather than reflect incident electromagnetic waves. The best way to achieve this is to use perfectly matched layers (PML), which requires a region of space which is lossy but is perfectly matched to the regular domains such that there are no reflections. An easier way is to use Mur's absorbing boundary conditions. In effect, the boundary conditions become

$$E_z^{t+\frac{1}{2}} = E_{z-1}^{t-\frac{1}{2}} + \frac{c\Delta t - \Delta z}{c\Delta t + \Delta z}\left(E_{z-1}^{t+\frac{1}{2}} - E_z^{t-\frac{1}{2}}\right) \tag{4.68}$$

and

$$H_{z+\frac{1}{2}}^{t+1} = H_{z-\frac{1}{2}}^{t} + \frac{c\Delta t - \Delta z}{c\Delta t + \Delta z}\left(H_{z-\frac{1}{2}}^{t+1} - H_{z+\frac{1}{2}}^{t}\right) \tag{4.69}$$

Try implementing these boundary conditions in the current model.

11. Include losses into your model. In particular, the electric and magnetic conductivities were removed from the above model. Can you add them again? A suitable discretization of Maxwell's equations for this might be of the form

$$\epsilon \frac{E_z^{t+\frac{1}{2}} - E_z^{t-\frac{1}{2}}}{\Delta t} + \sigma \frac{E_z^{t+\frac{1}{2}} + E_z^{t-\frac{1}{2}}}{2} = \frac{H_{z+\frac{1}{2}}^{t} - H_{z-\frac{1}{2}}^{t}}{\Delta z} \tag{4.70}$$

and

$$\mu \frac{H_{z+\frac{1}{2}}^{t+1} - H_{z+\frac{1}{2}}^{t}}{\Delta t} + \sigma_* \frac{H_{z+\frac{1}{2}}^{t+1} + H_{z+\frac{1}{2}}^{t}}{2} = -\frac{E_{z+1}^{t+\frac{1}{2}} - E_z^{t+\frac{1}{2}}}{\Delta z} \tag{4.71}$$

Additional Reading

Selberherr, S. (1984). Analysis and simulation of semiconductor devices. Wien; New York: Springer.

Taflove, A., & Hagness, S. C. (2000). Computational Electrodynamics: The Finite-Difference Time-Domain Method. Artech House, Norwood, MA.

5

Wind Power

CONTENTS

Wind power simply consists of extracting the kinetic energy from the wind and converting it to electrical energy. While the electronics and control system are an incredibly important (if in my opinion a little boring) aspect of modern wind turbines, here we'll focus more on the physics of the wind turbines themselves and the air flow around them.

5.1 Betz Limit

The Betz limit represents the theoretical maximum amount of power that can be extracted from the wind using a wind turbine. The fundamental assumption is that the fluid which goes through the turbine has to go somewhere. In other words, if the velocity of the air was reduced to zero as it went through the turbine (meaning that the turbine extracted all the kinetic energy from the air) then no other air would be able to pass though, without pushing this stationary air out of the way. In order for there to be a continuous flow of air through the turbine, some of the kinetic energy of the air cannot be extracted from the air as it passes though the turbine. The question is, how much energy can be extracted from the wind?

In determining the Betz limit we must first make a couple of assumptions. First, we idealize the wind turbine (don't we always?) and ignore any drag effects which might lower the idealized Betz limit in practice. Secondly, we assume that the flow through the turbine is axial and that we can conserve the mass of the air as it flows through the turbine. Lastly, we assume that the flow is incompressible, and density is constant, which is not strictly true for air flow. With these assumptions we can start by applying Bernoulli's equation to the problem. Bernoulli's equation is derived by conserving energy and consists of pressure (from a work term), gravitational potential energy, and kinetic energy. However, note that the fluid interacts with the rotor and, in particular, transfers energy from the fluid to the rotor. Therefore, we can't apply Bernoulli's equation as we cross the rotor.

Bernoulli's equation is often written in the form

$$\frac{1}{2}\rho v_1^2 + P_1 + \rho g h_1 = \frac{1}{2}\rho v_2^2 + P_2 + \rho g h_2 \tag{5.1}$$

where ρ is the density of the fluid (which is assumed to remain constant), and v_1 and v_2 are the velocities of the fluid along a streamline at two different locations in space (position 1 and position 2). P_1 and P_2 are the hydrostatic pressures at positions 1 and 2. g is the acceleration due to gravity, and h_1 and h_2 are the heights of the fluid at positions 1 and 2. Here, we take the heights h_1 and h_2 to be the same and so the gravitational potential energy terms cancel and we are left with

$$\frac{1}{2}\rho v_1^2 + P_1 = \frac{1}{2}\rho v_2^2 + P_2 \tag{5.2}$$

The slightly counter-intuitive upshot of this equation is that fluids flowing faster must have lower pressures (in order for the equation to balance).

An external force is provided on the air flow by the rotors (a consequence of Newton's third law as the air flow provides a force to the rotors to make them turn) which means that we can't apply Bernoulli's equation across the rotors. However, we can apply Bernoulli's equation to the air flow before the rotor, and separately to the air flow after the rotor. Let's look at the air flow before, or upstream from, the rotor. Bernoulli's equation can be used to compare the energy of the fluid at two locations in the fluid's streamline, and here we'll consider far upstream to a location just before the wind turbine rotors. Far upstream the wind velocity is v_∞ and the pressure is P_{atm}; whereas, just before the fluid stream interacts with the rotors the velocity is reduced to $v_\infty(1-a)$, where a is the axial induction factor, and the pressure at the rotors is $P + \Delta P$, where ΔP is the difference in pressure across the rotor plane. Bernoulli's equation, therefore, becomes:

$$\frac{1}{2}\rho v_\infty^2 + P_{atm} = \frac{1}{2}\rho \left(v_\infty(1-a)\right)^2 + P + \Delta P \tag{5.3}$$

Similarly, we can apply Bernoulli's equation to the fluid flow after it has passed through the rotors, downstream from the wind turbine. In particular, we compare two locations downstream from the wind turbine. The first location is just after the rotor plane of the turbine and the second location is far downstream from the wind turbine. We assume that there is no change in fluid velocity as the fluid flows through the actual rotor plane, but that the pressure drops from $P+\Delta P$ to just P, where recall that ΔP is defined as the change in pressure across the rotors. Far downstream the pressure in the fluid is assumed to have equilibrated with atmospheric pressure, but the velocity of the air flow is reduced to v_{wake}. In other words, applying Bernoulli's equation downstream from the wind turbine gives us the following equation

$$\frac{1}{2}\rho\left(v_\infty(1-a)\right)^2 + P = \frac{1}{2}\rho v_{wake}^2 + P_{atm} \tag{5.4}$$

Combining these two, and canceling terms, allows us to obtain an expression for the pressure difference across the rotor plane.

$$\Delta P = \frac{1}{2}\rho\left(v_\infty^2 - v_{wake}^2\right) \tag{5.5}$$

Now that we have an expression for the difference in pressure we can obtain an expression for the force in the streamwise direction, or thrust, acting on the rotors. In other words, we can find the force required to reduce the wind speed from v_∞ to v_{wake} by simply considering the definition of pressure as a force per unit area:

$$F = \Delta P A = \frac{1}{2}\rho\left(v_\infty^2 - v_{wake}^2\right) A \tag{5.6}$$

where $A = \pi R^2$ is the area swept out by the rotor blades, and R is the length of the rotor blades.

We can also take into consideration the fact that the force is the derivative of momentum with respect to time to obtain an alternative expression for the force. Note that force is the derivative of momentum with respect to time, which usually leads to Newton's $2^n d$ law, $F = ma$, but in this case it's a little different. For the fluid flow at the wind turbine, it is the mass which is changing as it is continuously flowing through the turbine and the velocity is constant (across the rotor plane). Therefore, we can write

$$F = \frac{dp}{dt} \tag{5.7}$$

where p is the pressure, and as the mass is flowing through the turbine this is equal to

$$F = \frac{dm}{dt}\left(v_\infty - v_{wake}\right) \tag{5.8}$$

where the derivative of mass, m, with respect to time, t, is the mass flow rate. The mass flow rate is the density, ρ, of the fluid times the flow rate (volume per second), Q, and the flow rate of the fluid is simply the area, A, multiplied by the velocity, here $(v_\infty(1-a))$, giving us the following expression for the force on the rotors

$$F = \rho A\left(v_\infty(1-a)\right)\left(v_\infty - v_{wake}\right) \tag{5.9}$$

Putting this expression for the force acting on the rotors in the streamwise direction, or thrust, equal to the earlier expression we obtained using Bernoulli's equation gives us an expression for the fluid velocity far downstream of the wind turbine

$$v_{wake} = (1 - 2a)\, v_\infty \tag{5.10}$$

where, as before, a is the axial induction factor which is defined as the ratio of the reduction in air velocity upon passing through the wind turbine, to that of the air velocity far away from the wind turbine

Let's go back to the force acting on the rotor. The power taken from the fluid is the force acting on the fluid multiplied by the velocity of the fluid, which given the above expressions can be written as

$$Fv = 2a(1 - a)^2 v_\infty^3 \rho A \tag{5.11}$$

where v_∞ is the velocity of the air far from the wind turbine, a is the axial induction factor, and A is the area of the rotors. One of the questions we might like to answer is, how much of the kinetic energy from the wind can we extract? Recall, that extracting 100% of the energy is silly: if the velocity of the wind was to go to zero, then air wouldn't be able to flow into the wind turbine as it would have nowhere to go. To characterize the fraction of the power extracted from the wind, we define the power coefficient as the ratio of the power extracted to the total power in the wind. The total power in the wind is of the form

$$\frac{1}{2}\rho A v_\infty^3 \tag{5.12}$$

and the power coefficient is given by

$$C_P = \frac{2a(1-a)^2 v_\infty^3 \rho A}{\frac{1}{2}\rho A v_\infty^3} = 4a(1-a)^2 \tag{5.13}$$

where the dimensionless coefficient of power C_P is a widely used measure of the performance of a wind turbine. Similarly, an expression for the coefficient of thrust, C_T, can also be obtained in the following form

$$C_T = \frac{F}{\rho A v_\infty^2} = \frac{2a(1-a)v_\infty^2 \rho A}{\rho A v_\infty^2} = 2a(1-a) \tag{5.14}$$

where the denominator in the above ratio is simply the force required to bring the fluid to a stop. A plot of both the power coefficient and the thrust coefficient as a function of the axial induction factor is depicted in Figure 5.1.

Our question remains: what is the maximum value of C_P using the Betz model? From Figure 5.1 we can clearly see that this occurs around $a = \frac{1}{3}$, but let's find this maximum by taking the derivative of C_P with respect to a and putting this to zero.

$$\frac{dC_P}{da} = 4(1-a)^2 - 8(1-a)a = 0 \tag{5.15}$$

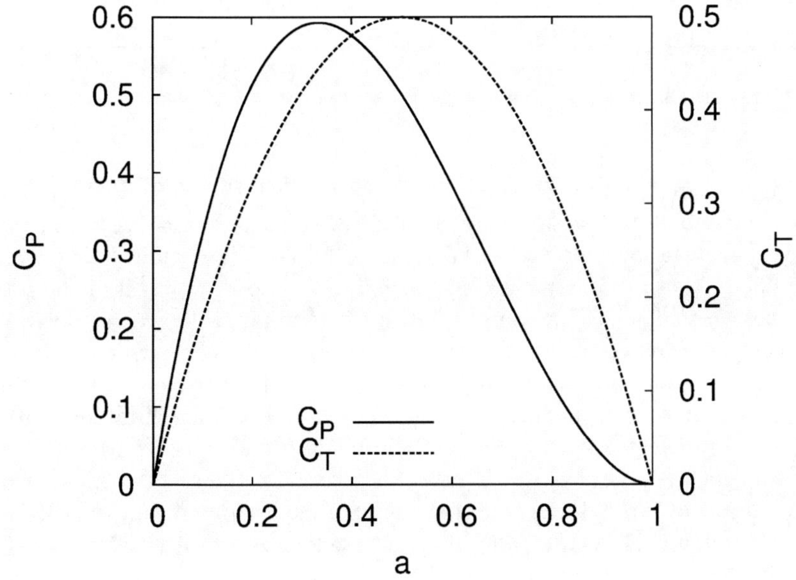

FIGURE 5.1
Plot of the power coefficient and thrust coefficient as a function of the axial induction factor.

From this we can clearly see that the maximum power coefficient occurs when $a = \frac{1}{3}$, and $C_{P,max} = \frac{16}{27}$. In other words, it is theoretically impossible for any wind turbine to extract more than 59% of the power in the wind. This is known as the Betz limit. Interestingly, this model predicts the flow velocity at the wind turbine to be two thirds of the wind speed, $\frac{2}{3}v_\infty$, and the wake velocity to be a third of the wind speed, $\frac{1}{3}v_\infty$. While in reality frictional losses make real wind turbines around 50% efficient, and it is possible to funnel more air into the stream to increase efficiencies beyond the Betz limit (impractical shrouded rotors), the Betz limit remains an important characteristic of wind turbines.

5.2 Blade Element Momentum Model

In the previous section we derived the Betz limit, where we considered how the change in momentum, as the rotor interacts with the air, and the motion of the air combine to give us a predicted ideal efficiency and a wake velocity. However, in the above analysis we didn't take into consideration the geometry of the blades that are interacting with the fluid. Instead, we treated the rotor plane in a very idealized way, and as such we were not able predict optimum blade configurations or characteristics. Blade element theory, which considers the forces on the blade as a result of air flow and blade geometry, can be combined with the arguments we made in the above Betz limit derivation to relate wind turbine performance to the blade geometry. In particular, the blade element momentum (BEM) model accounts for the angular momentum of the rotor (and through the conservation of angular momentum, the rotation of the air flow in the wake). The blade element momentum (BEM) model is not without simplifications, however. For example, the model ignores frictional drag and assumes radial velocities in the air are negligible. In particular, the wind turbine is considered in radial elements (or annular elements) where each element is solved independent of the other elements, and the force acting on the fluid from the blades in an annular element is assumed to be uniform (and not depend on the location of the blades as they spin around). Figure 5.2 depicts such an annular element. The radius of the area swept out by the blades is the length of the blades, R, the radius of the annular element is r, and the thickness of the annular element is dr. The entire area of the rotor can be broken up into a series of such elements, from $r = 0$ to $r = R$, in such a way that integrating over these elements and adding the power extracted from the wind in each element, we can ultimately find the total power extracted from the wind by the wind turbine.

Similar to our previous derivation of the Betz limit, we can consider the thrust (the force acting on the rotors in the streamwise direction) to be determined by rate of change in momentum. In particular, the thrust on the thin

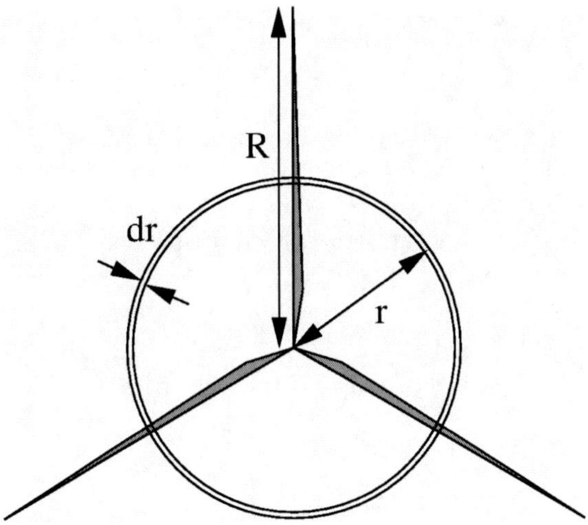

FIGURE 5.2
The blade element momentum model breaks the rotor plane into a series of annular elements, with each annular element solved independent of the others.

annular element is dT and is related to the velocities and the mass flow rate.

$$dT = (v_\infty - v_{wake})\frac{dm}{dt} = 2\pi r \rho v (v_\infty - v_{wake}) dr \tag{5.16}$$

where v_∞ the wind speed far from the wind turbine, v_{wake} is the velocity of the air in the wake of the wind turbine, m is the mass of air flowing through the wind turbine, and t is time. The mass flow rate can be written as the density multiplied by the flow rate, or density multiplied by the velocity and area of the air flow. The area of the annular element, which we are considering the air flowing through, is $2\pi r dr$ (circumference of the circle multiplied by the thickness). The velocity of the air flow through the wind turbine is v, and ρ is the density of the air as before (again assumed to be constant).

We can write a similar expression for the torque, dM, acting on the rotors where the rotational velocity upstream of the rotor is taken to be zero, while downstream it is v_{rotate} (recall, that if the wind turbine spins, then from conservation of angular momentum the air will have rotational motion also).

$$dM = r v_{rotate} \frac{dm}{dt} = 2\pi r^2 \rho v v_{rotate} dr \tag{5.17}$$

Note the velocity of the air, as it spins in the wake of the wind turbine, is v_{rotate}, and the units of torque are Newton meters as opposed to just Newtons for force. As before, we define axial induction factors where the axial induction factor, a, is the ratio of the reduction of wind speed at the rotor to the wind speed far way. Therefore, the velocity at the rotor, v, can be written in terms of the velocity far away from the wind turbine, v_∞, using the axial induction factor.

$$v = (1 - a)v_\infty \tag{5.18}$$

This time, however, we have an axial induction factor for the rotational components of the velocity, $\acute{a}$. The axial induction factor, $\acute{a}$, is really defined in the manner in which it relates the rotational velocity of the fluid in the annular element after the air flows through the wind turbine, v_{rotate}, and the angular velocity of the wind turbine blades, ω.

$$v_{rotate} = 2\acute{a}\omega r \tag{5.19}$$

Substituting these two definitions for the velocities into Equations 5.16 and 5.17, for the thrust and torque, respectively, acting on the rotor by the wind yields the following two expressions:

$$dT = 4\pi r \rho v_\infty^2 a(1 - a)dr \tag{5.20}$$

and

$$dM = 4\pi r^3 \rho v_\infty \omega(1 - a)\acute{a}dr \tag{5.21}$$

These equations describe the thrust and torque acting on the rotor blades, derived by considering the momentum in a discretized annular element. What we can do now is obtain alternative expressions for the thrust and torque on these annular elements, using the local air flow at the blades. In this manner we can equate the momentum conservation aspects of the fluid-structure interactions with the local air flow over and around the blades (which will incorporate elements of blade geometry).

Consider the geometry of the wind flowing onto a wind turbine blade in Figure 5.3. The cross section of the blade is depicted, along with the rotor plane (the plane in which the rotors are spinning). As the air interacts with the blade, its velocity is reduced in the direction normal to the rotor plane to $v_\infty(1 - a)$, and acquires a velocity tangential to the rotor plane of $\omega r(1 + \acute{a})$. The relative velocity, v_{rel}, therefore, can be obtained from these orthogonal components. The angle θ is the pitch of the blade, the turning of the blades with respect to the rotor plane (note this pitch will change as we consider different annular elements). The angle ϕ is the angle between the air flow at the rotor plane (note the air flow far away from the wind turbine, v_∞ is still perpendicular to the rotor plane) and the airfoil, or blade. The difference between these two angles is the angle of attack, $\alpha = \phi - \theta$, the angle of the wind speed in the rotor plane, and the plane in which the blade is oriented. From the above geometry, we can write an expression for ϕ

$$\tan\phi = \frac{(1 - a)v_\infty}{(1 + \acute{a})\omega r} \tag{5.22}$$

The lift per unit length on a blade is given by

$$L = \frac{1}{2}\rho v_{rel}^2 c C_L \tag{5.23}$$

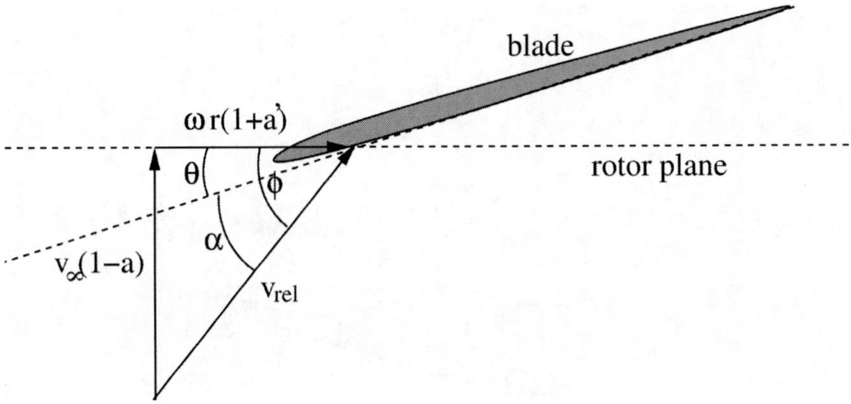

FIGURE 5.3
The geometry of air flow in relation to the blade orientation.

where ρ is the density of the air, and C_L is the lift coefficient which will depend on the angle of attack, α, of the air flow on the blade. The chord, c, of the blade is defined as the distance from the leading to the trailing edges of the airfoil, and will depend on the radius of the annular element as the chord length will change with distance along the blade. Similarly, the drag force per length is

$$D = \frac{1}{2}\rho v_{rel}^2 c C_D \tag{5.24}$$

where C_D is the drag coefficient, which also depends on the angle of attack, α, of the air flow on the blade. These lift and drag forces, however, are relative to the local direction of air flow at the wind turbine blades. In order to compare these forces with the thrust and torque calculated earlier using momentum conservation (Equations 5.20 and 5.21) we have to consider the components of these forces normal and tangential with the rotor plane.

Figure 5.4 depicts the orientation of the lift and drag forces, and the forces normal and tangential to the rotor plane, F_N and F_T, respectively. The force F_N will, in general, cause the blades to bend, while F_T will cause the blades to rotate. Projecting the lift and drag forces onto the normal and tangential directions of the rotor plane results in the following force components (or actually force per length)

$$F_N = L\cos\phi + D\sin\phi \tag{5.25}$$

and

$$F_T = L\sin\phi - D\cos\phi \tag{5.26}$$

where ϕ is the angle between the direction of the air flow at the wind turbine blade and the rotor plane.

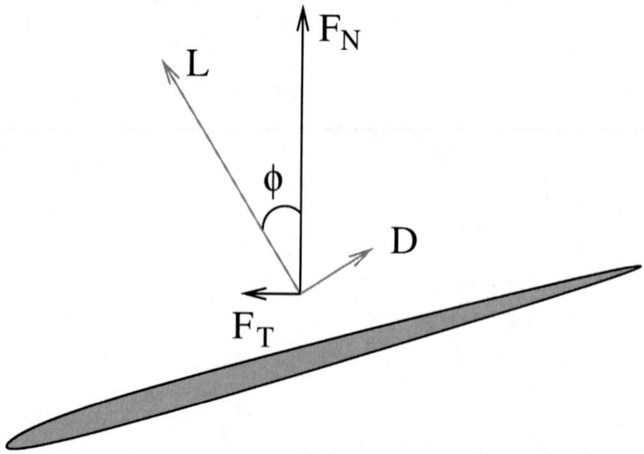

FIGURE 5.4
The lift and drag are perpendicular and parallel to the blade, whereas we need the forces acting normal and tangential to the rotor plane.

Similarly, we consider the coefficients of drag and lift to also be projected in the normal and tangential directions.

$$C_N = C_L \cos\phi + C_D \sin\phi \tag{5.27}$$

and

$$C_T = C_L \sin\phi - C_D \cos\phi \tag{5.28}$$

where these two equations will play an important role in solving the system and we'll come back to these a little later.

From the geometry depicted in Figure 5.3, we can see from trigonometry that

$$v_{rel} \sin\phi = v_\infty(1 - a) \tag{5.29}$$

and

$$v_{rel} \cos\phi = \omega r(1 + \acute{a}) \tag{5.30}$$

which gives us expressions containing the sine and cosine of the pitch angle, and we'll use this to help tidy up the math. We also have to consider the solidity of the turbine. The solidity is defined as

$$\sigma = \frac{cB}{2\pi r} \tag{5.31}$$

where c is the chord of the airfoil, and will depend on the radius of the annular element being considered, r. B is the number of blades (typically 3) that the wind turbine possesses.

Recall that F_N and F_T, the forces in the normal and tangential directions acting on the blades are actually forces per unit length. We can use these forces to calculate the thrust, dT, and torque, dM, acting on the wind turbine (and, hence, the wind as calculated earlier in Equations 5.20 and 5.21). In particular,

$$dT = BF_N dr \tag{5.32}$$

and

$$dM = rBF_T dr \tag{5.33}$$

where B is the number of blades, and we consider the total thrust and torque due to all blades. The normal and tangential projections of the coefficients of lift and drag can be written as

$$C_N = \frac{F_N}{\frac{1}{2}\rho v_{rel}^2 c} \tag{5.34}$$

and

$$C_T = \frac{F_T}{\frac{1}{2}\rho v_{rel}^2 c} \tag{5.35}$$

where the above relationships are simply the same as the relationships between lift and drag, and the coefficients of lift and drag. Putting all this together and combining the above equations we can show that the thrust and torque on the rotor blades are

$$dT = \frac{1}{2}\rho B \frac{v_\infty^2 (1-a)^2}{\sin^2\phi} c C_N dr \tag{5.36}$$

and

$$dM = \frac{1}{2}\rho B \frac{v_\infty (1-a)\omega r(1+\acute{a})}{\sin\phi\cos\phi} c C_T r dr \tag{5.37}$$

In other words, we now have two different expressions for thrust and torque, obtained from completely different means. Equations 5.20 and 5.21 were derived by considering the change in momentum of the fluid, while the above two expressions were obtained by looking at the forces acting on the rotor blades due to the fluid flow. Equating these two different expressions, we can obtain expressions for the axial induction coefficients

$$a = \frac{1}{\dfrac{4\sin^2\phi}{\sigma C_N} + 1} \tag{5.38}$$

and

$$\acute{a} = \frac{1}{\dfrac{4\sin\phi\cos\phi}{\sigma C_T} - 1} \tag{5.39}$$

Once we have the axial induction coefficients we can calculate the forces acting on the blades, and the power being extracted from the wind by the wind turbine. Therefore, acquiring the axial induction coefficients is the name of the game to solving this system. However, we are not done yet as the equations aren't quite complete.

At the beginning of this section, when we started to introduce the beam element momentum (BEM) model we outlined the main assumptions of the model; namely, that the annular elements don't interact (no radial fluid flow) and that the forces from the blades are assumed to be uniform throughout the annular element. Obviously these assumptions aren't entirely true. To correct for the latter assumption, that the forces from the blades are uniform throughout the annular element, Prandtl's tip loss factor can be incorporated. This is an empirical factor which factors in vorticity effects near the blade tips and other phenomena not accounted for in the model. The correction factor that Prandtl proposed modifies the momentum equations

$$dT = 4\pi r \rho v_\infty^2 a(1-a)K dr \tag{5.40}$$

and

$$dM = 4\pi r^3 \rho v_\infty \omega (1-a)\acute{a} K dr \tag{5.41}$$

where the factor K is added. This factor is empirically defined as

$$K = \frac{2}{\pi}\cos^{-1}\left(e^{-k}\right) \tag{5.42}$$

where

$$k = \frac{B}{2}\frac{R-r}{r\sin\phi} \tag{5.43}$$

recall B is the number of blades, R is the length of the rotor blades (and the radius of the area swept out by the blades), and r is the radius of the current annular element. Factoring in this change to the above analysis results in slightly modified expressions for the axial induction factors:

$$a = \frac{1}{\dfrac{4K\sin^2\phi}{\sigma C_N} + 1} \tag{5.44}$$

and

$$\acute{a} = \frac{1}{\dfrac{4K\sin\phi\cos\phi}{\sigma C_T} - 1} \tag{5.45}$$

where the addition of K is the only modification. There are other loss models, however, which we won't consider here, including the hub loss model (similar implementation to the tip loss model), the turbulent wake state, and the Glauert correction. The Glauert correction is necessary at higher values of a (note that for axial induction factors, a, greater than 0.5 the above model would predict negative thrust values. Therefore, a modified projection of the lift and drag coefficients in the tangential direction can be used.

$$C_T = \begin{cases} 4a(1-a)K & a \le \frac{1}{3} \\ 4a(1-\frac{1}{4}(5-3a)a)K & a > \frac{1}{3} \end{cases} \tag{5.46}$$

However it is worth noting that you may find that your values of a might never go above $\frac{1}{3}$, depending on the parameters in your model.

Once we have obtained the value for a we can determine the energy extracted from the wind over a given time frame, t.

$$E = t\int \left[2\rho v_\infty^3 a(1-a)^2 A\right] f(v_\infty)\, dv_\infty \tag{5.47}$$

where the contents of the square parenthesis is the power extracted from the wind and $f(v_\infty)$ is the probability density function for wind velocity far from the wind turbine (obviously location specific). Note that by considering a distribution of wind velocities we can calculate the average energy over an extended period of time, as long the probability density function for wind speeds are accurate over this same time frame.

Obtaining the axial induction factors are crucial to this calculation. a is defined as the fractional decrease in axial wind velocity between the wind speed far from the wind turbine and the rotor plane. $\acute{a}$ is defined as the fractional increase in tangential wind speed due to the air flows interaction with the rotating blades. Therefore, while these quantities are physically interesting in their own right, we are mainly concerned here with solving for other quantities like the extracted power. However, we can't obtain the axial induction factors directly because they are both functions of the angle between relative air speed and the rotor plane. Furthermore, this angle, in turn, depends on the axial induction factors. A numerical iteration is required to solve this problem, and a common procedure for solving this problem is listed below:

- loop through different annular segments at distances, r.
- pick initial a and $\acute{a}$.
- compute the flow angle using

$$\tan\theta = \frac{(1-a)v_\infty}{(1+\acute{a})\omega r} \tag{5.48}$$

- calculate the local angle of attack, $\alpha = \phi - \theta$.
- obtain the coefficients of drag and lift (from tables or fitted curves).
- rotate coefficients of drag and lift to get coefficients normal and tangential to the rotor plane

$$C_N = C_L\cos\phi + C_D\sin\phi \tag{5.49}$$

$$C_T = C_L\sin\phi - C_D\cos\phi \tag{5.50}$$

- calculate a and $\acute{a}$ from

$$a = \frac{1}{\dfrac{4K\sin^2\phi}{\sigma C_N} + 1} \tag{5.51}$$

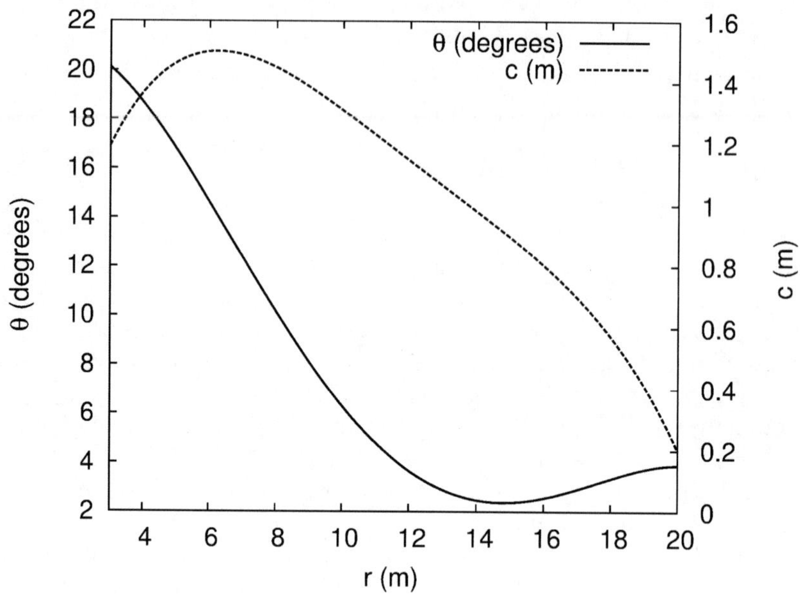

FIGURE 5.5
Plot of the pitch and chord of the rotor as a function of distance along the rotor.

$$\acute{a} = \frac{1}{\dfrac{4K \sin\phi \cos\phi}{\sigma C_T} - 1} \tag{5.52}$$

- iteratively calculate a and $\acute{a}$ until solutions converge.

The local pitch and chord for the airfoil are the main geometrical parameters for the wind turbine (although we'll come to coefficients of drag and lift in a moment). A plot of two functions used to describe the pitch, θ, and chord, c, as a function of distance from the center of the turbine, r, is depicted in Figure 5.5. The exact equations aren't important, for the present implementation, and these parameters are usually obtained from actual wind turbines. For illustration, Figure 5.5 shows typical functions, however, where the pitch angle is generally higher near the base and reduces as we move out towards the tip of the blade, and the chord increases initially and then tapers off as we approach the tip of the blade.

In contrast to the local pitch and chord, which are easily obtainable from the geometry of a wind turbine, one of the hardest aspects is determining the coefficients of drag and lift for a given airfoil geometry (as a function of α, the local angle of attack of the wind on the blade). In fact, it is more common to fit the results of an experimental study to the model, to obtain the coefficients of drag and lift (and of course, resulting in a model that appears

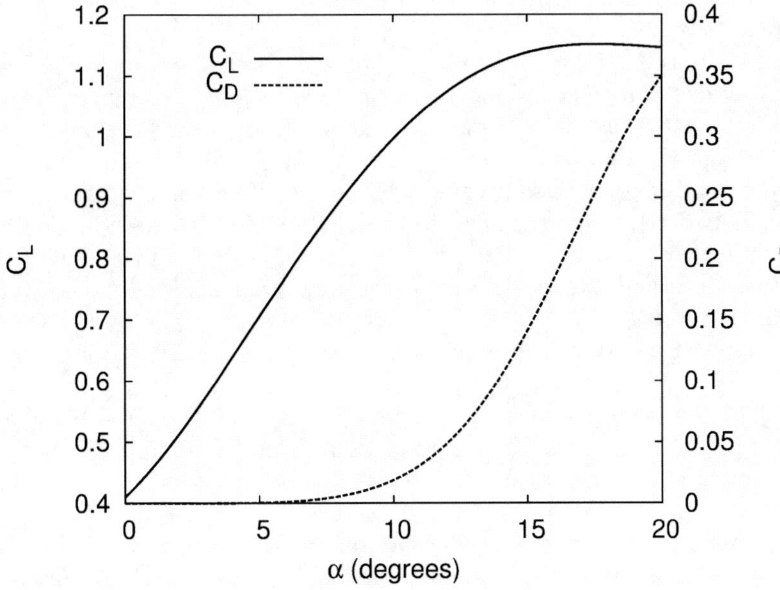

FIGURE 5.6
Plot of the coefficients of lift and drag as a function of the angle of attack of the wind.

to almost perfectly agree with the experimental results). Here, we'll make up some curves that have the correct general shape. The functions for coefficients of drag and lift, which are entirely made up functions, are

$$C_L = 2.39 \times 10^{-3}\alpha^3 - 0.0027\alpha^2 + 0.87\alpha + 0.37 \tag{5.53}$$

and

$$C_D = 0.403\left[1 - \exp\left(-(0.057\alpha)^{5.4}\right)\right] \tag{5.54}$$

and these curves are plotted in Figure 5.6. Both the coefficient of drag and the coefficient of lift tend to increase as we increase the local angle of attack of the wind on the blade.

Now that we have all the necessary information, we can solve the beam element momentum (BEM) model and obtain expressions for the axial induction coefficients and the power extracted from the wind. The spreadsheet implementation of the BEM model is depicted in Figures 5.7 and 5.8. At the top of the spreadsheet in Figure 5.7 are the constants; the length of the blades, the density of the air, the number of blades, and fixed rotational speed of the turbine. The velocity of the wind, far from the wind turbine, and radial distance r are also at the top of the spreadsheet (recall that we must solve for elements of different r and add them together at the end). As a function of the radial distance, r, the pitch and chord of the turbine blades are also determined.

Beneath these initial parameters, in Figure 5.7, is the solution of the beam element momentum (BEM) model. As we move across the columns in the spreadsheet, as depicted in Figure 5.7 and continuing into Figure 5.8, we move through the different variables; first starting with the axial induction factors, determining the coefficients of drag and lift (among other things), before finally determining the next iterative values of the axial induction factors. Subsequent iterations in the model are in the rows as we move down the spreadsheet, and after several rows we tend to find axial induction factors that are converged.

In cells A13 and B13 are the initial guesses at the axial induction factors (here both are taken to be 0, but the solution is not sensitive to this choice). In cell C13 the angle ϕ (between the direction of air flow at the blade and the rotor plane) is calculated and is a function of both axial induction factors. Subtracting the pitch gives us the angle of attack, the angle between the direction of air flow at the blade, and the orientation of the blade, in cell D13. As a function of α the coefficients of drag and lift are calculated in cells E13 and F13, although these might also have to be looked up from data tables (here we took a functional representation instead). The projection of these coefficients in the normal and tangential directions are calculated in cells G13 and H13. In cell I13 the solidity is calculated, although this is constant through all iterations as it only depends on the distance r and the chord, which in turn depends on r (in fact, we could have calculated this at the top of the spreadsheet instead of here). Finally the Prandtl's tip loss factor is calculated in cells J13 and K13, which depends on ϕ. Now we have everything required to calculate the next iteration's values of the axial induction factors, a and $\acute{a}$. The code in cell L13, calculates a

```
=1/(((4*K13*SIN(C13*PI()/180)*SIN(C13*PI()/180))/(I13*G13))+1)
```

while the radial axial induction factor, $\acute{a}$ is calculated in cell M13 using the code

```
=1/(((4*K13*SIN(C13*PI()/180)*COS(C13*PI()/180))/(I13*H13))-1)
```

Once we have the latest values, cells A14 and B14 are set equal to these updated values and the equations contained in row 13 are simply copied down for the subsequent iterations. After several iterations, the system converges and based on the axial induction factors we can calculate the power extracted from this portion of the wind turbine (across the area $2\pi r\,dr$). In particular, the code in cell O27 takes the axial induction factor, a, from cell L27 and calculates the power using the code

```
=2*$B$5*$F$4*$F$4*$F$4*L27*(1-L27)*(1-L27)*PI()*$B$4*$B$4
```

Figure 5.9 shows a surface plot generated from this model. The power as

Constants for the model, the wind velocity and distance along the rotor. As a function of the distance along the rotor, the pitch and chord (width) of the rotor are determined.

	A	B	C	D	E	F	G	H	I	J
1	Beam element momentum									
2										
3	Constants									
4	R	20			velocity	5				
5	rho	1.225			r	4				
6	B	3			pitch	18.67632				
7	omega	3			chord	1.3667456				
8										
9										
10										
11										
12	a	a'	phi	alpha	C L	C D	C N	C T	sigma	k
13	0	0	22.61986495	3.943544948	0.635558927	0.000127277	0.586718731	0.244328255	0.163143239	15.6
14	0.139241061	0.02887903	19.21767735	0.54135735	0.434327492	2.8044E-009	0.410124533	0.142962365	0.163143239	18.22832817
15	0.133740789	0.019118448	19.50262605	0.826306048	0.448313916	2.7516E-008	0.422592448	0.149669603	0.163143239	17.97213978
16	0.133930808	0.019781483	19.48694884	0.810628843	0.447524901	2.4812E-008	0.421889565	0.149290762	0.163143239	17.9860358
17	0.133917023	0.019744031	19.48789742	0.811577417	0.447572579	0.000000025	0.42193204	0.149313652	0.163143239	17.98519435
18	0.133917847	0.019746294	19.48784029	0.811520289	0.447569707	0.000000025	0.421929482	0.149312273	0.163143239	17.98524503
19	0.133917797	0.019746158	19.48784373	0.811523731	0.44756988	0.000000025	0.421929636	0.149312356	0.163143239	17.98524197
20	0.1339178	0.019746166	19.48784352	0.811523523	0.44756987	0.000000025	0.421929627	0.149312351	0.163143239	17.98524216
21	0.1339178	0.019746166	19.48784354	0.811523536	0.44756987	0.000000025	0.421929627	0.149312352	0.163143239	17.98524215
22	0.1339178	0.019746166	19.48784353	0.811523535	0.44756987	0.000000025	0.421929627	0.149312352	0.163143239	17.98524215
23	0.1339178	0.019746166	19.48784353	0.811523535	0.44756987	0.000000025	0.421929627	0.149312352	0.163143239	17.98524215
24	0.1339178	0.019746166	19.48784353	0.811523535	0.44756987	0.000000025	0.421929627	0.149312352	0.163143239	17.98524215
25	0.133917	0.019746166	19.48784353	0.811523535	0.44756987	0.00000002	0.421929627	0.149312352	0.163143239	17.98524215
26	0.133917	.019746166	19.48784353	0.811523535	0.44756987	0.0000000	.421929627	0.149312352	0.163143239	17.98524215
27	0.133917	0.019746166	19.48784353	0.811523535	0.44756987	0.00000002	0.421929627	0.149312352	0.163143239	17.98524215
28										

Cells A13 and B13 are the initial values of the axial induction factors, with iterative updates calculated in subsequent rows until the solution converges.

As a function of the axial induction factors the flow angle, local angle of attack, coefficients of lift and drag, and other parameters are calculated. As a function of these parameters updated axial induction factors are calculated in columns L and M (not shown), and copied down to the next row in columns A and B.

FIGURE 5.7
Spreadsheet implementation of the beam element momentum (BEM) model.

The updated values of the axial induction factors are in columns L and M.

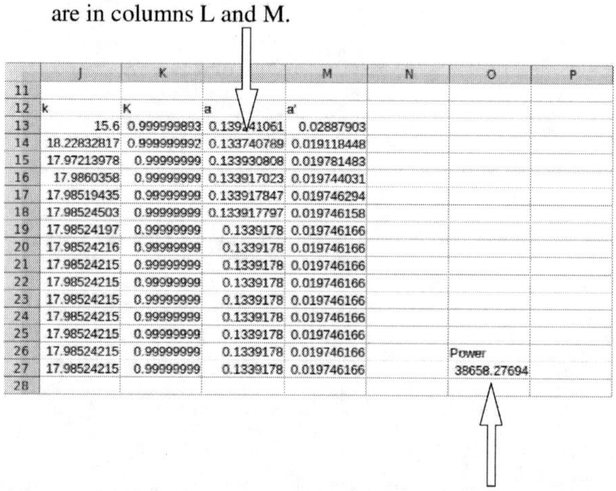

	J	K	L	M	N	O	P
11							
12	k	K	a	a'			
13	15.6	0.999999893	0.139[illegible]41061	0.02887903			
14	18.22832817	0.999999992	0.133740789	0.019118448			
15	17.97213978	0.99999999	0.133930808	0.019781483			
16	17.9860358	0.99999999	0.133917023	0.019744031			
17	17.98519435	0.99999999	0.133917847	0.019746294			
18	17.98524503	0.99999999	0.133917797	0.019746158			
19	17.98524197	0.99999999	0.1339178	0.019746166			
20	17.98524216	0.99999999	0.1339178	0.019746166			
21	17.98524215	0.99999999	0.1339178	0.019746166			
22	17.98524215	0.99999999	0.1339178	0.019746166			
23	17.98524215	0.99999999	0.1339178	0.019746166			
24	17.98524215	0.99999999	0.1339178	0.019746166			
25	17.98524215	0.99999999	0.1339178	0.019746166			
26	17.98524215	0.99999999	0.1339178	0.019746166		Power	
27	17.98524215	0.99999999	0.1339178	0.019746166		38658.27694	
28							

After several iterations the axial induction factor converges to a value, and based on this value we can calculate the power (for this annular section of the rotor) and the forces acting on the rotors.

FIGURE 5.8
Spreadsheet implementation of the beam element momentum (BEM) model (continued).

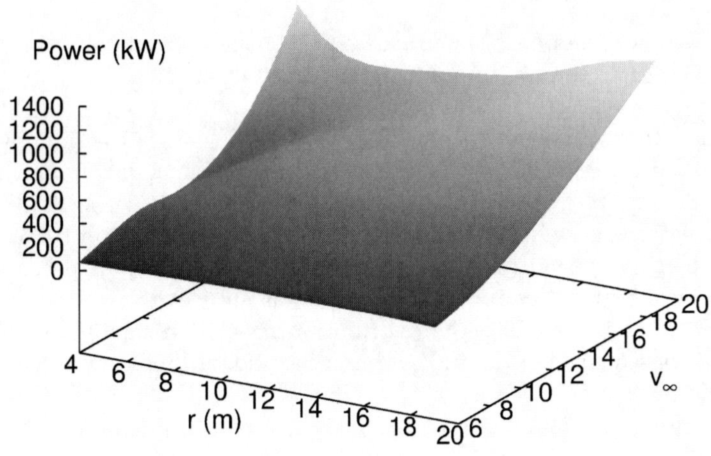

FIGURE 5.9
Plot of the power extracted from the wind, as a function of the velocity of the wind and the distance along the rotor.

a function of wind speed, v_∞, and the distance from the center of the rotor plane, r, are plotted. As expected increasing the wind speed causes the power to increase (at least given the variables here, although losses can also increase). The power extracted from different regions of the blade (different values of r) also varies. Integrating the power extracted over the length of the blade will result in the total power extracted from the wind turbine and this can be used to determine the power coefficient for the wind turbine.

The beam element momentum (BEM) model is a simple model, which is believed to provide relatively accurate results (although if the model was used as a curve fitting exercise, where experimental data is used to fit the model and obtain the coefficients of drag and lift, then such agreement would be expected). For such a simple model to provide a link between blade geometry and the performance of the wind turbine is really awesome. However, the model is limited and cannot account for all effects observed in wind turbines (for example, it can't capture the behavior of wind turbines not entirely turned into the wind).

5.3 Wake Models

Wake models are computer models used to consider the coordinated distribution of wind turbines within a given region. In particular, a wind turbine extracts energy from the wind, and in turn slows down the speed of the wind downstream from the wind turbine. The effects of the wind turbine will decrease with distance, and at a sufficiently large distance downstream from the wind turbine the velocity of the wind will return to its original strength. However, if a second wind turbine is placed too close to, and in the wake of, the first then the second turbine will experience reduced wind speeds and as a consequence extract less power than if it was subject to free-stream wind speeds. In order to optimize the distribution of wind turbines, the velocity fields as a consequence of the wakes from the wind turbines have to be taken into consideration.

Here, we will look at the Park model which was originally developed by N.O. Jensen and I. Katic, at the Risø National Laboratory in Denmark. While other more complicated models, often taking into consideration the full computational fluid dynamics and turbulence within the system, have been proposed, the Park wake model is a simple model (requiring the least computational effort) which still remains widely used today. The near field, just behind the wind turbine, is neglected in this model as the wind is generally quite tortuous and turbulent and the model is, therefore, only applicable in the far wake (a distance of a few rotor diameters away from the wind turbine). (Why would anyone want to put another wind turbine closer than this?) The basis of the model is to consider the conservation of momentum of the air flow in the axial direction of a wind turbine. In particular, the momentum of the wind at the wind turbine is written as $\pi r_o^2 v_o$ through the wind turbine, and at some distance downwind from the wind turbine it is $\pi r^2 v$. However, because the wake spreads out and gets larger (in radius) as the wake moves downwind $r > r_o$. Therefore, we include this extra region of space, at the wind turbine, where the wind around the wind turbine remains at its original strength

$$\pi r_o^2 v_o + \pi(r^2 - r_o^2)v_\infty = \pi r^2 v \tag{5.55}$$

where r_o is the radius of the rotor, v_o is the velocity of the wind through the turbine (see the Betz limit for this), $r = r_o + \alpha x$ is the expanded radius of the wake a distance x downstream, α is the wake decay constant (the entrainment constant), v_∞ is the free-stream wind speed far from the wind turbine, and v is the velocity in the wake of the wind turbine, a distance x downwind. The wake decay constant sets a linear rate of expansion for the radius of the wake (expanding out like a cone from the wind turbine) and depends on the turbulence in the wind (increasing with turbulence) that might arise, for example, if buildings and rough terrain were to be encountered in the surrounding region of the wind turbines. Note that inside the wake the

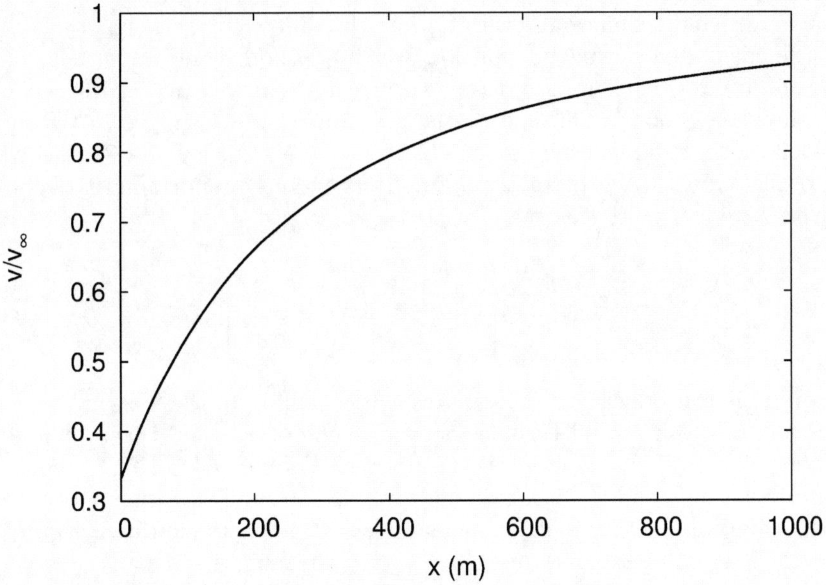

FIGURE 5.10
The velocity downstream from the wind turbine as a function of distance.

velocity is considered to be constant, which isn't particularly realistic, but is a reasonable simplification if we are trying to determine the power extracted from the wind turbines, and not calculate an accurate map of the velocity profiles. The constant velocity inside the wake of a wind turbine is of the form

$$v = v_\infty \begin{cases} 1 - 2a\left(\frac{r_0}{r_0+\alpha x}\right)^2 & \text{if } r < r_0 + \alpha x \\ 1 & \text{else} \end{cases} \tag{5.56}$$

where a is the axial induction factor, that the wind turbine is assumed to be operating under, and x and r represent the downwind direction and the radial distance, respectively, from the cylindrical coordinate system downwind. This constant velocity is entirely in the axial, x, direction and rotational velocities, for example, are neglected. α is the wake constant, and for flat landscapes (such as offshore environments) where turbulence in the wind is minimized, this is set equal to 0.04. The velocity downstream from a wind turbine as a function of distance in the axial direction, using Equation 5.56, is plotted in Figure 5.10. As you move further away from the wind turbine the velocity of the air can be seen to recover its original free-stream velocity, v_∞.

In order to capture the effects of multiple wakes overlapping, in the Park wake model, the deficits in kinetic energy are considered. In particular, the kinetic energy deficit is the fraction of energy lost from the wind because of the reduced velocity in the wind in the wake of the wind turbine. When two (or more) wind turbines all overlap, then the different kinetic energy deficits at the calculated downwind position are simply added together to calculate the total energy deficit. From this we can find the velocity of the wind as a consequence of the multiple overlapping wakes, and use this as the input velocity for a wind turbine that might be placed in this location (in the wake of other wind turbines). Because kinetic energy is proportional to velocity squared we can write this velocity as

$$v_i = v_\infty \left[1 - \sqrt{\sum_{k=1}^{k \neq i} \left(1 - \frac{v_k}{v_\infty} \right)^2} \right] \tag{5.57}$$

where v_i is the velocity of the wind that is incoming to the ith wind turbine, the sum of the kth wind turbines is over all the other wind turbines, and v_k is the velocity of the wind at the position of the ith wind turbine because of the presence of the kth wind turbine (using Equation 5.56). Interestingly, the same equation can be used to calculate the effects of the surface (ground or water, if the wind turbines are off shore) by placing a mirror image of the wind turbine beneath the real wind turbine (similar to how we handled the floor in the Gaussian Plume Model in Chapter 3). Finally, once we have the velocities at the turbines we can calculate the power coefficients relative to if the wind turbine was located in free-stream wind velocity, v_∞

$$C_P, i = 4 \frac{v_i^3}{v_\infty^3} a(1-a)^2 \tag{5.58}$$

where C_P, i is the power coefficient of the ith wind turbine.

The spreadsheet implementation of the Park wake model is considered in Figures 5.11 to 5.13. At the top of Figure 5.11 are the constants in the model; the wake decay constant, the wind speed far from the turbine, and the rotor radius. The remaining constants are the length of this square area, and the number of wind turbines in the area; in the current example we consider 10 turbines randomly placed in a square area 1 km × 1 km. Below the constants in Figure 5.11 is the solution to the Park wake model. In the first column is the rank, or reference number, of the ten wind turbines whose information is included in the rows, from row 12 to 21. The second and third columns contain randomly chosen x- and y-coordinates for the 10 wind turbines - the random positioning here is for demonstration purposes and the x- and y-coordinates can be chosen to be whatever we want (in fact it is these positions that we might seek to optimize). The fourth column contains the axial induction factors that the wind turbines are assumed to be operating at. Here, they are taken to be $\frac{1}{3}$, which produces the maximum efficiency for the wind turbine. Finally, Figure 5.11 determines if the wind turbine is in the wake of any of the other wind turbines in cells E12:N21. In particular, as we move down the rows we are considering the different wind turbines (whose input velocity and extracted power we hope to determine). As we move across from column E to N we go through the other wind turbines (with the reference number for these in cells E11:N11). If the wind turbine under consideration (going down the rows) is in the wake of another (going across the columns), then the cell will contain the number 1. If the wind turbine under consideration is not in the wake of another, then this number is 0. For example, the code in cell E15 is

```
=IF(INDEX($B$12:$B$21,$A15) > INDEX($B$12:$B$21, E$11),
(IF(((INDEX($C$12:$C$21,$A15)-INDEX($C$12:$C$21,E$11))^2) <
(($B$7+$B$5*(INDEX($B$12:$B$21,$A15)-
INDEX($B$12:$B$21,E$11)))^2), 1, 0)),0)
```

where the first IF statement is simply saying if the x-coordinate of the wind turbine under consideration (going down the rows) is greater than the x-coordinate of the second (going across the columns), then continue to see if the wind turbine is in the wake of the second. The x-coordinate of the first wind turbine (going down the rows) is determined using

```
INDEX($B$12:$B$21,$A15)
```

where cells B12:B21 contain the x-coordinates of all the wind turbines, and cell A15 is the cell containing the reference number for the first wind turbine (going down the rows). Note that as we copy this code to other cells, the x-coordinates (cells B12:B21) remain fixed and hence we put $'s before the column letter and row number, and the reference number in cell A15 will change as we go down, but as we go across the reference numbers are always going to be in column A so we only put a $ in front of the column letter. Next we check to see if the difference in y-coordinates between the two wind turbines (squared) is less than the size of the expanded wake (squared), to see if the wind turbine (going down the rows) is in the wake of another (going across the columns).

In Figure 5.12 we calculate the velocity of the second wind turbine's wake (going across the columns) at the location of the wind turbine under consideration (going down the rows), if the wind turbine is in the wake of another (if not we just put this equal to the free-stream wind velocity). For example, the code in cell O15 is

```
=IF(E15, ($B$6*(1 - 2*INDEX($D$12:$D$21,O$11)*
($B$7/($B$7+$B$5*(INDEX($B$12:$B$21,$A15) -
INDEX($B$12:$B$21,O$11))))^2)), $B$6)
```

Constants required in this model include the wake decay constant, the wind speed far from the wind turbine, the rotor radius, and the length of the square area and number of turbines in this area (see text).

	A	B	C	D	E	F	G	H	I	J
1	Park Wake Model									
2										
3	Constants									
4										
5	alpha	0.04								
6	v_infinity	10								
7	r	20								
8	L	1000								
9	N	10								
10					Turbine in wake?					
11	Index	x	y	a	1	2	3	4	5	6
12	1	767.7575694	534.6574765	0.33	0	0	0	1	0	0
13	2	154.1146003	204.3021782	0.33	0	0	0	0	0	0
14	3	527.5587873	587.3255557	0.33	0	0	0	0	0	0
15	4	209.055325	507.0615541	0.33	0	0	0	0	0	0
16	5	37.72367048	236.9802403	0.33	0	0	0	0	0	0
17	6	675.1503674	561.0521957	0.33	0	0	0	0	0	0
18	7	486.4305325	536.6563746	0.33	0	0	0	1	0	0
19	8	708.830886	346.4761598	0.33	0	0	0	0	0	0
20	9	68.0805866	104.2731015	0.33	0	0	0	0	0	0
21	10	568.5746875	488.5819708	0.33	0	0	0	1	0	0
22										

The first column is the index, the second and third columns are random numbers for positions, the first column is the axial induction factors, and in cells E12:N21 is a condition which is 0 if the turbine is not in the wake of another and 1 if it is. For example, the code in cell E15 is

```
=IF(INDEX($B$12:$B$21,$A15) > INDEX($B$12:$B$21, E$11),
(IF(((INDEX($C$12:$C$21,$A15)–INDEX($C$12:$C$21,E$11))^2) <
(($B$7+$B$5*(INDEX($B$12:$B$21,$A15)–INDEX($B$12:$B$21,E$11)))^2),
1, 0)),0)
```

FIGURE 5.11
Spreadsheet implementation of the Park wake model.

	O	P	Q	R	S	T	U	V	W	X
9										
10	Velocities									
11	1	2	3	4	5	6	7	8	9	10
12	10	10	10	8.527903512	10	10	7.297173063	10	10	10
13	10	10	10	10	10	10	10	10	10	10
14	10	10	10	10	10	10	10	10	10	10
15	10	10	10	10	10	10	10	10	10	10
16	10	10	10	10	10	10	10	10	10	10
17	10	10	10	10	10	10	6.521445894	10	10	10
18	10	10	10	7.269623278	10	10	10	10	10	10
19	10	10	10	10	10	10	10	10	10	10
20	10	10	10	10	10	10	10	10	10	10
21	10	10	10	7.766568613	10	10	10	10	10	10
22										

Different turbines are considered in each row as you move down in the spreadsheet, but just as we calculated whether or not a turbine is in the wake of another, we also need to calculate the velocity at the turbine as a consequence of the others (ranked from 1 to 10 going across). The code in cell O15 is

```
=IF(E15, ($B$6*(1 – 2*INDEX($D$12:$D$21,O$11)*
($B$7/($B$7+$B$5*(INDEX($B$12:$B$21,$A15) –
INDEX($B$12:$B$21,O$11))))^2)), $B$6)
```

FIGURE 5.12
Spreadsheet implementation of the Park wake model (continued).

where cell E15 is the corresponding flag (from Figure 5.11) to determine whether or not it was in the wake. And the equation used to calculate this velocity is simply Equation 5.56.

Figure 5.13 shows the solution to the model in the final depiction of the spreadsheet model. In particular, cells Y12:AH21 calculate the term

$$1 - \frac{v_k}{v_\infty} \tag{5.59}$$

from the velocities in cells O12:X21. In column AI we sum the above terms and then calculate the input velocity into the wind turbine, as a possible consequence of the wake from other wind turbines. In particular, the code in cell AJ12 is

```
=$B$6*(1-SQRT(AI12*AI12))
```

and once we have the input velocity, we can calculate the coefficient of power in relation to the wind turbine in free stream conditions. The code in cell Ak12 is

```
=4*(AJ12*AJ12*AJ12/($B$6*$B$6*$B$6))*D12*(1-D12)*(1-D12)
```

and the value of the coefficient (which doesn't take into consideration frictional losses) is 59% if the wind turbine is not in the wake of other turbines, but drops to lower values if the wind turbine is positioned in the wake of other wind turbines.

To summarize, the Park wake model is a very simple and computationally

We calculated the velocity in cells O12:X12, now we calculate in Y12:AH12. $\left(1 - \frac{v_k}{v_\infty}\right)$

	AB	AC	AD	AE	AF	AG	AH	AI	AJ	AK
9										
10										
11	4	5	6	7	8	9	10	Summation	v	Power Coeff.
12	0.021670681	0	0	0.073052735	0	0	0	0.094723415	9.052765848	0.439609834
13	0	0	0	0	0	0	0	0	10	0.592548
14	0	0	0	0	0	0	0	0	10	0.592548
15	0	0	0	0	0	0	0	0	10	0.592548
16	0	0	0	0	0	0	0	0	10	0.592548
17	0	0	0	0.121003387	0	0	0	0.121003387	8.789966133	0.402425175
18	0.07454957	0	0	0	0	0	0	0.07454957	9.254504296	0.469659402
19	0	0	0	0	0	0	0	0	10	0.592548
20	0	0	0	0	0	0	0	0	10	0.592548
21	0.049882158	0	0	0	0	0	0	0.049882158	9.501178424	0.508224922
22										

We sum the rows from colum Y to AH, and calculate the incoming velocity to the wind turbines, and from this calculate the power coefficient (idealized, so that the maximum for a turbine is 59%).

FIGURE 5.13
Spreadsheet implementation of the Park wake model (continued).

efficient way to determine the power extracted from the wind turbine farm, and the co-operative effects from the coordinated arrangement of the wind turbines. This is crucial to optimizing power extraction from wind farms. However, the model is limited and does not include the effects of turbulence or rotational flow in the downstream wake. That said, the performance of this simple model often stands up well in comparison to much more complicated numerical solutions.

5.4 Elastic Deformation of a Wind Turbine Blade

While it is common to simulate wind turbine blade elasticity using more complicated methods, a blade can be modeled as a simple beam. In particular, commercial packages which solve finite element methods are typically used to discretize the wind turbine blades into lots of small elements and solve the elastic (or plastic) deformation of each of the elements. In this way, the complete deformation of the wind turbine blade can be obtained as a function of external loads. Furthermore, finite element methods can be combined with computational fluid dynamics to simulate a complete aeroelastic solution with the wind applying a force on the wind turbine, causing it to move, and in turn the movement of the wind turbine blade feeding directly back into the solution of the wind velocities and forces. Such numerical methods are beyond this simple introduction, and as mentioned usually solved using commercial packages rather than using spreadsheet models. (It should be noted that alternative models can be used, such as the lattice spring model, but you'll probably never encounter such models as they are used very rarely.) Therefore, we'll limit our analysis of the elastic deformation of wind turbine blades to simple beam theory.

In the beam theory we will assume that we have access to the elasticity of the wind turbine blade at different locations along the blade, and using the forces acting on the blade from blade element momentum (BEM) theory, determine the deflections of the blade. In particular, we assume that we have the bending stiffness, which is defined as the product of the elastic modulus with the moment of inertia of the beam (obviously depending on the axis of rotation). If we take the forces acting in the normal and tangential directions to be known (we can get this from blade element momentum theory) then we can calculate the shearing forces as

$$\frac{dT_N}{dx} = -F_T(x) \tag{5.60}$$

and

$$\frac{dT_T}{dx} = -F_N(x) \tag{5.61}$$

Also the bending moments, or torques, can be found from these shearing forces as

$$\frac{dM_N}{dx} = T_T \tag{5.62}$$

and

$$\frac{dM_T}{dx} = -T_N \tag{5.63}$$

where the above equations can be derived using Newton's second law on an infinitesimal piece of the beam, and x is the direction along the length of the beam.

The bending moments, or torques, can now be transformed to the principal axes of the beam.

$$M_1 = M_N \cos(\beta) - M_N \sin(\beta) \tag{5.64}$$

and the second principal axis is perpendicular to the first principal axis

$$M_2 = M_T \sin(\beta) + M_N \cos(\beta) \tag{5.65}$$

where β is the angle between the tangential forces, from blade element momentum theory, and the first principal axis for the beam. The curvatures about the principal axes are from simple beam theory and equal to

$$\kappa_1 = \frac{M_1}{EI_1} \tag{5.66}$$

and

$$\kappa_2 = \frac{M_2}{EI_2} \tag{5.67}$$

The curvature is mathematically the second derivative of the deflection of the beam with respect to the axis of the beam (here the x-direction). These curvatures are then transformed back to the normal and tangential-axis as

$$\kappa_N = -\kappa_1 \sin(\beta) + \kappa_2 \cos(\beta) \tag{5.68}$$

and

$$\kappa_T = \kappa_1 \cos(\beta) + \kappa_2 \sin(\beta) \tag{5.69}$$

The angular deformations are now calculated from

$$\frac{d\Theta_T}{dx} = \kappa_T \tag{5.70}$$

and

$$\frac{d\Theta_N}{dx} = \kappa_N \tag{5.71}$$

and from a further integration we can obtain the deformations of the beam in the normal and tangential directions, u_N and u_T, respectively

$$\frac{du_T}{dx} = -\Theta_T \tag{5.72}$$

and

$$\frac{du_N}{dx} = -\Theta_N \tag{5.73}$$

Note that in the above model we must first acquire the angle β, the angle between the tangential forces, from blade element momentum theory, and the first principal axis for the beam, and the two bending stiffnesses, EI_1 and EI_2, for bending moments in the first and second principal axis, respectively. Figure 5.14 depicts an example of these quantities that will be used in the model. These values are taken (loosely) from "Aerodynamics of Wind Turbines" by Martin O. L. Hansen. The bending stiffness about both principal axis is higher near the base of the wind turbine blade and decreases as we approach the thinner tip. Furthermore, the angle between the tangential forces, from blade element momentum theory, and the first principal axis for the beam is also shown. This decreases to zero, almost linearly, as we move from the base of the wind turbine blade to the tip.

The spreadsheet implementation of this model is shown in Figures 5.15 to 5.17. In Figure 5.15, the forces acting on the wind turbine are obtained from beam element momentum (BEM) theory. In particular, the axial induction factors and the normal and tangential coefficients are in columns B and C, and D and E, respectively. In column F the relative velocity is calculated from the axial induction factors using the equation

$$v_{rel} = \sqrt{(v_\infty(1-a))^2 + (\omega r(1+\acute{a}))^2} \tag{5.74}$$

which is implemented, for example, in cell F6 with the code

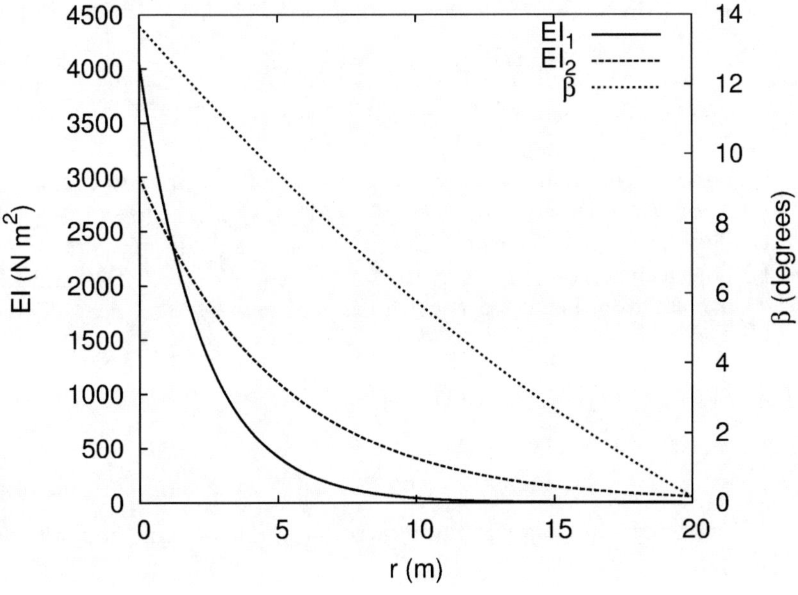

FIGURE 5.14
Elastic parameters of the wind turbine blade, required by the beam theory for capturing elastic deformations.

```
=SQRT((10*(1-B6))^2 + (3*A6*(1+C6))^2)
```

The forces are calculated in columns H and I, using the equations

$$F_N = C_N \frac{1}{2} \rho v_{rel}^2 c \tag{5.75}$$

and

$$F_T = C_T \frac{1}{2} \rho v_{rel}^2 c \tag{5.76}$$

where this is, by definition, how we arrived at the normal and tangential coefficients in the beam element momentum (BEM) theory. ρ is the density of air and c is the chord length (which depends on the position along the blade). The code in cell H6, for example, is

```
=D6*0.5*1.225*F6*F6*G6
```

Once we have the forces acting on the blade, we can calculate the deformation of the blade as a consequence of these forces.

In Figure 5.16 the moments about the principal axis are calculated. The forces from the beam element momentum (BEM) theory are copied to columns B and C. The shear forces are calculated from these applied forces, in columns D and E. The shear forces are zero at the tip of the blades (cells D40 and E40) and the shear forces are calculated using the derivatives

$$\frac{dT_N}{dx} = -F_T(x) \tag{5.77}$$

and

$$\frac{dT_T}{dx} = -F_N(x) \tag{5.78}$$

which we can integrate from the tip back towards the base of the blade. For example, the code in cell D25 is

```
=D26+0.5*(C26+C25)
```

and in cell E25

```
=E26+0.5*(B25+B26)
```

The resulting moments (or torques) about the normal and tangential directions is calculated in columns F and G. Again, the moments are zero at the tip of the blade and we can integrate back to get the momentums throughout the blade, from the tip back to the base of the turbine blade. The code in cell F25, for example, is

```
=F26-E26 - (B25/6 + B26/3)
```

Once we have the moments in the normal and tangential directions we can rotate these to obtain the moments in the two principal axis directions. In particular, the code in cell I25 is

To start with, it is necessary to obtain the forces acting on the wind turbine blade from the beam element momentum (BEM) theory. The following results, therefore, were obtained earlier in this chapter.

	A	B	C	D	E	F	G	H	I
1	Elasticity of Wind Turbine Blade								
2									
3	Results from BEM model								
4									
5	r	a	a'	CN	CT	Vrel	c	F_N	F_T
6	4.00E+000	8.91E-002	8.19E-002	7.91E-001	8.73E-001	15.85992892	1.3667456	166.5822116	166.3813633
7	5.00E+000	1.19E-001	5.04E-002	9.18E-001	5.85E-001	18.05406034	1.46375	268.183339	205.9759614
8	6.00E+000	1.32E-001	3.62E-002	9.07E-001	4.47E-001	20.57229956	1.4998496	352.4972625	217.3334263
9	7.00E+000	1.42E-001	2.80E-002	8.90E-001	3.68E-001	23.22928769	1.4904476	438.2412328	219.1424553
10	8.00E+000	1.53E-001	2.27E-002	8.87E-001	3.17E-001	25.96705836	1.4487296	530.4929212	216.1257805
11	9.00E+000	1.63E-001	1.90E-002	8.95E-001	2.83E-001	28.75971406	1.3856636	628.0239348	208.7052302
12	1.00E+001	1.71E-001	1.62E-002	9.07E-001	2.57E-001	31.59229089	1.31	726.3861446	196.5044105
13	1.10E+001	1.77E-001	1.39E-002	9.18E-001	2.37E-001	34.4545941	1.2282716	819.8427364	179.6384631
14	1.20E+001	1.80E-001	1.19E-002	9.23E-001	2.19E-001	37.33898747	1.1447936	902.4672146	158.8176734
15	1.30E+001	1.79E-001	1.01E-002	9.20E-001	2.02E-001	40.23983473	1.0616636	968.3159007	135.1125086
16	1.40E+001	1.74E-001	8.48E-003	9.06E-001	1.85E-001	43.153503	0.9787616	1010.910914	109.763138
17	1.50E+001	1.66E-001	7.04E-003	8.80E-001	1.68E-001	46.07803808	0.89375	1022.663517	84.06953906
18	1.60E+001	1.55E-001	5.78E-003	8.43E-001	1.51E-001	49.01230903	0.8020736	994.9438052	59.37047664
19	1.70E+001	1.41E-001	4.70E-003	7.98E-001	1.36E-001	51.95513469	0.6969596	918.9889092	37.13243317
20	1.80E+001	1.26E-001	3.80E-003	7.48E-001	1.22E-001	54.9049416	0.5694176	786.5662246	18.9985744
21	1.90E+001	1.15E-001	3.13E-003	7.01E-001	1.09E-001	57.8596982[illegible]	0.4082396	587.0775491	6.536970969

From BEM theory we calculate the relative velocity using

$$v_{rel} = \sqrt{(v_\infty(1-a))^2 + (\omega r(1+\acute{a}))^2}$$

and the forces using

$$F_N = C_N \frac{1}{2} \rho v_{rel}^2 c \quad \text{and} \quad F_T = C_T \frac{1}{2} \rho v_{rel}^2 c$$

where c is the chord.

FIGURE 5.15
The spreadsheet implementation of beam theory for capturing the elastic deformation of a wind turbine blade.

The forces in columns B and C are obtained from BEM theory. The shear forces in columns D and E and the moments in columns F and G are calculated from these forces. In particular, D40 is set to zero (no shear at tip) and the code in cell D25 is
=D26+0.5*(C26+C25)
Similarly, the moment in cell F40 is set to zero and the code in cell F25 is
=F26–E26 – (B25/6 + B26/3)

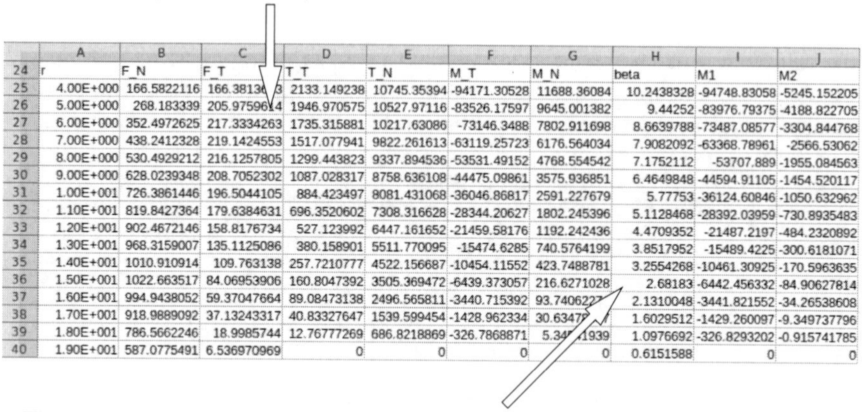

	A	B	C	D	E	F	G	H	I	J
24	r	F_N	F_T	T_T	T_N	M_T	M_N	beta	M1	M2
25	4.00E+000	166.5822116	166.38136[illegible]3	2133.149238	10745.35394	-94171.30528	11688.36084	10.2438328	-94748.83058	-5245.152205
26	5.00E+000	268.183339	205.97596[illegible]4	1946.970575	10527.97116	-83526.17597	9645.001382	9.44252	-83976.79375	-4188.822705
27	6.00E+000	352.4972625	217.3334263	1735.315881	10217.63086	-73146.3488	7802.911698	8.6639788	-73487.08577	-3304.844768
28	7.00E+000	438.2412328	219.1424553	1517.077941	9822.261613	-63119.25723	6176.564034	7.9082092	-63368.78961	-2566.53062
29	8.00E+000	530.4929212	216.1257805	1299.443823	9337.894536	-53531.49152	4768.554542	7.1752112	-53707.889	-1955.084563
30	9.00E+000	628.0239348	208.7052302	1087.028317	8758.636108	-44475.09861	3575.936851	6.4649848	-44594.91105	-1454.520117
31	1.00E+001	726.3861446	196.5044105	884.423497	8081.431068	-36046.86817	2591.227679	5.77753	-36124.60846	-1050.632962
32	1.10E+001	819.8427364	179.6384631	696.3520602	7308.316628	-28344.20627	1802.245396	5.1128468	-28392.03959	-730.8935483
33	1.20E+001	902.4672146	158.8176734	527.123992	6447.161652	-21459.58176	1192.242436	4.4709352	-21487.2197	-484.2320892
34	1.30E+001	968.3159007	135.1125086	380.158901	5511.770095	-15474.6285	740.5764199	3.8517952	-15489.4225	-300.6181071
35	1.40E+001	1010.910914	109.763138	257.7210777	4522.156687	-10454.11552	423.7488781	3.2554268	-10461.30925	-170.5963635
36	1.50E+001	1022.663517	84.06953906	160.8047392	3505.369472	-6439.373057	216.6271028	2.68183	-6442.456332	-84.90627814
37	1.60E+001	994.9438052	59.37047664	89.08473138	2496.565811	-3440.715392	93.74062[illegible]	2.1310048	-3441.821552	-34.26538608
38	1.70E+001	918.9889092	37.13243317	40.83327647	1539.599454	-1428.962334	30.6347[illegible]	1.6029512	-1429.260097	-9.349737796
39	1.80E+001	786.5662246	18.9985744	12.76777269	686.8218869	-326.7868871	5.34[illegible]1939	1.0976692	-326.8293202	-0.915741785
40	1.90E+001	587.0775491	6.536970969	0	0	0	0	0.6151588	0	0

The moments in the first and second principal axis are obtained by rotating the moments in the normal and tangential directions. The code in cell I25 is
=F25*COS(H25*0.01745329) – G25*SIN(H25*0.01745329)
(note 0.01745329 converts degrees to radians).

FIGURE 5.16
The spreadsheet implementation of beam theory for capturing the elastic deformation of a wind turbine blade (continued).

```
=F25*COS(H25*0.01745329) - G25*SIN(H25*0.01745329)
```

and in cell J25

```
=F25*SIN(H25*0.01745329) + G25*COS(H25*0.01745329)
```

Once we have the moments in the first and second principal axis we can calculate the deformation.

In Figure 5.17 the bending stiffnesses are calculated in columns K and L, although these are typically obtained from actual wind turbine blades. The curvature is then calculated in columns M and N, where the curvature is the second derivative of the displacement. Before we integrate twice to obtain the displacements, we rotate this back to the normal and tangential directions. In columns O and P we calculate the curvature in the tangential and normal directions, respectively. For example, the code in cell O25 is

```
=M25*COS(H25*0.01745329) +N25*SIN(H25*0.01745329)
```

and in cell P25

The bending stiffnesses are determined from the actual wind turbine blade, and this is a functional approximation

Finally, given the curvature, we can numerically integrate this twice to obtain the displacements. For example, Q25 is zero and the code in cell Q26 is
=Q25+0.5*(O26+O25)

	K	L	M	N	O	P	Q	R	S	T
24	EI1	EI2	k1	k2	k_T	k_N	theta_T	theta_N	u_T	u_N
25	66178885[illegible]	1349814535	-0.000143171	-3.8858E-006	-0.00014158	2.1637E-005	0	[illegible]	0	0
26	42054292[illegible]3	1105762566	-0.000199687	-3.7882E-006	-0.000197602	2.9023E-005	-0.000169591	2.5330E-005	1.2050E-005	0.000080127
27	267239844.8	905836186.9	-0.000274986	-3.6484E-006	-0.000272397	3.7817E-005	-0.000404591	5.8750E-005	5.3357E-005	0.000360985
28	169821273.3	742057312.4	-0.00037315	-3.4587E-006	-0.000370077	4.7915E-005	-0.000725828	0.000101616	0.000132699	0.000918055
29	107915288.1	607890325.9	-0.000497686	-3.2162E-006	-0.00049419	5.8972E-005	-0.001157961	0.000155059	0.000260115	0.001849607
30	68576269.57	497981277.3	-0.000650297	-2.9208E-006	-0.00064649	7.0319E-005	-0.001728301	0.000219705	0.000446552	0.003280046
31	43577743.52	407944232.7	-0.000828969	-2.5754E-006	-0.000825018	8.0887E-005	-0.002464055	0.000295307	0.000703177	0.005361347
32	27692082.73	334186252.8	-0.001025276	-2.1871E-006	-0.001021392	8.9192E-005	-0.00338726	0.000380347	0.001040312	0.00827064
33	17597318.81	273764016.2	-0.001221051	-1.7688E-006	-0.001217473	9.3422E-005	-0.004506692	0.000471653	0.001465959	0.012201277
34	11182460.79	224266366.3	-0.001385153	-1.3405E-006	-0.001382114	9.1711E-005	-0.005806486	0.00056422	0.001984038	0.017344146
35	7106050.116	183718093.2	-0.001472169	-9.2858E-007	-0.001469846	8.2674E-005	-0.007232467	0.000651412	0.002592607	0.023856311
36	4515638.301	150501113.2	-0.001426699	-5.6416E-007	-0.001425163	6.6191E-005	-0.008679971	0.000725845	0.003282609	0.031816254
37	2869525.113	123289898.5	-0.001199439	-2.7793E-007	-0.00119862	4.4323E-005	-0.009991863	0.000781102	0.004037905	0.041171049
38	1823479.612	100998582.3	-0.000783809	-9.2573E-0[illegible]8	-0.000783505	2.1833E-005	-0.010982925	0.00081418	0.00483742	0.051693036
39	1158755.46	82737627.01	-0.000282052	-1.1068E-[illegible]	-0.000282	5.3921E-006	-0.011515678	0.000827792	0.005659775	0.06298413
40	736347.2608	67778326.83	0	[illegible]	0	0	-0.011656678	0.000830488	0.006489365	0.074593808

The curvature is simply the moment divided by bending stiffness, and once we have the curvature along the principal axis we can rotate this back to the normal and tangential directions. The code in cells O25 and P25 is
=M25*COS(H25*0.01745329) +N25*SIN(H25*0.01745329)
and
=–M25*SIN(H25*0.01745329) + N25*COS(H25*0.01745329)

FIGURE 5.17
The spreadsheet implementation of beam theory for capturing the elastic deformation of a wind turbine blade (continued).

```
=-M25*SIN(H25*0.01745329) + N25*COS(H25*0.01745329)
```

Finally, we integrate this twice to obtain the displacement. First we integrate to obtain the angular deformations in columns Q and R. For example, cell Q25 is zero and the code in cell Q26 is

```
=Q25+0.5*(O26+O25)
```

and cell R25 is zero and the code in cell R26 is

```
=R25+0.5*(P26+P25)
```

Finally we integrate again to obtain the displacements in columns S and T. For example, cell S25 is zero and the code in cell S26 is

```
=S25+R25+(P26/6 + P25/3)
```

and cell T25 is zero and the code in cell T26 is

```
=T25-Q25-(O26/6+O25/3)
```

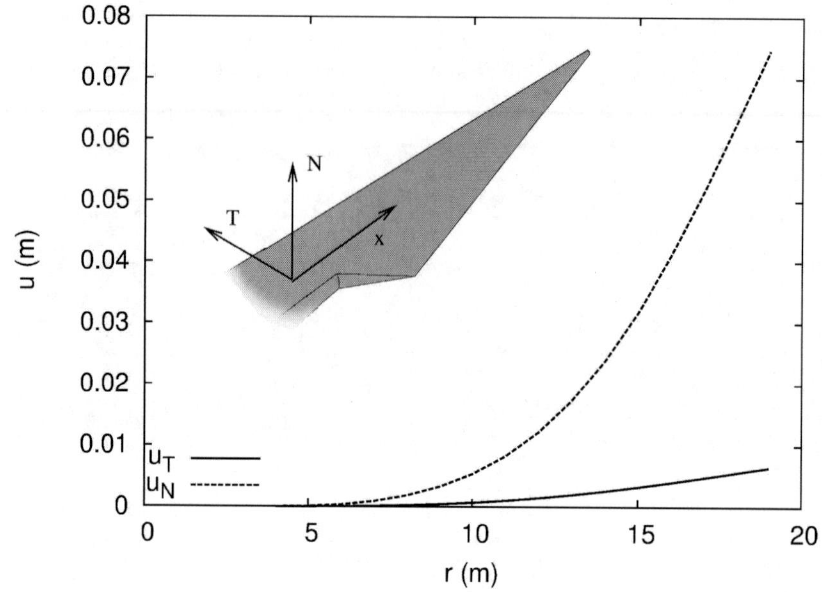

FIGURE 5.18
The deformation of the wind turbine blade. Deformations in the normal and tangential directions are plotted as a function of distance along the blade.

Figure 5.18 depicts the deformation of the wind turbine blade as a function of distance along the wind turbine blade. The inset depicts the orientation of the blade with respect to the normal and tangential directions. The deformation of the blade is obviously greater as we approach the tip of the wind turbine blade, and the deformation is greater in the normal direction than in the tangential direction.

5.5 Exercises

1. Implement the beam element model (BEM) and use it to calculate the power coefficient for the turbine. How does this compare to the Betz limit?

2. The geometry of blade is incorporated in the beam element model (BEM) through the pitch and the cord of the rotor, which are specified along the length of the rotor. Derive your own expressions (either as an equation, or as discrete data points) and investigate the effects that the blade shape can have

on the power coefficient of the turbine.

3. Write a spreadsheet model which calculates the velocity fields as a function of position assuming the Park wake model. For a random distribution of wind turbines generate a contour plot of wind velocity as a function of position.

4. Implement the Park wake model. Imagine you are tasked with positioning 10 wind turbines within a circular region, to create a new wind farm. Find the optimum configuration of wind turbines and discuss how the size of the circle might affect the performance of the wind farm.

5. Implement the Park wake model and simulate the random distribution of 10 wind turbines in a rectangular region (similar to the example presented earlier). Now, however, design a Monte Carlo model to optimize the placement of the wind turbines. In particular, the power coefficients of the wind turbines should be maximized by your Monte Carlo model, which will require the Metropolis algorithm.

6. The forces acting on a wind turbine blade, and the subsequent deformation of the blade depend on the pitch of the blade. Systematically study the pitch of the blade, varying the angle at different locations along the blade, to predict the optimum shape of the blade which would maximize the wind turbines power coefficient, whilst simultaneously limiting the deformation of the blade.

Additional Reading

Hansen, M. O. (2013). Aerodynamics of wind turbines. Earthscan, London.

6

Biofuels

CONTENTS

Biofuels are an important contribution to our energy supply, and have historically made up the majority of our energy requirements (prior to fossil fuels). Legislative mandates are being enacted with the purpose of reducing greenhouse gas emissions from transportation fuels, and while there are many ways in which this might be achieved, it certainly encourages the substitution of fossil fuels with biofuels. This raises questions about the possible implications of substituting biofuels for petroleum fossil fuels. In particular, biofuels are often touted as a method of saving our planet from a climate catastrophe. Furthermore, in the US, 30% of the ethanol is produced in Iowa and, as the Iowa caucuses are the first electoral event in the process of a politician running for the presidency of the US, ethanol is widely supported politically. However, there are concerns about biofuels. The amount of water required to produce biofuels is often debated with advocates arguing that ethanol production requires less water than gasoline, while opponents note that a shortage of water throughout the world is already having a detrimental effect on agricultural productivity.

One of the main criticisms against biofuels is that the production of biodiesels may be energetically inefficient. In other words, the production of biofuels might require more energy than can be obtained from burning the fuel. Recently it has been claimed that producing ethanol from corn, switchgrass, and wood biomass, or biodiesel using soybeans and sunflower, requires the input of more energy from fossil fuels than would actually be gained from the biofuels that are produced. Such inefficiencies in the production of biofuels arise as a consequence of the low efficiency of photosynthesis, the slow growth of land-based crops, and the energy-intensive oil extraction process. However, the lower estimates are often considered misleading because such low estimates neglect the energy value of by-products, and once these by-products are considered the biofuels may be more economically viable (especially when government subsidies help lower the direct cost to consumers, at the expense of indirect costs in the form of taxes).

In a world where 14% of our ever-increasing population is undernourished it might be considered dangerous to remove agricultural land from necessary food production. If we are to continue increasing biofuel production it might seem inevitable that we would continue to impinge upon land needed for growing food (or to destroy forests which would only further contribute to global warming by releasing carbon sequestered in woodland). This poses ethical constraints on the production of such biofuels and raises concerns about potential food shortages and food price inflation.

A viable alternative to using land-based crops for producing biofuels is to use algae, or microalgae (photosynthetic microorganisms), as the source of the biofuel. The ability to grow on non-agricultural land, using untreated water (potentially treating it in the process), and consuming industrial CO_2 in the process, could help us develop renewable energy in a sustainable and responsible way. Microalgae has attracted a lot of attention because under conditions of nitrogen deprivation, they produce high amounts of lipids (oil) with dry weights reported to be as high as 60%. This opens up the possibility of using algae for biodiesel. The fact that they yield significantly more biomass when compared to terrestrial plants has only added to the recent hype. However, under conditions of high stress, the nitrogen deficiency required to produce high lipid contents, the yields are drastically reduced. Optimizing the growth rate and oil production for different algal species is an important area of research.

6.1 Droop Model

The Droop model is a mathematical model capable of predicting growth rates of algae whilst accounting for deficiencies in nutrient, such as nitrogen or phosphorus. Nutrient uptake (nitrogen, phosphorus, vitamins, etc...) and biomass growth are known to be uncoupled processes for microalgae and the biomass is observed to continue growing for a few days after the nutrients have been exhausted. Therefore, the Droop model accounts for this by having a variable which describes the internal concentration of a nutrient. Furthermore, the growth rate of the biomass depends on this internal concentration. Typically nitrogen is considered to be the limiting nutrient, because nitrogen limitation induces lipid formation and produces more oil-rich algae.

The concentration of the dissolved nitrogen is given by s, the dilution rate (the flow rate into the system divided by the volume of the system) is given by D, and influent inorganic nitrogen concentration is given by s_{in}. The rate of change in dissolved nitrogen is described by the following equation

$$\frac{ds}{dt} = Ds_{in} - \rho(s)x - Ds \qquad (6.1)$$

where, in this model, the absorption rate, $\rho(s)$, is represented by Michaelis-Menten kinetics

$$\rho(s) = \rho_m \frac{s}{s + K_s} \tag{6.2}$$

where K_s is the half saturation constant for substrate uptake, associated with the maximum uptake rate ρ_m.

The internal nitrogen cell quota, q, is defined as the amount of nitrogen per biomass unit, and the rate of change of internal nitrogen cell quota is given by

$$\frac{dq}{dt} = \rho(s) - \mu(q)q \tag{6.3}$$

where $\mu(q)$ is the growth rate of the biomass, x, which is evolved using the following equation

$$\frac{dx}{dt} = \mu(q)x - Dx \tag{6.4}$$

The growth rate, $\mu(q)$, is related to the internal concentration of the limiting element and given by the Droop function

$$\mu(q) = \bar{\mu}\left(1 - \frac{Q_o}{q}\right) \tag{6.5}$$

where $\bar{\mu}$ is defined as the growth rate at hypothetical infinite quota, while Q_0 is the minimal cell quota, for which no algal growth can take place.

Given the above model we can capture the growth of algae and account for limited nutrient supply. The spreadsheet implementation of the Droop model is depicted in Figure 6.1. At the top of Figure 6.1 are the constants; the dilution rate $D = 0.02$, the influent inorganic nitrogen concentration $s_{in} = 0.5$, the maximum uptake rate $\rho_m = 0.05$, the half saturation constant for substrate uptake $K_s = 0.01$, the growth rate at hypothetical infinite quota $\bar{\mu} = 0.1$, and the minimal cell quota $Q_o = 0.05$. These constants are specific to the system being studied. Also at the top of the spreadsheet is the time constant for the model. This can be varied to ensure stability (too large a time step and the model will become unstable).

Below the constants in the spreadsheet is the solution to the model. The different variables are stored in the columns and as we move down the rows of the spreadsheet we move forward in the simulation time. Column A is simply the iteration number and column B is time (the time step multiplied by the iteration number). Column C updates the concentration of the dissolved nitrogen. In particular, the code in cell C16 is

```
=C15+$E$5*($B$5*$B$6 - F15*E15 - $B$5*C15)
```

which is the equation from above,

$$\frac{ds}{dt} = Ds_{in} - \rho(s)x - Ds \tag{6.6}$$

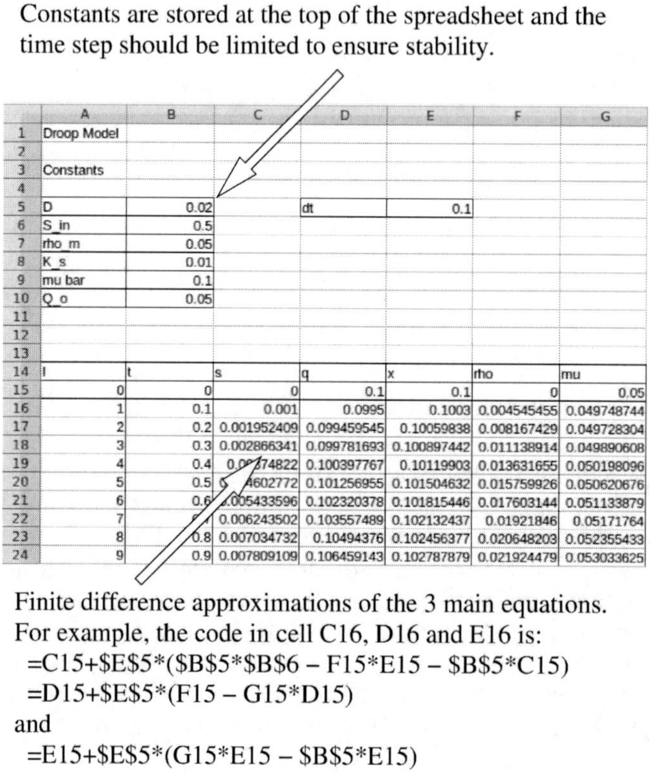

	A	B	C	D	E	F	G
1	Droop Model						
2							
3	Constants						
4							
5	D	0.02		dt	0.1		
6	S_in	0.5					
7	rho_m	0.05					
8	K_s	0.01					
9	mu bar	0.1					
10	Q_o	0.05					
11							
12							
13							
14	I	t	s	q	x	rho	mu
15	0	0	0	0.1	0.1	0	0.05
16	1	0.1	0.001	0.0995	0.1003	0.004545455	0.049748744
17	2	0.2	0.001952409	0.099459545	0.10059838	0.008167429	0.049728304
18	3	0.3	0.002866341	0.099781693	0.100897442	0.011138914	0.049890608
19	4	0.4	0.0[illegible]74822	0.100397767	0.10119903	0.013631655	0.050198096
20	5	0.5	[illegible]4602772	0.101256955	0.101504632	0.015759926	0.050620676
21	6	0.6	[illegible].005433596	0.102320378	0.101815446	0.017603144	0.051133879
22	7	[illegible]	0.006243502	0.103557489	0.102132437	0.01921846	0.05171764
23	8	0.8	0.007034732	0.10494376	0.102456377	0.020648203	0.052355433
24	9	0.9	0.007809109	0.106459143	0.102787879	0.021924479	0.053033625

FIGURE 6.1
The spreadsheet implementation of the Droop model of algae growth.

However, in order to solve this numerically we solve the finite difference approximation of this equation

$$s^{t+1} = s^t + \Delta t \left[Ds_{in} - \rho(s^t)x^t - Ds^t \right] \tag{6.7}$$

where the superscripts represent the time step; the concentration of the dissolved nitrogen at time $t + 1$ is set equal to variables from the previous time step, at time t.

Similarly, Column D updates the internal nitrogen cell quota, q, from the equation

$$\frac{dq}{dt} = \rho(s) - \mu(q)q \tag{6.8}$$

The code in cell D16 is, therefore,

```
=D15+$E$5*(F15 - G15*D15)
```

Again, a finite difference approximation is used to numerically capture the differential equation

$$q^{t+1} = q^t + \Delta t \left[\rho(s^t) - \mu(q^t)q^t \right] \tag{6.9}$$

where again the superscripts represent the time.

Finally, once we have determined how much nitrogen is stored internally in the algae we can update the concentration of algae biomass (in column E). In particular, the code in cell E16 is

```
=E15+$E$5*(G15*E15 - $B$5*E15)
```

which again is the finite difference approximation of

$$\frac{dx}{dt} = \mu(q)x - Dx \tag{6.10}$$

This finite difference approximation is of the form

$$x^{t+1} = x^t + \Delta t \left[\mu(q^t)x^t - Dx^t \right] \tag{6.11}$$

where, again, the superscripts represent time. Note that because the differential equations are updated using finite differences, the size of these differences (in other words, the time step) must be kept small to appropriately approximate the differential equation. If the time step is too large, the solution will either oscillate between high and low numbers, or increase rapidly towards infinite. Therefore, if the system becomes unstable it is usually very obvious.

As part of the differential equations we needed the absorption rate, $\rho(s)$, and the growth rate, $\mu(q)$, which are calculated in columns F and G, respectively. In particular, the code in cell F16 is

```
=$B$7*(C16/(C16+$B$8))
```

And the code in cell G16 is

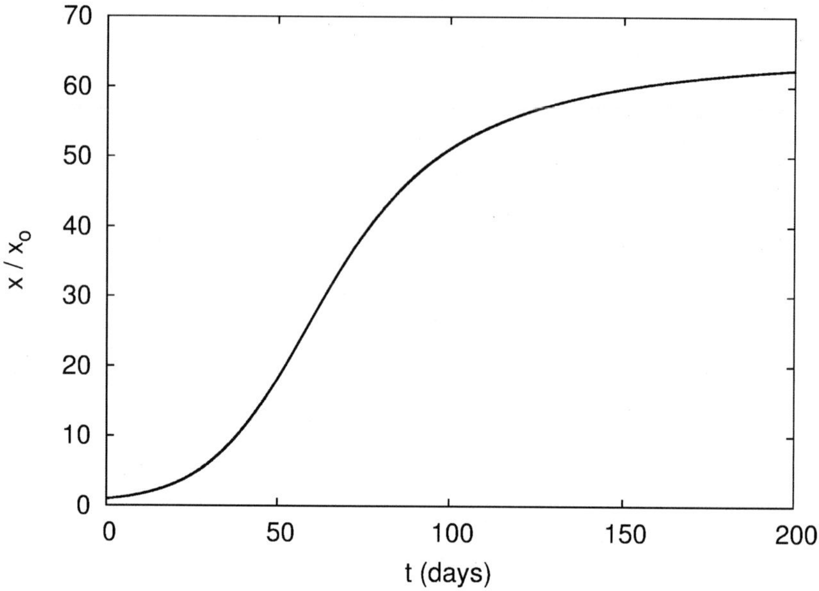

FIGURE 6.2
Plot of the relative biomass concentration as a function of time calculated using the Droop model.

```
=$B$9*(1 - ($B$10/D16))
```

This represents the full solution of the simple Droop model. The solution, given the parameters considered here, is depicted in Figures 6.2 and 6.3.

Figure 6.2 depicts the relative concentration of algae biomass as a function of time. The concentration of algae increases, as the initial concentration of biomass has access to sufficient nutrients, but then saturates later on as nutrients (here nitrogen) become more limited.

The concentration of dissolved nitrogen, relative to the influent inorganic nitrogen concentration, and the internal nitrogen cell quota, relative to the minimal cell quota, are plotted as a function of time in Figure 6.3. Both the relative concentration of dissolved nitrogen, s/s_{in}, and the relative internal nitrogen cell quota, q/Q_o, increase initially as nutrients are introduced into the system through s_{in}. The biomass grows in response to this abundance of nutrients, and eventually the nutrients are limited and so becomes the growth of the algae biomass.

The above model represents a very simple but powerful technique for predicting the growth rates of algae. However, algae can also become photoinhibited, whereby the algae growth is inhibited by too much light. Therefore, it is often required to "flash" a light source periodically such that the algae has a chance to relax back to a resting state. The system can become even

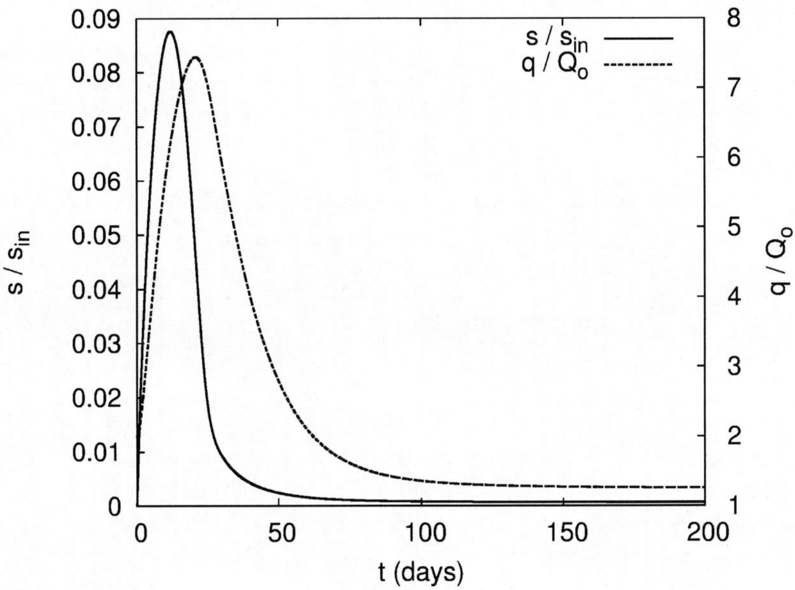

FIGURE 6.3
Plot of the internal nitrogen cell quota and the concentration of the limiting dissolved inorganic nitrogen as a function of time calculated using the Droop model.

more complicated when nutrient deficiency and variations in light are coupled, because when the nitrogen is limited, the pigment of the algae can change. Let's look at a model, the photosynthetic factory model, which takes into consideration the effects of varying light intensity, but for a system with more than sufficient nutrients (perhaps more applicable if all one cares about is the quantity of biomass).

6.2 Photosynthetic Factory of Algae Growth

The photosynthetic factory model is a three-state model that captures both the photosynthesis and photoinhibition of algae when subject to light. The photosynthetic factory model was initially developed by Eilers and Peeters (Ecol. Modell., 42, 199, 1988), but was improved by Wu and Merchuk (Chem. Eng. Sci., 56, 3527, 2001) who applied the model to capturing the growth of *Porphyridium sp.* To capture the dynamics of algae, the algae is considered to exist in three different states; a resting state, an activated state, and an inhibited state. The fraction of cells in the resting, active, and inhibited states are x_1, x_2 and x_3, respectively. Upon absorbing energy from light a cell can switch from being in the resting state (where it does not grow) to the activated state (where the algae does grow). The activated cell can then either receive additional energy from light, which will switch it temporarily to an inhibited state (where it will not grow), or participate in the photosynthesis of the algae and then return to the resting state. The crucial aspect of this model, in contrast to the Droop model considered in the previous section, is that nutrients are assumed to be in ample supply, and the light source (or its attenuation) is the only variable in the system. The reaction rate for each transition that involves the capture of energy from light is assumed to be of first-order with respect to the intensity of light. The remaining processes are assumed to be of zero-order. The kinetic equation that describes the evolution of the resting state is given by

$$\frac{dx_1}{dt} = -\alpha I x_1 + \gamma x_2 + \delta x_3 \tag{6.12}$$

where α, γ, and δ are the rate constants, and I is the light intensity which can vary as a function of position in the system (as light becomes attenuated by the dark water). The fraction of algae in the activated state is given by

$$\frac{dx_2}{dt} = \alpha I x_1 - \gamma x_2 - \beta I x_2 \tag{6.13}$$

where β is another rate constant. Because x_1, x_2, and x_3 are fractions they must sum up to one, and so the final equation for the fraction of algae in the photoinhibited state is simply

$$x_3 = 1 - x_1 - x_2 \tag{6.14}$$

Note that the above variables are fractions and not concentrations. To calculate the concentration of algae we must calculate the growth rate.

Once we have the fraction of algae in the different states we can determine the growth rate, which only depends on the fraction of algae in the activated state. The growth rate of algae, μ, is given by

$$\mu = \kappa\gamma x_2 - M \tag{6.15}$$

where κ is the photosynthetic yield, and M is a maintenance term.

The implementation of the above model is depicted in Figure 6.4. Here we will adopt the same values for the rate constants as used by Wu and Merchuk to successfully capture the growth of *Porphyridium sp.* In addition, in the current model, the light intensity is assumed to be a sinusoidal function of time. Therefore at the top of the spreadsheet, next to the variables required for the model, is the time step, the maximum light intensity and the angular frequency with which the light intensity varies. Varying the light source is important, as the algae would otherwise occupy a photoinhibited state. However, it should be noted that the varying light intensity could be as a consequence of algae circulating in the water through areas close to the light source, and areas further from from the light source. In areas further from the light source, the light can be significantly attenuated and the algae is essentially in the dark.

Below the constants is the solution of the photosynthetic factory model using finite difference approximations for the derivatives. The iteration number is included in column A, and the time (iteration number multiplied with the time step) is calculated in column B. The light intensity is calculated in column C. For example, the code in cell C16

```
=0.5*(SIN(B16*$E$7) + 1)*$E$6
```

which ensures the light intensity varies sinusoidally from 0 to the maximum light intensity stored in cell E6.

The fraction of light in the resting state is calculated from the finite difference approximation of the equation

$$\frac{dx_1}{dt} = -\alpha I x_1 + \gamma x_2 + \delta x_3 \tag{6.16}$$

which takes the form

$$x_1^{t+1} = x_1^t + \Delta t \left(-\alpha I x_1^t + \gamma x_2^t + \delta x_3^t\right) \tag{6.17}$$

The code required to calculate this is in column D, with the code in cell D16 being

```
=D15 + $E$5*(-$B$5*C15*D15 +$B$7*E15 +$B$8*F15)
```

Note that as we move down the rows we move forward in time, and so the fraction of algae in the resting state in row 16 depends on variables from the previous time step in row 15.

Rates and other constants are stored at the top of the spreadsheet. Note that the intensity of light and the angular frequency with which it varies are important variables for algae growth.

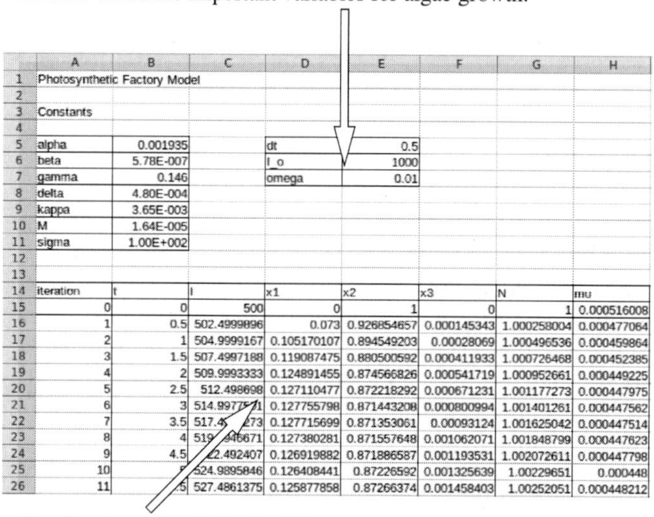

	A	B	C	D	E	F	G	H
1	Photosynthetic Factory Model							
2								
3	Constants							
4								
5	alpha	0.001935		dt	0.5			
6	beta	5.78E-007		I_o	1000			
7	gamma	0.146		omega	0.01			
8	delta	4.80E-004						
9	kappa	3.65E-003						
10	M	1.64E-005						
11	sigma	1.00E+002						
12								
13								
14	iteration	t	I	x1	x2	x3	N	mu
15	0	0	500	0	1	0	1	0.000516008
16	1	0.5	502.4999896	0.073	0.926854657	0.000145343	1.000258004	0.000477064
17	2	1	504.9999167	0.105170107	0.894549203	0.00028069	1.000496536	0.000459864
18	3	1.5	507.4997188	0.119087475	0.880500592	0.000411933	1.000726468	0.000452385
19	4	2	509.9993333	0.124891455	0.874566826	0.000541719	1.000952661	0.000449225
20	5	2.5	512.498698	0.127110477	0.872218292	0.000671231	1.001177273	0.000447975
21	6	3	514.9977[illegible]1	0.127755798	0.871443208	0.000800994	1.001401261	0.000447562
22	7	3.5	517.4[illegible]273	0.127715699	0.871353061	0.00093124	1.001625042	0.000447514
23	8	4	519[illegible]46671	0.127380281	0.871557648	0.001062071	1.001848799	0.000447623
24	9	4.5	[illegible]22.492407	0.126919882	0.871886587	0.001193531	1.002072611	0.000447798
25	10	[illegible]	524.9895846	0.126408441	0.87226592	0.001325639	1.00229651	0.000448
26	11	[illegible]5	527.4861375	0.125877858	0.87266374	0.001458403	1.00252051	0.000448212

The fractions of cells in the different states are calculated as a function of varying light intensity. Only cells in the active state will result in an algae growth rate, which is used to update the concentration of algae.

FIGURE 6.4
The spreadsheet implementation of the photosynthetic factory model of algae growth.

Similarly, the fraction of algae in the activated state is calculated using the finite difference approximation of the equation

$$\frac{dx_2}{dt} = \alpha I x_1 - \gamma x_2 - \beta I x_2 \tag{6.18}$$

which takes the form

$$x_2^{t+1} = x_2^t + \Delta t \left(\alpha I x_1 - \gamma x_2 - \beta I x_2\right) \tag{6.19}$$

This is implemented in column E with the code in cell E16 being

```
=E15+$E$5*($B$5*C15*D15 -$B$7*E15 - $B$6*C16*E15)
```

The fraction of the algae in the inhibited state, calculated in column F, is simply one minus the other fractions. The code in cell F16 is

```
=1 - D16-E16
```

Now that we have calculated the fraction of algae in different states, we can calculate the concentration of algae using the growth rate.

Column G calculates the concentration of algae and column H calculates the growth rate. The update of the algae concentration, N, is achieved using the equation

$$\frac{dN}{dt} = \mu N \tag{6.20}$$

or in finite difference form

$$N^{t+1} = N^t + \Delta t \mu N^t \tag{6.21}$$

where the superscripts represent time and μ is the growth rate. This is implemented in column G with the code in cell G16 being

```
=G15+$E$5*H15*G15
```

The growth rate is simply calculated from

$$\mu = \kappa \gamma x_2 - M \tag{6.22}$$

with the code in cell H16 being

```
=$B$9*$B$7*E16-$B$10
```

The above set of equations take the system variables (the fractions of algae in different states, the growth rate, and the concentration of algae) at a given time step and calculate those same variables at the next time step, subject to variations in light intensity. Therefore, we can evolve the system and calculate the evolution of this algal system. The results from the simulations are depicted in Figures 6.5 through 6.7.

In Figure 6.5 the fraction of algae cells in the resting state is plotted as

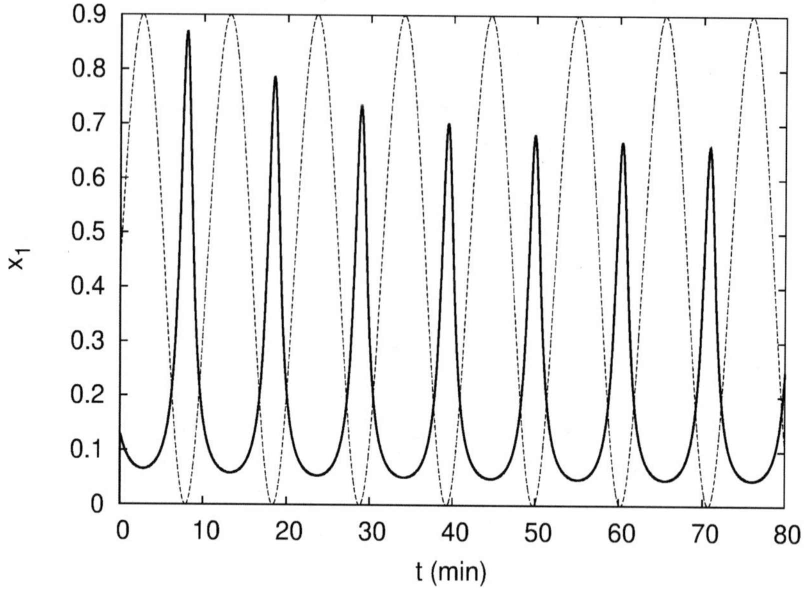

FIGURE 6.5
Plot of the fraction of algae cells in the resting state as a function of time. The variation of light intensity as a function of time is indicated in the background as a dashed line.

a function of time. Note that the fraction oscillates as a function of time, as the light concentration oscillates in the current model. Therefore, plotted in the background is the light intensity as a function of time (this varies from 0 to the maximum value in cell E6). As the light intensity reduces to low values, the fraction of algae in the resting state dramatically increases. This is why varying the light intensity can be so important to the growth of algae, as times of darkness allow the algae to return to this resting state and reduce the number in the unproductive photoinhibited state.

The fraction of algae in the active state is plotted as a function of time in Figure 6.6. Note that in contrast to the fraction of algae in the resting state, which shot up when the light intensity dropped to low values, the fraction of algae in the activated state increases dramatically when the light intensity increases. This is a consequence of the light exciting algae from the resting state to the activated state.

The fraction of algae in the photoinhibited state are plotted as a function of time in Figure 6.7, where again the variation in light intensity is plotted in the background as a dashed line. The fraction of algae in this photoinhibited state is limited because periods of low light intensity allow algae to return to the resting state. Interestingly, the fraction of algae in the photoinhibited state

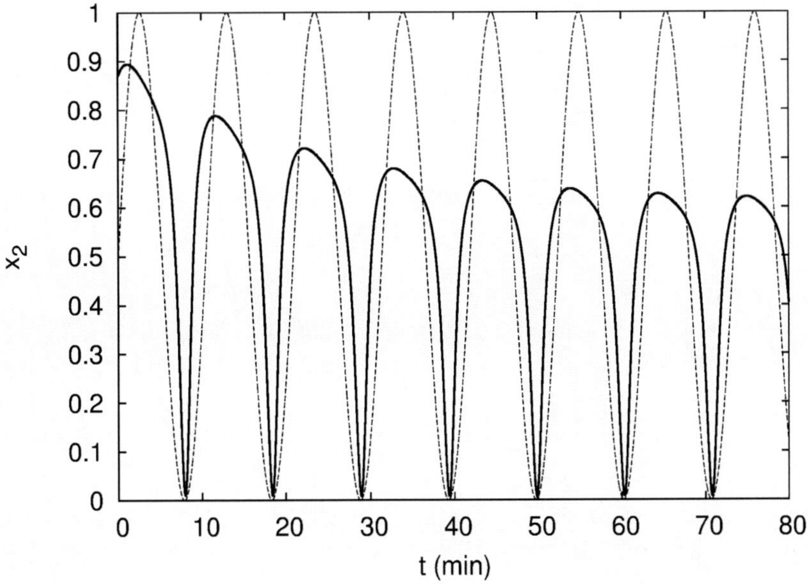

FIGURE 6.6
Plot of the fraction of algae cells in the activated state as a function of time. The variation of light intensity as a function of time is indicated in the background as a dashed line.

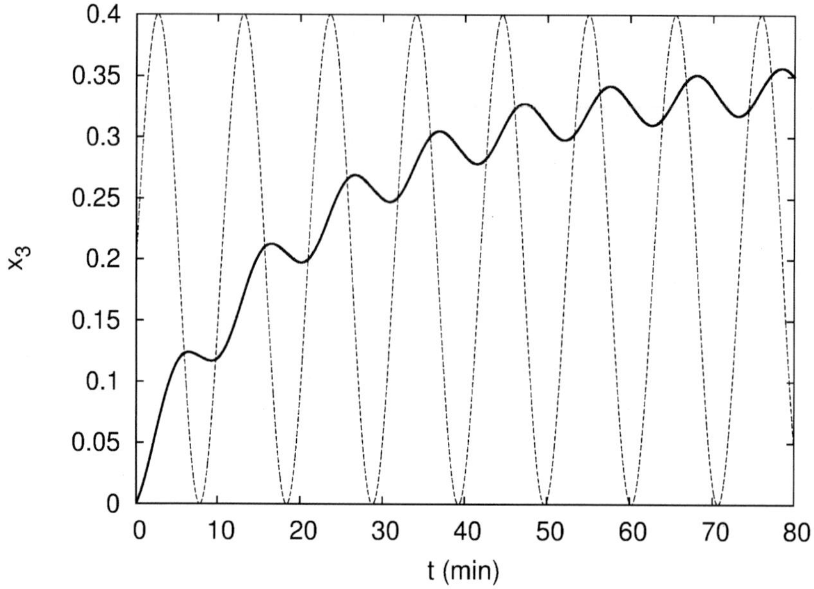

FIGURE 6.7
Plot of the fraction of algae cells in the photoinhibited state as a function of time. The variation of light intensity as a function of time is indicated in the background as a dashed line.

doesn't vary as dramatically as the fraction of algae in both the resting and activated states. Note, however, that the variation in light intensity is chosen somewhat haphazardly, and this could be optimized to ensure the fraction of algae in the activated state is maximized. Furthermore, the light attenuation might be increased as the algae growths (increasing the turbidity, or haziness, of the water). Therefore, the sinusoidal function might not be the best function to capture the variation in light intensity as a function of time.

The above photosynthetic factory model wonderfully captures the strong coupling between biology (microalgae growth) and physics (light absorption and hydrodynamics) at play in the growth of algae. A number of models try to capture the hydrodynamics of the algae growth medium (usually photobioreactors, although open pond systems are more economical) and feed the location of the algae as a function of time into the photosynthetic factory model. However, the model might not be too sensitive to the exact hydrodynamics and handling the variation of light intensity in a functional form is fine (although purely sinusoidal functions, as considered here, are too simplistic). One of the main limitations of this model is the need to parameterize the model using experimental data. One wonders what the predictive capabilities of a model are, if you need to do the experiment beforehand to parameterize the model.

6.3 Cellular Automata Model of Wood Combustion

An important, if a little out-of-fashion, class of simulations is cellular automata. Cellular automata were particularly popular in the 1980's because they are simple to implement and computationally efficient. At the heart of cellular automata is the idea that by making the computer follow very simple rules in local areas, or cells, quite rich and complicated behavior can emerge. For example, the equations that capture fluid dynamics (Navier Stokes equations) can be quite complicated to solve numerically, yet lattice gas cellular automata can capture fluid dynamics by locally propagating and colliding discrete particles of fluid on a lattice. The correct fluid dynamics can emerge from these simple rules, and avoid the complexities of having to solve differential equations. However, as computers have become increasingly faster, solving complex systems of equations is now rather routine and more efficient cellular automata models are no longer a necessity. Furthermore, cellular automata models can be more difficult to parameterize. That said, cellular automata models are still an important class of models that may well come back into fashion in the future, and remain very computationally efficient. This makes cellular automata models perfect for our considerations here, as we can implement such models very easily in a spreadsheet.

The cellular automata model considered here captures the combustion of a log of wood (or other solid fuel) through a model which is similar in many aspects to the traditional forest fire models. The forest fire models consider regions in space to be populated with trees or other vegetation, which can change state to a region which is on fire. The probability of a region, or cell, being ignited depends on whether or not neighboring cells are also on fire. Once a cell is on fire, there is a probability that the cell will burn out and the site is considered to be empty. This is a classic model, and has been used to capture the dynamics of disease spreading and other phenomena, besides that of the spread of forest fires. Here, however, we will apply the model to burning solid fuel. In contrast to traditional forest fire models, oxygen supply is important to burning wood, where wood will char without oxygen but not burn, and heat can be transferred through radiation in a non-local manner. While the model considered here is much simpler than models which incorporate differential equations which couple mass transport, chemical reactions, heat transfer, and fluid dynamics, the cellular automata model can capture the basic behavior of these systems.

The burning wood consists of a series of states. First we start with wet wood, and then progress through two states of drying wood, a state of dry wood, three states of char, a state of burning char, before finally ending up with a state of ash. Although, after ash the site can be transitioned to being empty (or occupied with air). For each possible transition, from one state to the next, there is a minimum amount of energy which is required for the transition to occur. Once a transition occurs, energy is either consumed (considered a negative energy contribution) or released (considered a positive energy contribution), and this energy is stored locally within the cells. Finally, for a transition to occur a stochastic condition must be met, and there is a probability of a given transition occurring within the model. Some transitions may also require an neighboring empty cell (if oxygen is needed for the underlying process).

The table below lists the index for a given state (which simply helps define which state we are in), the minimum energy required to transition to the next state ($E^{transition}_{min}$), the amount of energy which will be released if the transition occurs ($E^{released}_{k \leftarrow k+1}$), and the probability of a transition occurring to the next state ($p^{transition}_{k \leftarrow k+1}$).

Type of cell	index	$E^{transition}_{min}$	$E^{released}_{k \leftarrow k+1}$	$p^{transition}_{k \leftarrow k+1}$
air/empty	0	-	-	-
wet wood	1	1	-1	1/2
drying wood	2	1	-1	1/2
drying wood	3	1	-1	1/2
dry wood	4	2	-1	1/2
char	5	2	2	1
char	6	2	2	1
char	7	3	2	1
burning char	8	0	4	1
ash	9	0	0	1/5

This is a very new model, and as such most of the parameters are rough estimates, but if the model is to describe an exothermic reaction which can sustain itself then during the entire process the sum of energy released (negative if energy is consumed) must be greater than zero. Furthermore, in the original model energy "packages" were considered to be mobile, which enables the fire to spread, and each "package" (an energy of 1) was subject to random walk statistics. In other words, during each iteration there is a probability of the local energy moving in discrete packets in different directions. Here, to make the model more tractable to spreadsheet modeling, the energy diffuses out continuously using a moving average approach.

The spreadsheet implementation of the cellular automata model is depicted in Figures 6.8 through 6.15. In Figure 6.8 the above table is written at the top of the spreadsheet, and referenced in the cells below when required. Also at the top is a measure of the porosity of the system (the fraction of the system which initially consists of empty space). A section to the right of the area of the spreadsheet shown in Figure 6.8 (not shown) is used to calculate the initial configuration using code of the form

The parameters for the model include the index, which indictaes the state of a cell, the minimum energy required to transition to the next state, the energy released when transitioning to the next state and, finally, the probability of undergoing the transition in a given time step.

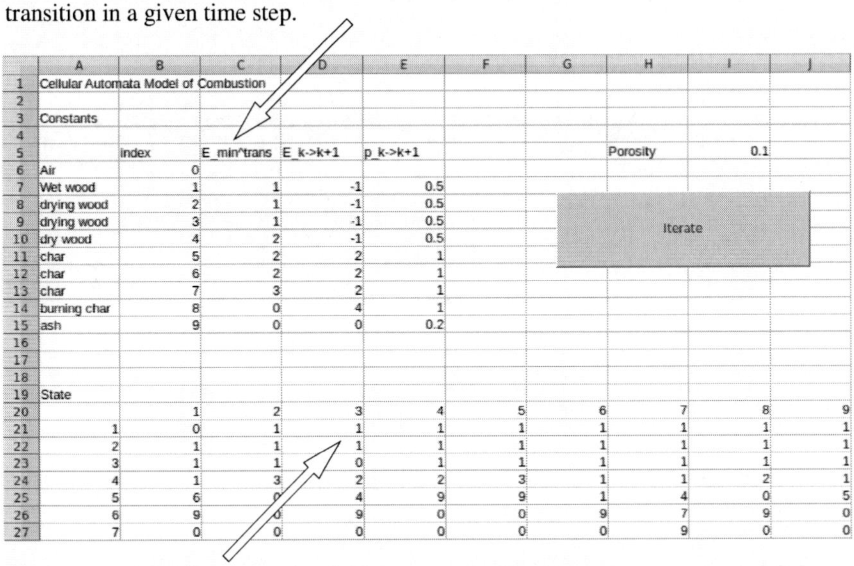

	A	B	C	D	E	F	G	H	I	J
1	Cellular Automata Model of Combustion									
2										
3	Constants									
4										
5		index	E_min^trans	E_k->k+1	p_k->k+1			Porosity	0.1	
6	Air	0								
7	Wet wood	1	1	-1	0.5					
8	drying wood	2	1	-1	0.5					
9	drying wood	3	1	-1	0.5			Iterate		
10	dry wood	4	2	-1	0.5					
11	char	5	2	2	1					
12	char	6	2	2	1					
13	char	7	3	2	1					
14	burning char	8	0	4	1					
15	ash	9	0	0	0.2					
16										
17										
18										
19	State									
20		1	2	3	4	5	6	7	8	9
21	1	0	1	1	1	1	1	1	1	1
22	2	1	1	1	1	1	1	1	1	1
23	3	1	1	0	1	1	1	1	1	1
24	4	1	3	2	2	3	1	1	2	1
25	5	6	0	4	9	9	1	4	0	5
26	6	9	0	9	0	0	9	7	9	0
27	7	0	0	0	0	0	0	9	0	0

In cells B21:AO60 the current state of the system is contained – with the integers in the cells indicating the local state as defined by the indices.

FIGURE 6.8
The spreadsheet implementation of a cellular automata model of wood combustion. System parameters, a button and the state of the system are depicted.

```
=IF($I$5>RAND(), 0, 1)
```

Once the initial configuration is established it is possible to copy and paste this across to the actual cells which contain the state variables (the index which characterizes what state the burning wood is in). The state variables are stored in cells B21:AO60 (these cells just contain numerical values and not any code) and shown at the bottom of Figure 6.8. Also in Figure 6.8 is a button which runs a macro necessary to update the simulation (copy the calculated solution for the next time step over the numerical values representing the current time step). We'll come back to this macro later.

Figure 6.9 depicts the region of the spreadsheet containing the energy stored in the system (cells B67:AO106). Recall that in the current model this energy variable represents energy (for example, in the form of heat transfer) and temperature as well. This energy variable is updated locally as the fire burns the wood, but also spreads out spatially which we'll discuss in a moment.

In cells B112:AO151 we check to see if a site in the simulation is neigh-

Cells B67:AO106 contain the energy in the system, which in this simple model captures both energy and temperature.

	A	B	C	D	E	F	G	H	I	J
65										
66		1	2	3	4	5	6	7	8	9
67	1	0.235953163	0.218327078	0.199560644	0.183937447	0.174020087	0.150797116	0.132161421	0.116900689	0.09580261
68	2	0.312606826	0.322901783	0.271637687	0.286811298	0.2599731	0.234683963	0.205934	0.192726222	0.158311457
69	3	0.454158847	0.528266574	0.483359546	0.6218722	0.572211445	0.453768154	0.440817587	0.292166873	0.380806572
70	4	1.340333643	1.045591487	1.336658371	2.37845083	2.246272982	1.020374922	0.900387033	0.748877227	0.732421782
71	5	3.834559097	3.28699747	4.312839918	6.464667979	5.866733335	3.046037928	2.36916638	2.413541752	2.306664825
72	6	5.722652015	5.388365227	5.931539119	6.375636012	5.688678329	4.198497328	4.30484526	4.797614611	5.240949707
73	7	5.476464142	5.388687355	5.758924968	5.62535266	5.381618332	5.209875496	5.279914223	5.581090983	5.929899981
74	8	5.122279124	5.08556159	5.350947811	5.385847985	5.483577324	5.508705844	5.511644804	5.571727849	5.693973395
75	9	5.143878525	5.191662566	5.28952341	5.395949121	5.50853433	5.554438867	5.535058874	5.48979204	5.461063928
76	10	5.292747032	5.321791365	5.375625154	5.451955992	5.529184121	5.565757514	5.546649592	5.481077496	5.392247066
77	11	5.419619573	5.435603658	5.470432394	5.520260276	5.567545653	5.588681917	5.568082535	5.503076828	5.405133909
78	12	5.55561548	5.562559216	5.582488203	5.609488991	5.630782917	5.631400226	5.600163998	5.534244132	5.440083119
79	13	5.73110926	5.7296748	5.732604011	5.734795621	5.727711569	5.701937635	5.650924111	5.574419442	5.478294517
80	14	5.954243804	5.944085514	5.927897576	5.902832635	5.864349593	5.807582457	5.729810924	5.633069008	5.52410799
81	15	6.220910838	6.201885732	6.16608487	6.113138441	6.042145248	5.952158701	5.843646336	5.7204115	5.590131563
82	16	6.523768482	6.496270522	6.441588669	6.361589794	6.258523587	6.134868594	5.993950058	5.841197146	5.684948546
83	17	6.854720365	6.819517348	6.747238059	6.641630382	6.507606722	6.350655018	6.176810858	5.993266364	5.809019646
84	18	7.204248469	7.162264254	7.073824181	6.944226059	6.780617901	6.591127273	6.384368713	6.169485981	5.956405281

FIGURE 6.9
The spreadsheet implementation of a cellular automata model of wood combustion. Local energy in the system.

boring an empty site. This is depicted in Figure 6.10. For example, the code in cell C113 is

```
=IF((C21=0) OR (B22=0) OR (C23=0) OR (D22=0), 1, 0)
```

This is important as a couple of the reactions require air or empty space in order to occur. In particular, the reactions of char to burning char (index 7 to 8) requires oxygen and the removal of ash requires empty space. In other words, for either of these reactions to occur the site has to be adjacent to an empty space.

In cells B158:AO197 we check to see if any of the sites occupied by wet wood (index 1) are going to undergo a reaction to dry wood (index 2). For example, the code in cell B158 is

```
=IF(AND((B21=1), (B67>$C$7),($E$7>RAND())), 1, 0)
```

where the first part of the AND statement is just checking that the index is 1 (`B21=1`), the second part is checking that the energy at this site (cell B67) is greater than the minimum required (cell C7) and, finally, that the probability that this reaction is to occur has been met. If all these conditions are met, then the cell is filled with a number 1, to indicate that the reactions will occur, else it is filled with a number 0. Similar calculations are required for the other reaction in the system in an almost identical fashion. Except for the reactions which require oxygen or empty space, which have an additional condition. For example, the reaction from char to burning char requires oxygen and so the check to see if a reaction from state 7 to state 8 is to occur is of the form calculated in cell B434

For two reactions, from state 7 to 8 and 9 to 0, oxygen or air is required and so these reactions can only occur if the site neighbors an empty site, with index 0.

	A	B	C	D	E	F	G	H	I	J
109	Is neighbour air?									
110										
111		1	2	3	4	5	6	7	8	9
112	1	1	1	0	0	0	0	0	0	0
113	2	1	0	1	0	0	0	0	0	0
114	3	0	1	0	1	0	0	0	0	0
115	4	0	1	1	0	0	0	0	1	0
116	5	1	1	1	1	1	0	1	0	1
117	6	1	1	1	1	1	1	0	1	1
118	7	1	1	1	1	1	1	1	1	1
119	8	1	1	1	1	1	1	1	1	1
120	9	1	1	1	1	1	1	1	1	1
121	10	1	1	1	1	1	1	1	1	1
122	11	1	1	1	1	1	1	1	1	1
123	12	1	1	1	1	1	1	1	1	1
124	13	1	1	1	1	1	1	1	1	1
125	14	1	1	1	1	1	1	1	1	1
126	15	1	1	1	1	1	1	1	1	1
127	16	1	1	1	1	1	1	1	1	1
128	17	1	1	1	1	1	1	1	1	1

The code in cell C113 is

```
=IF((C21=0)   OR   (B22=0)   OR   (C23=0)   OR   (D22=0), 1, 0)
```

FIGURE 6.10
The spreadsheet implementation of a cellular automata model of wood combustion. Check if a site has access to oxygen.

	A	B	C	D	E	F	G	H	I	J
154	Is transition going to occur? #1									
155										
156										
157		1	2	3	4	5	6	7	8	9
158	1	0	0	0	0	0	0	0	0	0
159	2	0	0	0	0	0	0	0	0	0
160	3	0	0	0	0	0	0	0	0	0
161	4	1	0	0	0	0	1	0	0	0
162	5	0	0	0	0	0	1	0	0	0
163	6	0	0	0	0	0	0	0	0	0
164	7	0	0	0	0	0	0	0	0	0
165	8	0	0	0	0	0	0	0	0	0
166	9	0	0	0	0	0	0	0	0	0
167	10	0	0	0	0	0	0	0	0	0
168	11	0	0	0	[illegible]	0	0	0	0	0
169	12	0	0	0	[illegible]	0	0	0	0	0
170	13	0	0	0	[illegible]	0	0	0	0	0
171	14	0	0	0	[illegible]	0	0	0	0	0

In cells B158:AO197 we check to see if a reaction from state 1 to state 2 is to occur.
The code in cell B158 is

```
=IF(AND((B21=1), (B67>$C$7),($E$7>RAND())), 1, 0)
```

Similar calculations are required to check if other reactions are to occur.

FIGURE 6.11
The spreadsheet implementation of a cellular automata model of wood combustion. Determination of whether a transition from state 1 to state is to occur.

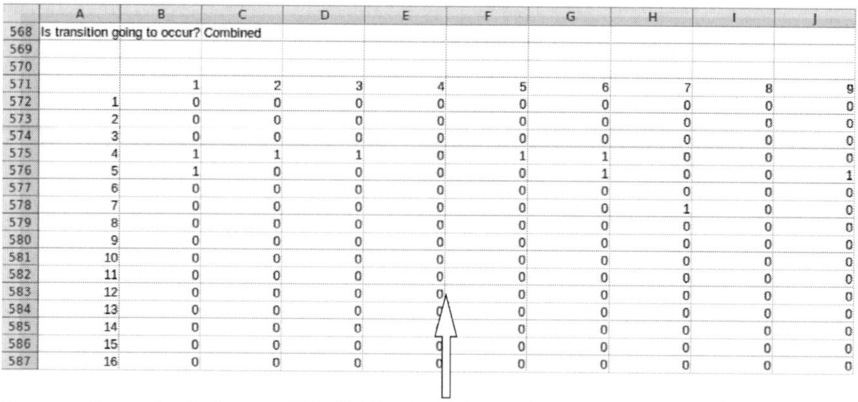

	A	B	C	D	E	F	G	H	I	J
568	Is transition going to occur? Combined									
569										
570										
571		1	2	3	4	5	6	7	8	9
572	1	0	0	0	0	0	0	0	0	0
573	2	0	0	0	0	0	0	0	0	0
574	3	0	0	0	0	0	0	0	0	0
575	4	1	1	1	0	1	1	0	0	0
576	5	1	0	0	0	0	1	0	0	1
577	6	0	0	0	0	0	0	0	0	0
578	7	0	0	0	0	0	0	1	0	0
579	8	0	0	0	0	0	0	0	0	0
580	9	0	0	0	0	0	0	0	0	0
581	10	0	0	0	0	0	0	0	0	0
582	11	0	0	0	0	0	0	0	0	0
583	12	0	0	0	0	0	0	0	0	0
584	13	0	0	0	0	0	0	0	0	0
585	14	0	0	0	[illegible]	0	0	0	0	0
586	15	0	0	0	0	0	0	0	0	0
587	16	0	0	0	0	0	0	0	0	0

Once we have checked to see if individual reactions might occur, we combine these to see if any reaction is going to occur. For example, the code in cell B572 is
=SUM(B158,B204,B250,B296,B342,B388,B434,B480,B526)

FIGURE 6.12
The spreadsheet implementation of a cellular automata model of wood combustion. Determination if any reaction is to occur.

```
=IF(AND((B21=7), (B67>$C$13),(B112 = 1),($E$13>RAND())), 1, 0)
```

where the additional check (B112 = 1) simply looks at the above determination as to whether or not a site is neighbored by an empty site. This is depicted in Figure 6.11.

We also have to determine if any reaction is going to occur (regardless of which reaction it is) and so in cells B572:AO611 we simply add up all the numbers which checked if any of the reactions were to occur. See Figure 6.12. For example, the code in cell B572 is

```
=SUM(B158,B204,B250,B296,B342,B388,B434,B480,B526)
```

Of course, only one reaction can occur within a given cell and so the result of this is still an integer which is 1 if a reaction is to occur (any reaction this time) and 0 if nothing is to occur.

Based upon the above determination of any reaction occurring we need to transition the state of the variables. This is depicted in Figure 6.13, which simply shows the area of the spreadsheet where we check if a transition occurs and if it has we add 1 to the current state, such that it transitions to the next state. For example, in cell B618 we include the code

```
=IF(B572=1, B21+1, B21)
```

Note, however, that if a site is occupied by ash then it will react to index 10 (which doesn't exist) and so we need to loop this back around to make any sites with index of 10 equal to 0. For example, the code in cell B664 (not shown) is

	A	B	C	D	E	F	G	H	I	J
614	Update state if in transition									
615										
616										
617		1	2	3	4	5	6	7	8	9
618	1	0	1	1	1	1	1	1	1	1
619	2	1	1	1	1	1	1	1	1	1
620	3	1	1	0	1	1	1	1	1	1
621	4	2	4	3	2	4	2	1	2	1
622	5	7	0	4	9	9	2	4	0	6
623	6	9	0	9	0	0	9	7	9	0
624	7	0	0	0	0	0	0	10	0	0
625	8	0	0	0	0	0	0	0	0	0
626	9	0	0	0	0	0	0	0	0	0
627	10	0	0	0	0	0	0	0	0	0
628	11	0	0	0	0	0	0	0	0	0
629	12	0	0	0	0	0	0	0	0	0
630	13	0	0	0	0	0	0	0	0	0
631	14	0	0	0	0	0	0	0	0	0
632	15	0	0	0	0	0	0	0	0	0
633	16	0	0	0	0	0	0	0	0	0

If a transition occurs then simply update the index. The code in cell B618 is =IF(B572=1, B21+1, B21)

FIGURE 6.13
The spreadsheet implementation of a cellular automata model of wood combustion. Index is updated if a transition occurs.

```
=IF(B618=10, 0, B618)
```

which performs this calculation and creates the final updated state variables of the system (just waiting to be copied and pasted up to the top of the simulation for an iteration in the model). However, we still have to update the energy in the system.

If a reaction occurs then energy can be released (or if the reaction requires energy then a negative amount is released). To determine how much energy is released we need to check if any of the reactions occurred. This is depicted in Figure 6.14. For example, the code in cell B709 is

```
=IF(B158=1, $D$7,0) + IF(B204=1, $D$8,0) + IF(B250=1,  $D$9,0) +
IF(B296=1, $D$10,0) + IF(B342=1, $D$11,0) + IF(B388=1, $D$12,0)
+ IF(B434=1, $D$13,0) + IF(B480=1, $D$14,0) +
IF(B526=1, $D$15,0) + B67
```

which simply adds the energy from the reactions to the original energy (in cell B67). For example, in the above code, `IF(B158=1, $D$7,0)` simply states that if a reaction from index 1 to 2 occurs (which was calculated in cell B158) then add the energy released during this reaction (stored in the table at the top of spreadsheet in cell D7) but if no reaction occurred then add nothing.

Figure 6.15 depicts the region of the spreadsheet where we take this updated energy and allow it to spread out to some extent. Recall that in the original model there was a random walk assigned to "units" of energy (with

	A	B	C	D	E	F	G	H	I	J
705	Update energy if in transition.									
706										
707										
708		1	2	3	4	5	6	7	8	9
709	1	0.235953163	0.218327078	0.199560644	0.183937447	0.174020087	0.150797116	0.132161421	0.116900689	0.09580261
710	2	0.312606826	0.322901783	0.271637687	0.286811298	0.2599731	0.234683963	0.205934	0.192726222	0.158311457
711	3	0.454158847	0.528266574	0.483359546	0.6218722	0.572211445	0.453768154	0.440817587	0.292166873	0.380806572
712	4	0.340333643	0.045591487	0.336658371	2.37845083	1.246272982	0.020374922	0.900387033	0.748877227	0.732421782
713	5	5.834559097	3.28699747	4.312839918	6.464667979	5.866733335	2.046037928	2.36916638	2.413541752	4.306664825
714	6	5.722652015	5.388365227	5.931539119	6.375636012	5.688678329	4.198497328	4.30484526	4.797614611	5.240949707
715	7	5.476464142	5.388687355	5.758924968	5.62535266	5.381618332	5.209875496	5.279914223	5.581090983	5.929899981
716	8	5.122279124	5.08556159	5.350947811	5.385847985	5.483577324	5.508705844	5.511644804	5.571727849	5.693973395
717	9	5.143878525	5.191662566	5.28952341	5.395949121	5.50853433	5.554438867	5.535058874	5.48979204	5.461063928
718	10	5.292747032	5.321791365	5.375625154	5.451955992	5.529184121	5.565757514	5.546649592	5.481077496	5.392247066
719	11	5.419619573	5.435603658	5.470432394	5.520260276	5.567545653	5.588681917	5.568082535	5.503076828	5.405133909
720	12	5.55561548	5.562559216	5.582488203	5.609488991	5.630782917	5.631400226	5.600163998	5.534244132	5.440083119
721	13	5.73110926	5.7296748	5.732604011	5.734795621	5.727711569	5.701937635	5.650924111	5.574419442	5.478294517
722	14	5.954243804	5.944085514	5.927897576	5.90283263	5.864349593	5.807582457	5.729810924	5.633069008	5.52410799
723	15	6.220910838	6.201885732	6.16608487	6.113138441	6.042145248	5.952158701	5.843646336	5.7204115	5.590131563
724	16	6.523768482	6.496270522	6.441588669	6.361589794	6.258523587	6.134868594	5.993950058	5.841197146	5.684948546

We add together the energies that are released from the reactions to the original energy to get the new energy. The code in cell B709 is

=IF(B158=1, $D$7,0) + IF(B204=1, $D$8,0) + IF(B250=1, $D$9,0) + IF(B296=1, $D$10,0) + IF(B342=1, $D$11,0) + IF(B388=1, $D$12,0) + IF(B434=1, $D$13,0) + IF(B480=1, $D$14,0) + IF(B526=1, $D$15,0) + B67

FIGURE 6.14

The spreadsheet implementation of a cellular automata model of wood combustion. Calculation of new energy after transition occurs.

an energy value of 1) and these units randomly wandered around. To simplify the implementation in a spreadsheet (without I believe substantively changing the model) we will just apply a moving average to the energy and allow the energy to diffuse out.

For example, the code in cell C756 is

```
=0.2*(2*C710+0.75*(C709+D710+B710+C711))
```

This mimics the probability of "units" of energy moving to a neighboring direction (either up, down, left or right) being 0.15 and the probability of the energy not moving being 0.4 (although here we don't have the probability of "units" of energy moving around, but in effect split the "units" up and allow these proportions to move instead). To maintain similar probabilities at the boundaries (with less directions for the energy to move in) we modify the above code. For example, in cell B755 the code is

```
=0.2857*(2*B709+0.75*(C709+B710))
```

and in cell C755 the code is

```
=0.23529*(2*C709+0.75*(D709+B709+C710))
```

Although it might be better to allow the probability of the energy moving

	A	B	C	D	E	F	G	H	I	J
751	Energy diffuses out									
752										
753										
754		1	2	3	4	5	6	7	8	9
755	1	0.2485895	0.236576059	0.212831086	0.203095014	0.186836947	0.166407285	0.145773142	0.129249283	0.106675457
756	2	0.325870532	0.328786438	0.302550066	0.315337584	0.294148259	0.254444441	0.232431979	0.193087442	0.181075485
757	3	0.422162724	0.407208379	0.457109043	0.806873848	0.616167543	0.371720449	0.454165444	0.38135089	0.364946515
758	4	1.277953955	0.692075003	1.217699615	2.251801062	1.824174773	0.705119883	0.897040231	0.950328507	1.20775716
759	5	4.395594968	3.652002347	4.128115408	5.425916206	4.663541917	2.686630966	2.397388348	2.798765157	3.308819595
760	6	5.63986492	5.204827485	5.647980566	6.106790118	5.548844083	4.266814483	4.218716985	4.55011	5.731484905
761	7	5.441820568	5.411872331	5.648049029	5.685445159	5.453769904	5.139260557	5.20308417	5.469309893	5.77947746
762	8	5.18202377	5.192261165	5.368357818	5.432713232	5.461136903	5.467412811	5.48896894	5.570166323	5.698646026
763	9	5.174679698	5.20277826	5.312937062	5.403758906	5.497886147	5.539484031	5.529402345	5.503256038	5.487829256
764	10	5.293898217	5.323062308	5.380305536	5.453935198	5.525742672	5.55914618	5.5411563	5.482195827	5.401580217
765	11	5.423955379	5.440401846	5.475269551	5.523017565	5.567354646	5.585390656	5.563018864	5.499511442	5.404589592
766	12	5.563713692	5.570531008	5.589257973	5.614044649	5.632735133	5.630795061	5.59776325	5.531359161	5.437380198
767	13	5.739162877	5.73742362	5.739270034	5.739813829	730864493	5.703417808	5.651319444	5.574247542	5.477303607
768	14	5.960029066	5.949689492	5.933000085	5.907160239	867780624	5.810271511	5.732207656	5.635540081	5.526308931
769	15	6.223831126	6.204857055	6.169110511	6.116153258	045083647	5.955099876	5.846908212	5.724371208	5.594707784
770	16	6.52375954	6.496522244	6.442312955	6.36286808	260340989	6.137240542	5.997058464	5.845365328	5.690318635

In the current model the combustion spreads spatially through the diffusion of the energy. The code in cell C756 is

=0.2*(2*C710+0.75*(C709+D710+B710+C711))

FIGURE 6.15

The spreadsheet implementation of a cellular automata model of wood combustion. Allow energies to diffuse to neighboring regions (required for fire to spread).

to be less if this is an actual physical barrier (such as the edge of the burning sample), and include energy being released to its environment (which is essentially the entire point point of burning the fossil fuel).

Once we have the updated states and the updated energy calculated, to finish the iteration we simply have to copy these up to the top of the simulation to replace the old values. Upon doing this, of course, the spreadsheet will repopulate itself with new values and another iteration will have occurred. However, if we copy the new states first then the energy will update based on these states (and not on the current states as it should do) or vice versa, if we copy the new energies then the states will be recalculated based on these new energies before we've had a chance to copy the states. Therefore, we copy the energy and states and paste (using paste special to paste only the numbers) to a separate portion of the spreadsheet (not shown). Then copy these values to the top of the spreadsheet. This avoids updating the state or energy separately. The above copy and pasting is recorded in a macro, such that running the macro allows an iteration to occur. Once we have this macro, we can create a button which when pressed will perform this macro and update the system to the next iteration.

Some preliminary results from this simulation are shown in Figure 6.16. Three snapshots are shown after 12, 24, and 36 iterations. The shades of gray correspond to the state of the burning sample, with an index of 0 representing empty space. An index of 1 represents wet wood and as it burns the state transitions through to ash which has an index of 9. In the initial conditions some of the sites are considered empty, while most of the sites are considered to be wet wood. The initial energy is 0 everywhere in the system, except for a small region of the simulation on the left boundary of the simulation which is subject to a localized high energy input (ignition). As a consequence the wood burns in this localized region, releasing more energy in the process, while also spreading out. As can be seen, the fire spreads out and propagates through the sample, and would eventually consume all of the biofuel. It should be noted that while this model is a wonderfully simple model that correctly captures the qualitative behavior of the burning of fossil fuels, it is very hard to parameterize such models. Perhaps this model could be mapped on to a more complicated model (incorporating, for example, mass transport, chemical reactions, heat transfer, and fluid dynamics) which might allow the above parameters in this simple cellular automata model to be parameterized.

6.4 Exercises

1. Systematically vary the parameters in the Droop model and determine which parameters are most important when optimizing biomass yield.

2. Systematically change the frequency with which the light intensity varies in the photosynthetic factory model. Is there a frequency which maximizes the algae growth rate? Plot the maximum algae growth rate as a function of frequency.

3. Simulate the growth of algae using the photosynthetic factory model in a photobioreactor. Note that the variation in light intensity, in a real photobioreactor, is likely to depend on the amount of algae present in the water. If the water is free of algae then the light intensity will be constantly high, as the water is clear. However, if there is a substantial amount of algae in the water then it may be quite opaque and only the algae at the surface of the photobioreactor tube will receive any light. Using the Beer-Lambert law modify the photosynthetic factory model such that the variations in light intensity are due to the different levels of light experienced by the algae as they circulate around the photobioreactor tube. The Beer-Lambert law is given by

$$I = I_0 e^{-x\sigma N} \tag{6.23}$$

where I is the light intensity, I_0 is the light intensity outside the photobioreactor, x is the distance through the water which the light passes, N is the algae concentration (calculated in the photosynthetic factory model), and σ is an attenuation coefficient for the algae in the water.

4. Simulate a photobioreactor subject to the optimum amount of light using

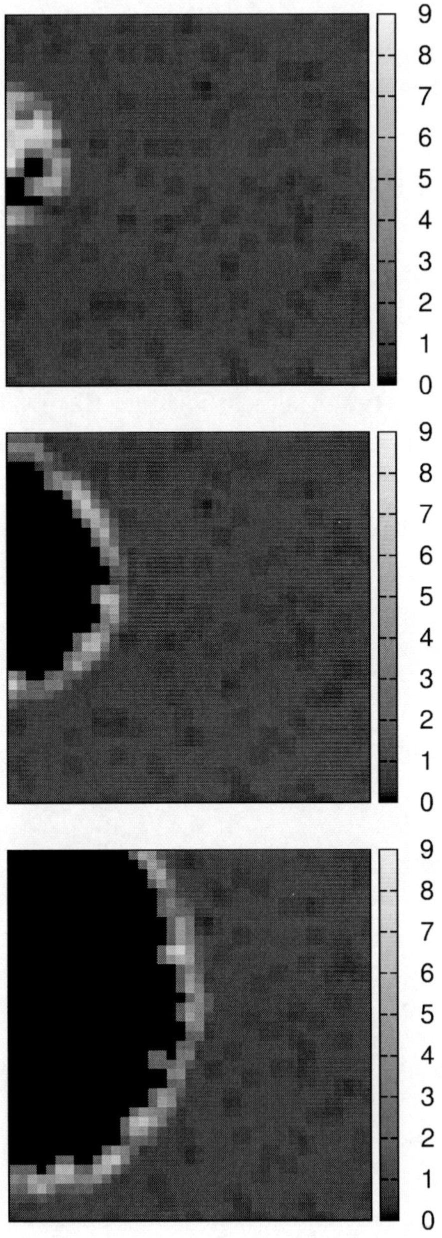

FIGURE 6.16
Example of combustion of a piece of wood (two-dimensional square in current model) using a cellular automata model. The index varies from 0 (empty space), through wet wood (1) all the way up to 9 (ash).

the photosynthetic model. In particular, when the algae concentrations are small the entire photobioreactor is subject to light, and the optimum amount of light for growing algae is likely to be small. However, as the algae grows and and the water becomes increasingly turbid, the optimum intensity of light is likely to increase.

5. Rather than have the energy diffuse out as a continuous function in the cellular automata model of wood burning, incorporate a scheme where packets of energy are able to randomly move around in the form of a random walk. Should this random walk be guided, or biased in any way?

6. To capture gravitational effects in the cellular automata model of wood burning, enable filled sites to "fall" down and occupy empty sites. In other words, if there is an empty site in the model then there should be a probability that the solid material above it would be unstable enough to fall down and occupy this empty space. Obviously this wouldn't happen all the time, as the burning wood still maintains some structural stability, but over time we might expect to find a pile of ash on the floor and not a solid block of ash filled with air pockets (as would occur currently in the model).

Additional Reading

Barsanti, L., & Gualtieri, P. (2014). Algae: Anatomy, Biochemistry, and Biotechnology. CRC Press, Taylor & Francis Group, Boca Raton, FL.

7

Hydrogen Production and Fuel Cells

CONTENTS

Hydrogen is a way in which to store energy, and could have an especially huge impact in automotive or residential applications. The process is typically a two-stage process consisting of first using energy to produce the hydrogen and then, second, of extracting the energy from hydrogen in another form (typically electrical energy using fuel cells).

The production of hydrogen is currently dominated by steam reforming. This involves the direct production of hydrogen from hydrocarbons. The source of these hydrocarbons is mainly fossil fuels. For example, hydrogen can be generated from natural gas with a high degree of efficiency. At high temperatures (generated by burning the natural gas) steam reacts with the natural gas (methane) in an endothermic reaction creating syngas in a process known as steam reforming. In other words, the first stage involves the reaction

$$CH_4 + H_2O \rightarrow CO + 3H_2 \tag{7.1}$$

Additional hydrogen is also generated via a water gas shift reaction, where

$$CO + H_2O \rightarrow CO_2 + H_2 \tag{7.2}$$

where the important consequence is the generation of carbon dioxide (CO_2), a greenhouse gas which we would preferably like to avoid emitting. While it may be possible to sequester the carbon dioxide from the steam reforming process, there are other potentially less polluting methods of generating hydrogen. For example, other (less commonly used at present) methods include electrolysis (using electrical current to drive the reaction), thermolysis (using heat to drive the reaction), or fermentative hydrogen production (using bacteria to convert organic matter to biohydrogen). A particularly interesting extension of the electrolysis method is the use of an "artificial leaf", which couples the current and voltage from a photovoltaic cell with that of the electrochemical cell, which facilitates water splitting through the introduction of electrical energy.

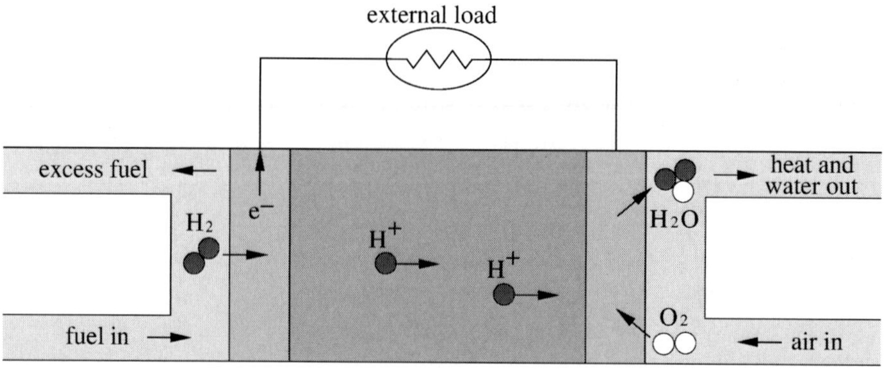

FIGURE 7.1
A schematic of a fuel cell.

In this manner, solar energy can be directly converted to hydrogen fuel in a clean and relatively inexpensive manner. Once we have the hydrogen, however, it is necessary to extract the energy for practical use; this is typically achieved using fuel cells.

Fuel cells are, in may ways, similar to batteries; the main difference is that fuel cells work continuously. In particular, fuel cells create an electric current through an electrochemical reaction. Typically, hydrogen is the fuel used by the device, which is taken into the anode of the fuel cell. Meanwhile, oxygen is taken into the cathode of the fuel cell. The chemical reaction produces electricity, heat, and vaporized water. The main types of fuel cells are polymer electrolyte membrane fuel cells (PEMFC), also known as proton exchange membrane fuel cells, where a polymer membrane with high proton conductivity serves as the electrolyte which separates the anode and the cathode. This allows protons to travel from the anode to the cathode within the electrolyte, while the electrons move through an external circuit creating a current.

For typical fuel cells, with a proton conducting electrolyte, hydrogen interacts with the anode, creating two electrons and two protons via

$$H_2 \rightarrow 2H^+ + 2e^- \tag{7.3}$$

The protons enter the electrolyte and are transported to the cathode. At the cathode the supplied oxygen reacts according to

$$O_2 + 4e^- \rightarrow 2O^{2-} \tag{7.4}$$

creating oxygen ions. During the reactions, the electrons travel from the anode to the cathode via an external circuit. The oxygen ions recombine with the protons which have drifted through the electrolyte to form water

$$O_2^- + 2H^+ \rightarrow H_2O \tag{7.5}$$

This means that the fuel cells are clean and do not pollute the environment, at least while they are in operation.

Polymer electrolyte membrane fuel cells (PEMFCs) are best known for their applications in transportation, with fuel cell vehicles becoming increasingly commercial. The benefits of PEM fuel cells for automotive applications are that they are robust and stable devices that operate at relatively low temperatures (on the order of 80^oC) which enables fast start-up times. Furthermore, the fuel cells do not emit carbon monoxide, carbon dioxide, nitrogen oxides or fine particulate matter, pollutants which plague traditional fossil-fuel based automobiles. Fuel cells could also help to reduce America's dependence on foreign imports of fossil fuels, as the hydrogen fuel could be produced more locally. Of course, the hydrogen production would require energy from some source (hopefully, clean technologies such as nuclear or alternative energies) but the fuel cells are much more efficient than combustion engines (and they operate at peak efficiencies throughout their use, rather than combustion engines which operate more inefficiently in urban or suburban driving conditions). Furthermore, if the energy for hydrogen production were to come from irregular power sources (such as wind or solar), then the need to correlate power availability with demand would no longer be necessary, as hydrogen fuel can be produced at any time. The use of clean technologies would, of course, mean that the fuel cell vehicles would not emit any pollution either during their operation, or during the production of hydrogen fuels to run the vehicles.

The general structure and operation of a PEM fuel cell is depicted in Figure 7.1. The device itself consists of an electrolyte layer in the middle (through which the protons travel), sandwiched between the two porous electrodes. Hydrogen fuel is continuously fed into the cell from one side and interacts with the anode, while simultaneously the oxidant (in general the oxygen in the air) is continuously fed into the other side of the cell and interacts with the cathode. In Figure 7.1 the hydrogen atoms are depicted as solid circles while the oxygen atoms are depicted as white circles. The anode and cathodes are porous to enable them to be permeable to the hydrogen fuel and air, respectively. Meanwhile, the electrolyte would ideally possess a gas permeability as low as possible. However, there remain many challenges (such as reducing costs, while improving efficiencies).

To characterize the performance of a fuel cell it is quite common to look at the polarization curve, otherwise known as the voltage-current (V-I) curve. An example of such a curve is shown in Figure 7.2. The polarization curve can be separated into three different regions, where different mechanisms influence the losses in the system. To the left of the curve, at low currents, the behavior of the fuel cell is dominated by activation or kinetic losses. In other words, the reactions taking place at the surface of the electrodes are not instantaneous and the kinetics of the electrochemical reactions cause a decrease in the voltage of the cell. For example, the slow oxygen reduction kinetics at the cathodes of polymer electrolyte fuel cells can cause such losses. As the current is de-

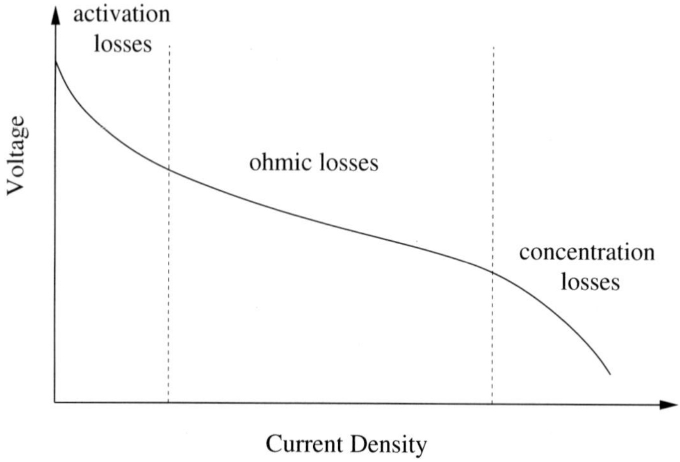

FIGURE 7.2
The different regions of the polarization (V-I) curve corresponding to activation, ohmic, and concentration losses.

creased through the cell, ohmic losses become an increasingly important factor in the lowering of the voltage. This arises simply because of the resistance of the cell electrolyte and electrodes. At high currents, mass-transport limitations become increasingly important. Concentrations of reactants at catalyst sites are consequently reduced, or depleted, faster than the reactants can be transported to the catalyst sites. This results in concentration losses as seen in Figure 7.2 for high current densities. The polarization curve is often used to both experimentally characterize fuel cell performances and gauge the performance of models and simulations, and we will look at a couple of computer models used to capture the behavior and physics of PEM fuel cells. First we will turn our attention to simple electrochemical models which are similar to the equivalent circuit models of solar cells encountered in Chapter 4, before considering a continuum macroscopic model of fuel cell concentrations.

7.1 Photovoltaic and Electrochemical Equivalent Circuit Model

As a society we are desperately in need of clean energy sources, but furthermore, we are desperately in need of portable energy sources which can be used in automotive applications. The clean and renewable aspects of using photovoltaic devices to absorb sunlight, when coupled with the use of electrochemical reactions to produce hydrogen fuel for mobile applications, is an attractive solution to our long term energy problems. The combination of photovoltaic and electrochemical cells has recently attracted a lot of attention in what is commonly referred to as an "artificial leaf".

Leaves, and other photosynthetic organisms, directly convert the energy of sunlight into hydrogen fuel by splitting water to produce oxygen and hydrogen. Subsequently, this chemical fuel is fixed through their combination with carbon dioxide to produce carbohydrates. This last step, however, does not change the energy of the system, and is just a convenient way for nature to store energy and the main extraction of energy from sunlight is in the creation of hydrogen fuel. In other words, the energy storage in photosynthesis is primarily due to water splitting, and carbohydrate production is simply a method of storing the hydrogen. The solar water-splitting cells, or artificial leaves, consist of a silicon photovoltaic interfaced to hydrogen- and oxygen-evolving catalysts. This offers a direct method of converting solar energy directly into fuel, in a manner similar to that of real leaves. In order to drive the electrochemical process of water-splitting we must provide enough energy to overcome both the energetic barrier of water oxidation (1.23 V) and any overpotentials needed to drive the catalysis. Typically, solar cells will not have a large enough voltage to drive water splitting, and a number of solar cells in series will be required. Therefore, multi-junction or tandem solar cell designs are required for these kinds of devices.

To model these photovoltaic-electrochemical systems we can use an equivalent circuit model, capable of predicting the performance of the system and, crucially, the efficiency of the system. Note that the equivalent circuit approach to solar cells is well established, as we looked at the use of equivalent circuit models of solar cells in Chapter 4. To recap, the light absorption by the solar cell produces a current source of excited electrons and holes. Some of this current is lost due to internal recombination, and this recombination is represented by a diode in the equivalent circuit model, I_D. Mechanical defects and material dislocations also cause losses within the solar cell, and these are captured through the insertion of the shunt resistance, R_{SH}. Internal electrical resistances, particularly at the interface of the semiconductor and the metal contacts, can also result in losses and this is captured by the insertion of a series resistor, R_S. These are depicted on the left-hand side of the equivalent circuit in Figure 7.3. On the right-hand side of the circuit is the electrochemical cell which shares both the same voltage and current as the photovoltaic cell. The electrochemical reaction can only occur if the voltage produced is greater than the thermodynamic potential ($\mu_{th} = 1.23$ V), the overpotential for each reaction, and any losses due to the resistance of the solution between the electrodes. The overpotential for the two reactions, both the oxygen and hydrogen reactions, are captured using diodes in the equivalent circuit (which interestingly enough is consistent with Tafel's law, relating the rate of an electrochemical reaction to the overpotential), while the resistance of the solution is a simple resistor. We can now solve this equivalent circuit model and find the operating current in the photovoltaic-electrochemical system.

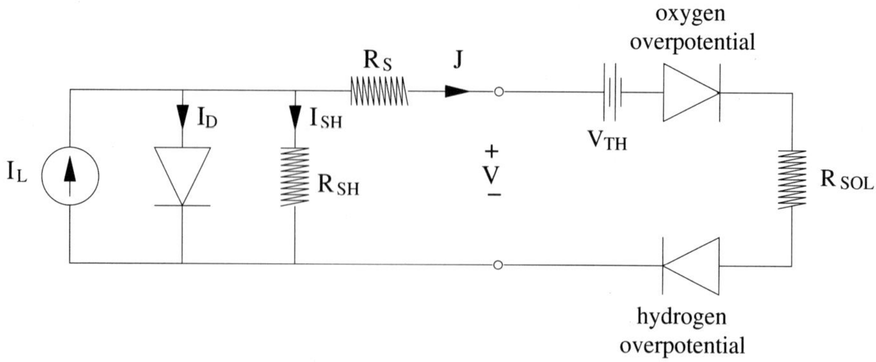

FIGURE 7.3
The equivalent circuit diagram for a photovoltaic electrochemical system (a.k.a. an artificial leaf). Note the solar cell is on the left of the circuit and the electrochemical cell is on the right.

The equivalent circuit of a solar cell has already been covered in Chapter 4, but here it is a little different. Not only do we have the electrochemical part, which we'll come to in a minute, but we also have to combine more than one solar cell in series. A single solar cell would be unable to generate the required voltage to drive water splitting, and cannot be used for direct solar to fuel conversion systems. However, tandem or multi-junction cells are a single photovoltaic device made up of multiple p-n junctions, of different semiconductor materials, where the voltage across the multi-junction device is the addition of the voltage across each individual p-n junction. Furthermore, the p-n junctions, and the different semiconductor materials used for each p-n junction, absorb different regions of the spectrum of sunlight which further increases the efficiency. It should be noted that currently the cost of such multi-junction solar cells is prohibitively high for all but the most specialized applications; it is unclear if or when this technology might become more economically viable. The current density from a multi-junction solar cell is the same as for a single junction solar cell, except we introduce the number of cells (here, junctions) in series, N:

$$J = J_L - J_o \left\{ \exp\left[\frac{q\,(V + JNR_s)}{nNkT} \right] - 1 \right\} - \frac{V + JNR_s}{NR_{sh}} \tag{7.6}$$

where J_L is the photogenerated current density, J_o is the reverse saturation current density, q is the elementary charge, V is the voltage across the cell, N is the number of cells (or junctions) in series, R_s is is the series specific resistance, n is the diode ideality factor, k is the Boltzmann constant, T is the cell temperature, and R_{sh} is the shunt specific resistance. On the right-side of the circuit is the electrochemical cell, where the voltage required to drive the electrochemical reaction must exceed the thermodynamic potential, μ_{th}, of the water splitting reaction by an amount sufficient for the reaction kinetics to occur. To drive the reaction kinetics the overpotential for each reaction (given by Tafel's law) must be met

$$V_O = \tau_O \ln\left(\frac{J}{J_o^O}\right) \tag{7.7}$$

for the oxygen reaction, and

$$V_H = \tau_H \ln\left(\frac{J}{J_o^H}\right) \tag{7.8}$$

for the hydrogen reaction, where τ_O is the Tafel slope for the oxygen reaction, J is the current density from the solar cell, J_o^O is the exchange current density for the oxygen reaction, τ_H is the Tafel slope for the hydrogen reaction, and J_o^H is the exchange current density for the hydrogen reaction. We add to this the resistance of the solution, and obtain the required voltage to drive water splitting as

$$V = \mu_{th} + \tau_O \ln\left(\frac{J}{J_o^O}\right) + \tau_H \ln\left(\frac{J}{J_o^H}\right) + JR_{sol} \tag{7.9}$$

We can now couple the two equations, Equation 7.6 for the photovoltaic cell and Equation 7.9 for the electrochemical cell, to obtain a single solution for the current density in the coupled photovotaic-electrochemical system

$$\begin{aligned} J = J_L - & \\ J_o \left\{ \exp\left[\frac{q\left(\mu_{th} + \tau_O \ln\left(\frac{J}{J_o^O}\right) + \tau_H \ln\left(\frac{J}{J_o^H}\right) + J(R_{sol} + NR_s)\right)}{nNkT}\right] - 1\right\} & \\ - \frac{\mu_{th} + \tau_O \ln\left(\frac{J}{J_o^O}\right) + \tau_H \ln\left(\frac{J}{J_o^H}\right) + J(R_{sol} + NR_s)}{NR_{sh}} & \end{aligned} \tag{7.10}$$

In particular, the above equation simply matches the current and voltage output from the photovoltaic device to the current and voltage input to the electrochemical cell. In this steady-state configuration it can be shown that the efficiency of the solar power to fuel conversion is of the form

$$\eta_{sf} = \frac{\mu_{th} J}{P_{sun}} \tag{7.11}$$

where $\mu_{th} J$ represents the intensity (power per unit area, Wcm^{-2}) devoted to the water splitting process, and P_{sun} represents the solar intensity.

In order to solve the above transcendental equation we must use a numerical method. In other words, above we have one equation and one unknown

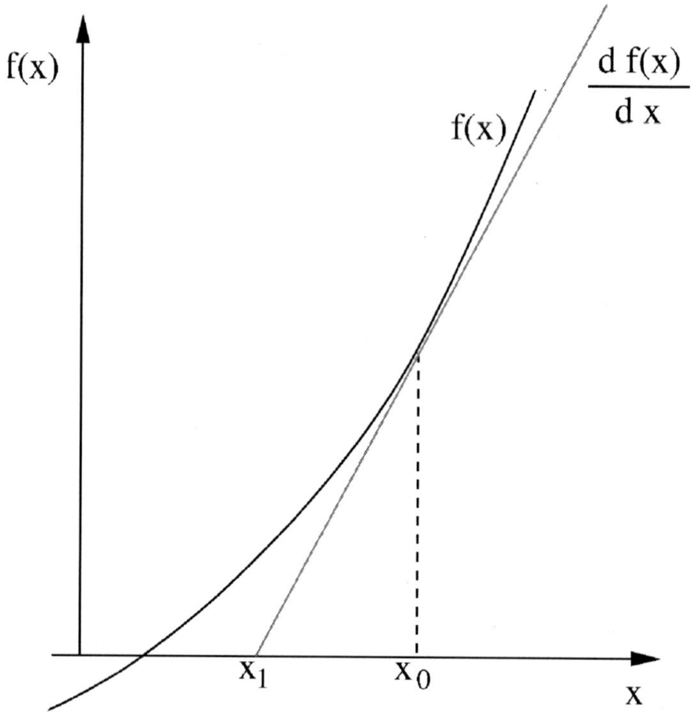

FIGURE 7.4
Schematic of how the Newton-Raphson method works.

(the current density), but the solution is not obvious. However, we solved a similar equation in chapter 4 when looking at equivalent circuits. In that case we used the bisection method which is a very reliable and accurate method of solving such problems. Here we'll use a much faster, although arguably less reliable, method called the Newton-Raphson method. The Newton-Raphson method is depicted graphically in Figure 7.4, where the function we are trying to solve is the dark black curve and we are trying to find where the function crosses the x-axis, and is equal to zero. We start with an initial guess x_o which is not the correct solution, and find the value of the function, $f(x_o)$, and its derivative at this point. Note the derivative is represented by the straight line. The next approximation to the solution, x_1 is obtained by finding where the slope crosses the x-axis. This is repeated, with each step yielding successively better approximations, until an acceptable solution is found. Mathematically, this is often written as the solution of the function

$$f(x) = 0 \tag{7.12}$$

and the equation for finding successively better approximations is of the form

$$x_{n+1} = x_n - \frac{f(x_n)}{f(x_n)'} \tag{7.13}$$

In terms of our system, therefore, we need not only the function, $f(x_n)$, but also the derivative of the function, $f(x_n)'$. The function we are solving is simply Equation 7.10, which upon putting the current density on the right-hand side to make the left-hand side zero gives us the function

$$f(J) = J_L - J_o \left\{ \exp\left[\frac{q\left(\mu_{th} + \tau_O \ln\left(\frac{J}{J_o^O}\right) + \tau_H \ln\left(\frac{J}{J_o^H}\right) + J(R_{sol} + NR_s)\right)}{nNkT} \right] - 1 \right\} - \frac{\mu_{th} + \tau_O \ln\left(\frac{J}{J_o^O}\right) + \tau_H \ln\left(\frac{J}{J_o^H}\right) + J(R_{sol} + NR_s)}{NR_{sh}} - J \tag{7.14}$$

The derivative of this function is

$$\frac{df(x_n)}{dx} = -\frac{J_o q\left[\frac{\tau_H}{J} + \frac{\tau_O}{J} + R_{sol} + NR_s\right]}{nNkT} \times \exp\left\{ \frac{q\left(\tau_H \ln\left(\frac{J}{J_o^H}\right) + \tau_O \ln\left(\frac{J}{J_o^O}\right) + (R_{sol} + NR_s)J + \mu_{th}\right)}{nNkT} \right\} - \frac{\frac{\tau_H}{J} + \frac{\tau_O}{J} + R_{sol} + NR_s}{NR_s h} - 1 \tag{7.15}$$

and we can now enter this into a spreadsheet and iteratively solve for the current density in our photovoltaic-electrochemical system.

The spreadsheet implementation of the above model is depicted in Figure 7.5. The constants required for the mode, including the photovoltaic properties and the electrochemical properties, are included at the top of the spreadsheet. Note that these constants are taken from the literature, but different systems could have quite different parameters. Below the constants is the area of the spreadsheet devoted to solving the Newton-Raphson method. This consists of three columns. In the first column is the approximate solution of the current density for a given iteration (as we move down the rows in the spreadsheet we perform subsequent iterations). The second and third columns contain the function and the derivative of the function, respectively.

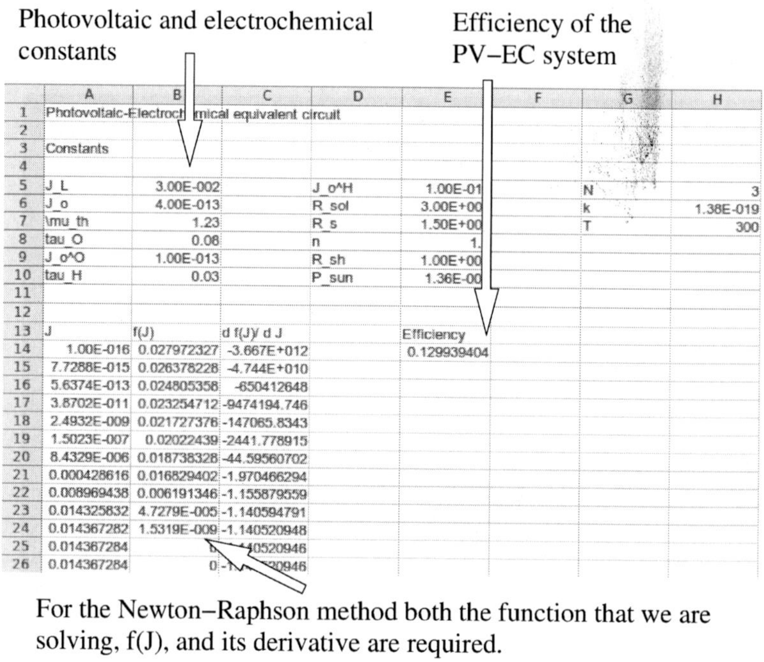

	A	B	C	D	E	F	G	H
1	Photovoltaic-Electroch[illegible]mical equivalent circuit							
2								
3	Constants							
4								
5	J_L	3.00E-002		J_o^H	1.00E-01		N	3
6	J_o	4.00E-013		R_sol	3.00E+00		k	1.38E-019
7	\mu_th	1.23		R_s	1.50E+00		T	300
8	tau_O	0.08		n	1.			
9	J_o^O	1.00E-013		R_sh	1.00E+00			
10	tau_H	0.03		P_sun	1.36E-00			
11								
12								
13	J	f(J)	d f(J)/ d J		Efficiency			
14	1.00E-016	0.027972327	-3.667E+012		0.129939404			
15	7.7288E-015	0.026378228	-4.744E+010					
16	5.6374E-013	0.024805358	-650412648					
17	3.8702E-011	0.023254712	-9474194.746					
18	2.4932E-009	0.021727376	-147065.8343					
19	1.5023E-007	0.02022439	-2441.778915					
20	8.4329E-006	0.018738328	-44.59560702					
21	0.000428616	0.016829402	-1.970466294					
22	0.008969438	0.006191346	-1.155879559					
23	0.014325832	4.7279E-005	-1.140594791					
24	0.014367282	1.5319E-009	-1.140520948					
25	0.014367284	[illegible]	[illegible]40520946					
26	0.014367284	0	-1[illegible]20946					

FIGURE 7.5
Spreadsheet implementation of the photovoltaic-electrochemical equivalent circuit model.

In cell A14 is an initial guess at a solution. Note that the Newton-Raphson method can be quite sensitive to this initial guess, and if the method becomes unstable (giving very large and silly numbers, for example, that continue to grow towards infinite) then you might want to try changing this initial guess. In cell B14 we calculate the function using the code

```
=$B$5 - $B$6*(EXP((1.6E-019*($B$7+$B$8*LN(A14/$B$9)
+ $B$10*LN(A14/$E$5) +
A14*($E$6+ $H$5*$E$7)))/($E$8*$H$5*$H$6*$H$7)) - 1) -
($B$7+$B$8*LN(A14/$B$9) + $B$10*LN(A14/$E$5) +
A14*($E$6+ $H$5*$E$7))/($H$5*$E$9) - A14
```

In cell C14 we calculate the derivative of the function, using the present iteration's estimate for the current density, using the code

```
=-($B$6*1.6E-019*($B$10/A14 + $B$8/A14 +
$E$6+$H$5*$E$7)*EXP(1.6E-019*($B$10*LN(A14/$E$5) +
$B$8*LN(A14/$B$9) +
A14*($E$6+$H$5*$E$7) + $B$7)/($E$8*$H$5*$H$6*$H$7)))/
($E$8*$H$5*$H$6*$H$7) -
($B$10/A14+$B$8/A14+$E$6+$H$5*$E$7)/($H$5*$E$9) - 1
```

The previous estimate is updated, to obtain successively better estimates using both the function and its derivative, and this is reflected in cell A15 which takes the initial guess in cell A14 and updates it using the code

```
=A14-B14/C14
```

After only a few iterations the Newton-Raphson method typically gives a very good solution to the transcendental equation (Equation 7.10). Once we have the current density we can obtain the predicted efficiency of the photovoltaic-electrochemical system using

$$\eta_{sf} = \frac{\mu_{th} J}{P_{sun}} \tag{7.16}$$

where $\mu_{th} J$ represents the intensity (power per unit area, Wcm^{-2}) devoted to the water splitting process, and P_{sun} represents the solar intensity. This is calculated in cell E14. In other words, this model can take fundamental variables from both a photovoltaic system and a separate electrochemical system, and ultimately predict the efficiency of a coupled photovoltaic electrochemical system.

7.2 Simple Electrochemical Model of Fuel Cells

The electrochemical models of fuel cells consider the maximum voltage possible (as a consequence of the chemical reaction within the fuel cell) and then subtract various current-dependent losses from this voltage. In particular, ohmic losses can occur due to the electrical resistance of the electrodes and electrolyte layer, activation losses can occur predominantly due to the kinetics at the electrodes, and concentration losses can occur as a consequence of the gas concentrations being reduced at the electrodes when consumed by the reaction. This result is an equation for the voltage across the fuel cell as a function of the current produced by the fuel cell. In this manner, the electrochemical fuel cell model is capable of capturing the behavior of the devices, but require the input of parameters that are experimentally obtained. The models, therefore, do not generally provide insights into how the properties of the fuel cells emerge from the device physics, but rather can be used to predict how the devices might behave in external (and potentially time-dependent) circuits.

We start by considering the open-circuit (or no-load) voltage. This is the maximum voltage that can be obtained in a fuel cell due to the reactions. In particular, the energy difference between the initial states of the reagents (H_2 and $\frac{1}{2}O_2$) and the final state of the reagents (H_2O) enable this voltage to be calculated through the Nernst expression. The maximum voltage is given by

$$E_N = E_0 + \frac{RT}{2F}\left\{\ln\left(P_{H_2}\right) + \frac{1}{2}\ln\left(P_{O_2}\right)\right\} \tag{7.17}$$

where E_0 is the maximum theoretical voltage (1.229 V), T is the temperature in Kelvin, R is the universal gas constant (8314 J K^{-1} mol^{-1}), and F is the Faraday constant (96487 C). The voltage depends on the hydrogen pressure (in atm), P_{H_2}, and the oxygen pressure (also in atm), P_{O_2} at the anode and cathode, respectively.

The activation losses in voltage arise due to the kinetics at the electrodes. The voltage losses at the anode, however, are much lower than the voltage losses at the cathode and for fuel cells fed with pure hydrogen the voltage losses at the anode can be neglected. The activation voltage losses occur at relatively low currents in the fuel cell. The activation losses can be calculated from the following equation:

$$V_{activ} = -\left[\xi_1 + \xi_2 T + \xi_3 T \ln\left(\frac{P_{O_2}}{5.08\times 10^6 e^{-\frac{498}{T}}}\right) + \xi_4 T\ln(I)\right] \tag{7.18}$$

where I is the current in the fuel cell, the constants ξ_i are phenomenological and, while they may be derived from equations that capture aspects of the fuel cells kinetic, thermodynamic and electrochemical behavior, they can also be obtained experimentally. The oxygen concentration is also included in the above equation as

$$C_{O_2} = \frac{P_{O_2}}{5.08 \times 10^6 e^{-\frac{498}{T}}}. \tag{7.19}$$

where P_{O_2} is the pressure of the oxygen and T is temperature.

As fuel is consumed at an electrode, the concentration of fuel neighboring the electrode will be reduced. It is not possible for the fuel cell concentration to be instantaneously replenished, and this can result in a voltage loss at high currents. In other words, at high currents the fuel is being consumed at the electrodes at a higher rate, and because it cannot be replenished as quickly there is a drop in voltage. The loss of voltage as a consequence of the concentration gradient near the electrodes is of the form

$$V_{conc} = -B \ln\left(1 - \frac{I}{I_L}\right) \tag{7.20}$$

where B is a constant, which can't be obtained from theoretical considerations, and it is more accurate to allow this constant to be fitted. The current, I_L, is the limit current. This represents the point at which the voltage will abruptly decline.

The last mechanism by which voltage can be lost in the fuel cell is through the ohmic resistance of the polymer electrolyte layer (to the ionic flux) and the electrical resistance of the electrodes. The voltage drop as a consequence of ohmic losses is

$$V_{ohm} = R_s I \tag{7.21}$$

where R_s is the combined membrane and electrodes resistance.

Combining the above equations results in the voltage across the cell as a function of the current through it. In other words, the fuel cell's voltage is

$$V_{cell} = n_c(E_N - V_{activ} - V_{conc} - V_{ohm}) \tag{7.22}$$

where n_c accounts for the number of cell in series. In other words, the fuel cell's voltage can now be calculated as a function of the open circuit voltage and the current-dependent voltage drops. We now calculate these voltages using a simple spreadsheet model.

Figure 7.6 depicts the spreadsheet implementation of an electrochemical model of a fuel cell. At the top of the spreadsheet is the constants required for the model (which are taken directly from the literature). The concentration of O_2 is calculated from the pressure of the oxygen, P_{O_2}, using the equation above. In column B is the current, which the fuel cell's voltage will depend upon. The maximum voltage, E_N, calculated from the Nernst expression, is included in column C and the code in cell C16 is

```
=$C$3 + (($C$4*$F$3)/(2*$C$5)) * (LN($F$4) + 0.5*LN($F$5))
```

This does not depend on the current through the fuel cell and is a constant as a function of current, going down the column. In Column D the spreadsheet calculates the voltage drop due to activation losses and the code in cell D16 is

```
=-($C$7 + $C$8*$F$3 + $C$9*$F$3*LN($F$7) + $C$10*$F$3*LN(B16))
```

where it is the last term which is current-dependent (referencing cell B16). The voltage drop due the concentration gradient at the electrodes is calculated in column E. The code in cell E16 is

```
=-$C$11*LN(1-(B16/$C$12))
```

which results in the voltage dropping off at high currents. In column F are the ohmic losses in voltage, which are simply proportional to the current and, therefore, the code in cell F16 is simply

```
=$C$6*B16
```

Finally, the voltage across the entire fuel cell is calculated in column G with the code in cell G16 being

```
=$F$8*(C16-D16-E16-F16)
```

and consisting of the open cell voltage minus the current-dependent losses.

A plot of voltage as a function of current for the fuel cell is depicted in Figure 7.7. In particular, the maximum open-circuit voltage is plotted across the top and depicted by the dashed line. Next, the open-circuit voltage minus the voltage drop arising from activation losses is plotted as the more spaced out dotted line. The effects of concentration losses are considered next as the the open-circuit voltage minus the voltage drop arising from activation losses and concentration losses are plotted as the closely spaced dotted line. Finally, the voltage across the entire fuel cell (open circuit voltage minus voltage drops from all losses) is plotted as a solid line. In this manner, the voltage drop due to the different loss mechanisms are clearly observable. In particular, the ohmic losses can be seen to linearly increase with current, while the concentration losses occur predominantly at higher currents. Therefore, the model can be seen to capture the behavior of the fuel cell and predict how the fuel cell will operate under variable conditions.

7.3 Continuum Mathematical Model of Fuel Cells

Mathematical modeling plays an important role in the design of proton exchange membrane fuel cells. In particular, there is a wide variety of continuum macroscopic models that are considered successful as they agree well with the

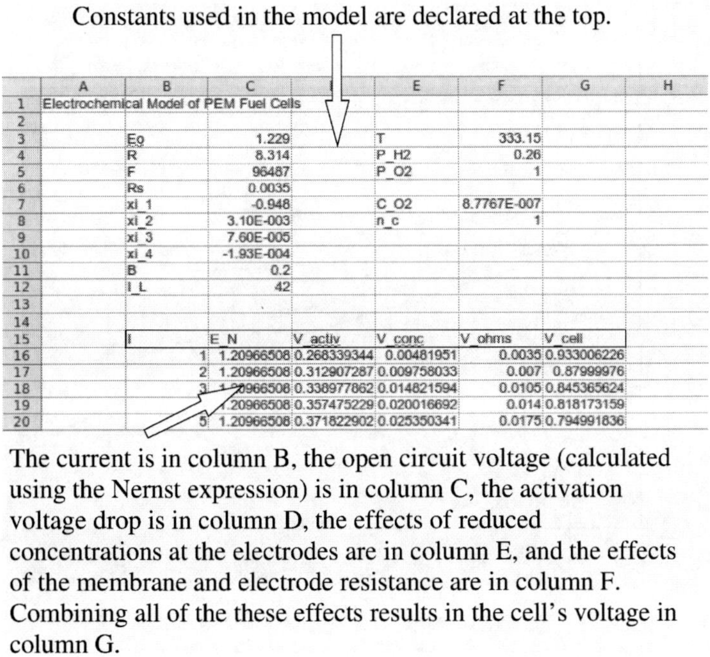

	A	B	C	D	E	F	G	H
1	Electrochemical Model of PEM Fuel Cells							
2								
3		Eo	1.229		T	333.15		
4		R	8.314		P_H2	0.26		
5		F	96487		P_O2	1		
6		Rs	0.0035					
7		xi_1	-0.948		C_O2	8.7767E-007		
8		xi_2	3.10E-003		n_c	1		
9		xi_3	7.60E-005					
10		xi_4	-1.93E-004					
11		B	0.2					
12		I_L	42					
13								
14								
15		I	E_N	V_activ	V_conc	V_ohms	V_cell	
16		1	1.20966508	0.268339344	0.00481951	0.0035	0.933006226	
17		2	1.20966508	0.312907287	0.009758033	0.007	0.87999976	
18		3	[illegible]	0.338977862	0.014821594	0.0105	0.845365624	
19		[illegible]	[illegible]	0.357475229	0.020016692	0.014	0.818173159	
20		5	1.20966508	0.371822902	0.025350341	0.0175	0.794991836	

FIGURE 7.6
Spreadsheet implementation of a simple electrochemical model of a fuel cell.

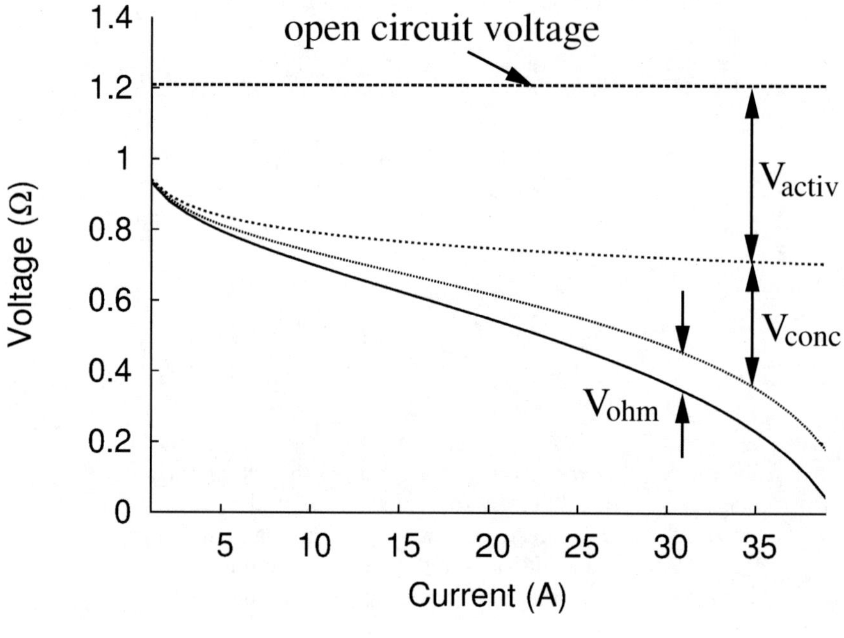

FIGURE 7.7
Current-voltage curve for a fuel cell calculated using the simple electrochemical model.

experimental data. (Although it should be noted that when a model or simulation has a number of free parameters, then agreement with experimental data is often either fortuitous or obtained through fitting the free parameters.) In general, the model or simulation is considered valid when the experimental and modeled polarization curves are in good agreement. However, it is often found that different models and simulations, that are significantly different from one another, are found to agree with the same experimental data. Increasingly complex models are continually being developed, incorporating full dynamic three-dimensional fuel cell models, multiple phase fluid flows, heat transfer and other processes, that capture the important physics of fuel cells. Rather than explore these increasingly complex models and simulations that are emerging within this field, however, we will look at an earlier model which is more suited to spreadsheet modeling.

The model we will consider was originally developed by Springer *et al* (J. Electrochemical Soc., 1991). The model is a simple one-dimensional model which is capable of capturing the fuel cell current-voltage performance, and was found to agree reasonably well with experimental data. This model is important, not only because it is the cornerstone from which later models were descended, but also because it emphasized the importance of keeping the membrane well hydrated, and the importance of water management. In particular, the model is steady-state, isothermal and one-dimensional, with the fuel cell consisting of five layers (membrane, diffusion media, and catalyst layers), and importantly this model considered the water content in the membrane (allowing for variable properties in the membrane such as conductivity and the water diffusion coefficient).

In the one-dimensional model all quantities are considered to vary only in the direction perpendicular to the anode and cathode interfaces. Recall that at the anode the hydrogen reacts to yield electrons to the platinum and protons to the membrane material

$$H_2 \rightarrow 2H^+ + 2e^- \tag{7.23}$$

At the cathode, the oxygen reacts with the protons coming through the membrane to form molecules of water

$$O_2^- + 2H^+ \rightarrow H_2O \tag{7.24}$$

The crux of this model, therefore, is to capture the water transport throughout the cell. This includes transport through the porous electrodes, and transport through the membrane electrolyte, to obtain the steady-state water distribution throughout the cell.

A schematic of how the model works is shown in Figure 7.8. The interfaces are labeled 1 through 4, as we go from the anode to the cathode, and the water concentrations are obtained at these outer interfaces first. Then the water content is calculated using differential equations (that we will cover shortly) starting from interface 1 and working our way through to interface 3,

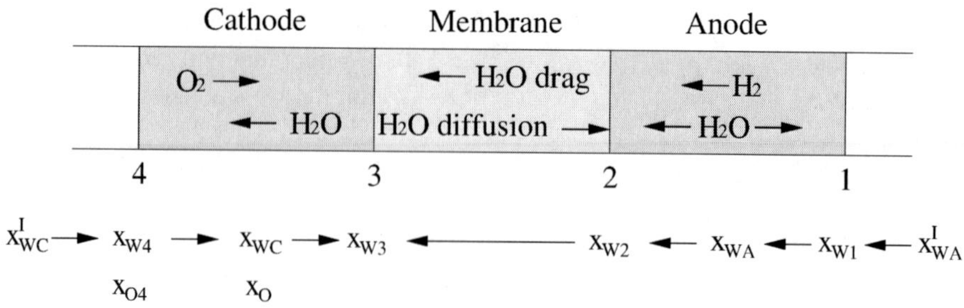

FIGURE 7.8
Schematic of the continuum mathematical model of a fuel cell.

while at the same time the water content is calculated starting from interface 4 and working our way through to interface 3 through the cathode. In this manner we obtain two separate solutions for the water content at interface 3. The ratio of water flux to the molar flux of H_2 through the anode is a free parameter, which can be varied until the two solutions agree, giving us a consistent distribution of water content throughout the fuel cell. Note that not only does water enter the fuel cell as humidified fuel and air streams, but it is also generated at the cathode by the oxygen and hydrogen reaction.

The cathode inlet stream is assumed to be saturated with water vapor, and the mole fraction of water is given by

$$x_{WC}^{I} = \frac{P_C^{SAT}}{P_C} \tag{7.25}$$

where the subscripts W and C represent water and cathode, respectively. P_C^{SAT} is the saturated pressure at the cathode and P_C is the pressure at the cathode. Essentially this sets up a Dirichlet boundary condition (where we assume that we know the water content at this boundary).

The mole fraction of water and oxygen at interface 4 depend on the mole fraction of water in the cathode inlet stream. In particular, the mole fraction of water at interface 4 is set up as a ratio of the water flow and the total flow.

$$x_{W4} = \frac{x_{WC}^{I}\nu_O + 2(1+\alpha)\left(1 - x_{WC}^{I}\right)x_{ON}}{\nu_O + (2\alpha+1)\left(1 - x_{WC}^{I}\right)x_{ON}} \tag{7.26}$$

and, similarly for oxygen

$$x_{O4} = \frac{(\nu_O - 1)\left(1 - x_{WC}^{I}\right)x_{ON}}{\nu_O + (2\alpha+1)\left(1 - x_{WC}^{I}\right)x_{ON}} \tag{7.27}$$

where ν_O is a stochiometric coefficient and the ratio of inlet flux to the total flux across interface 4, and x_{ON} is the inlet dry gas mole fraction of oxygen (and is assumed to be known). The above equations, in a relatively straightforward manner, give us the water and oxygen mole fractions at interface 4.

Note, however, that this depends on the free parameter, α, which is the ratio of water flux to the molar flux of H_2.

Once we have the mole fractions of oxygen and water at interface 4 we can obtain the mole fractions at interface 3 by integrating

$$\frac{dx_{WC}}{dz} = -\frac{RTI}{P_C}\left[\frac{(1 - x_{WC} - x_O)(1+\alpha)}{D_{WN}} + \frac{0.5x_{WC} + x_O(1+\alpha)}{D_{ON}}\right] \tag{7.28}$$

and

$$\frac{dx_O}{dz} = \frac{RTI}{P_C}\left[\frac{x_O(1-\alpha) + 0.5x_{WC}}{D_{WO}} + \frac{1 - x_{WC} - x_O}{D_{ON}}\right] \tag{7.29}$$

where x_{WC} and x_O are the mole fractions of water and oxygen in the cathode, respectively. z is the distance through the cathode, R is the gas constant, T is temperature, I is current density, P_C is the pressure at the cathode, and α is again the ratio of water flux to the molar flux of H_2. D_{WN} is the binary diffusion coefficient between water and nitrogen, D_{ON} is the binary diffusion coefficient between oxygen and nitrogen, and D_{WO} is the binary diffusion coefficient between water and oxygen. We can integrate the above equations using a simple finite difference approximation. In other words, we can write

$$x_{WC}^{i+1} = x_{WC}^{i} + \Delta z\left\{-\frac{RTI}{P_C}\left[\frac{(1 - x_{WC}^{i} - x_O^{i})(1+\alpha)}{D_{WN}} + \frac{0.5x_{WC}^{i} + x_O^{i}(1+\alpha)}{D_{ON}}\right]\right\} \tag{7.30}$$

and

$$x_O^{i+1} = x_O^{i} + \Delta z\left\{\frac{RTI}{P_C}\left[\frac{x_O^{i}(1-\alpha) + 0.5x_{WC}^{i}}{D_{WO}} + \frac{1 - x_{WC}^{i} - x_O^{i}}{D_{ON}}\right]\right\} \tag{7.31}$$

where i represents a discrete location in space and the spatial discretization step is Δz. This is solved all the way along to interface 3, such that the water content at interface 3 can be determined.

Turning our attention to the other side of the fuel cell, the inlet mole fraction of water at the anode is

$$x_{WA}^{I} = \frac{P_A^{SAT}}{P_A} \tag{7.32}$$

where the subscript A represents the anode. P_A^{SAT} is the saturated pressure at the anode and P_A is the pressure at the anode. And, similar to the cathode side of the fuel cell, this sets up another Dirichlet boundary condition.

The mole fraction of water at interface 1 can be obtained from the following equation

$$x_{W1} = \frac{\nu_H x_{WA}^{I} - \alpha\left(1 - x_{WA}^{I}\right)}{x_{WA}^{I} - \alpha\left(1 - x_{WA}^{I}\right) + \nu_H - 1} \tag{7.33}$$

where ν_H is a stochiometric coefficient and the ratio of inlet hydrogen flux to the total flux across interface 1.

The mole fraction of water at interface 2 can be obtained through integrating the following equation across the anode.

$$\frac{dx_{WA}}{dz} = \frac{RTI}{P_A D_{WH}} \left[x_{WA}(1+\alpha) - \alpha\right] \tag{7.34}$$

where x_{WA} is the water mole fraction in the anode and $D_{WH}$is the binary diffusion coefficient between water and hydrogen. This differential equation is much easier to solve than Equations 7.28 and 7.29, and a simple analytical equation can be obtained to give us the mole fraction of water at interface 2

$$x_{W2} = \left(x_{w1} - \frac{\alpha}{1+\alpha}\right) \exp\left(\frac{RTIt_A}{P_A D_{WH}}\right) + \frac{\alpha}{1+\alpha} \tag{7.35}$$

where t_A is the anode thickness.

Now we have obtained the water content at interface 2, we must obtain the water content at interface 3, which we will then be able to compare with the solution from earlier which involved the numerical integration of water content from interface 4 to interface 3. When the parameter α is chosen such that these two solutions agree then the simulation is solved.

To get from interface 2 to interface 3 we have to go through the membrane, and for that we must consider λ, the membrane water content or the local ratio H_2O/SO_3 in the membrane. This is calculated directly from the mole fraction of water content. First we calculate the water vapor activity using

$$a = \frac{x_W P}{P_{SAT}} \tag{7.36}$$

where x_W is the mole fraction of water, P is pressure and P_{SAT} is the saturation pressure. Next we calculate the water content using a phenomenological equation of the form

$$\lambda = \begin{cases} 0.043 + 17.81a - 39.85a^2 + 36a^3 & \text{if } a \leq 1; \\ 14 + 1.4(a-1) & \text{if } 1 < a < 3. \end{cases} \tag{7.37}$$

It is this water content which we will integrate across the membrane using the differential equation

$$\frac{d\lambda}{dz} = \left[2\eta_{drag} - \alpha\right] \frac{IM}{\rho_{dry} D(\lambda)} \tag{7.38}$$

where η_{drag} is the number of molecules per proton, M is the molecular weight of the membrane, ρ_{dry} is the density of dry membrane, and the diffusion coefficient (which depends on water content) is given by

$$D(\lambda) = 10^{-6} \exp\left[2416\left(\frac{1}{303} - \frac{1}{T}\right)\right] (2.563 - 0.33\lambda + 0.0264\lambda^2 - 0.000671\lambda^3) \tag{7.39}$$

From the above system of equations we can obtain the water content at interface 3, λ_3, both through the integration of water content from the anode and the cathode. We can vary α, the ratio of water flux to the molar flux of H_2, until these two solutions converge and we have a complete calculation of the water content throughout the fuel cell. From here we can then calculate the voltage across the fuel cell, and hence obtain the polarization curve.

The conductivity of the membrane as a function of water content, λ, has been experimentally determined to be of the form

$$\sigma = \exp\left[1268\left(\frac{1}{303} - \frac{1}{T}\right)\right](0.005139\lambda - 0.00326) \tag{7.40}$$

where T is temperature. From this conductivity, we can calculate the resistance of the membrane using

$$R = \int \frac{dz}{\sigma} \tag{7.41}$$

where the integration is over the entire membrane. To obtain the polarization curve (voltage-current curve) the model accounts for kinetic losses and ohmic losses in the membrane. To calculate kinetic losses, the Tafel equation is used, which relates the current density to the overpotential, η.

$$J = j_o P_C \frac{x_O 3}{1 - x_{liq}} \exp\left[\frac{0.5F}{RT}\eta\right] \tag{7.42}$$

where j_o is the exchange current density (referenced to pure oxygen at 1 atm at the open cell potential, V_{oc}). P_C is the pressure at the cathode, x_{liq} is the mole fraction of liquid which is generally very small (and often ignored), F is Faraday's constant, and η is given by

$$\eta = V_{oc} - V_{cell} - JR \tag{7.43}$$

where V_{oc} is the open cell voltage, and v_{cell} is the voltage across the cell, which we are trying to obtain in order to generate the polarization curve.

To summarize, the simulation involves calculating the water content at interface 3, λ_3, from both the anode and cathode sides of the fuel cell such that the two solutions agree. Then we can take the water content throughout the cell and calculate the resistance and voltage across the cell. One of the parameters used throughout these calculations is the current density. Therefore, we can obtain the voltage across the cell as a function of current density, and generate the polarization curve.

Let's turn our attention to the spreadsheet implementation of the model described above. Figure 7.9 depicts the top of the spreadsheet, where the many constants are contained. Most of these numbers are obtained directly from the literature, but the current density in cell B8 is changed to obtain the polarization (or voltage-current) curve. To the right of Figure 7.9 the

The model uses a large number of constants, most of which are obtained experimentally. The current density in cell B8 is the variable changed when obtaining the voltage–current curves.

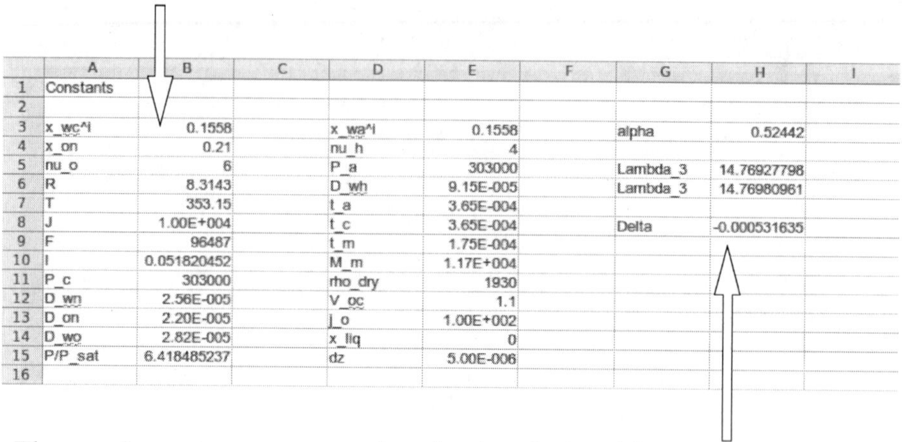

	A	B	C	D	E	F	G	H	I
1	Constants								
2									
3	x_wc^i	0.1558		x_wa^i	0.1558		alpha	0.52442	
4	x_on	0.21		nu_h	4				
5	nu_o	6		P_a	303000		Lambda_3	14.76927798	
6	R	8.3143		D_wh	9.15E-005		Lambda_3	14.76980961	
7	T	353.15		t_a	3.65E-004				
8	J	1.00E+004		t_c	3.65E-004		Delta	-0.000531635	
9	F	96487		t_m	1.75E-004				
10	I	0.051820452		M_m	1.17E+004				
11	P_c	303000		rho_dry	1930				
12	D_wn	2.56E-005		V_oc	1.1				
13	D_on	2.20E-005		i_o	1.00E+002				
14	D_wo	2.82E-005		x_liq	0				
15	P/P_sat	6.418485237		dz	5.00E-006				
16									

The membrane water content at location is calculated from the anode and from the cathode. When an α is found which results in these two values being close together then the simulation is considered solved.

FIGURE 7.9
Spreadsheet implementation of the continuum mathematical model of a fuel cell.

variable α is determined (α is stored in cell H3). As a function of this variable, the water content at interface 3, λ_3, is obtained twice (as described in the description of the model above). These two λ's are copied from elsewhere in the spreadsheet model (which we will come back to later) in to the cells H5 and H6. The difference between these two solutions is calculated in H8. The objective, therefore, is to find a value of α which would minimize this difference. While this could be done numerically, using for example the bisection method, in the current implementation this is done by hand (although it doesn't take too many iterations to converge).

The calculation of the two λ_3's, and ultimately the voltage across the cell, is depicted in Figure 7.10. In cells A20 and A23 the mole fractions of water and oxygen, respectively, are calculated at interface 4. In terms of these concentrations, the mole fractions of water and oxygen are calculated by integrating over distance. These integrations (the equations of which are covered above) occur in cells D20:BY20 and D23:BY23, for the mole fraction of water and oxygen, respectively. For example, the code contained in cell E20 is of the form

```
=D20+$E$15*($B$6*$B$7*$B$10/$B$11) *
```

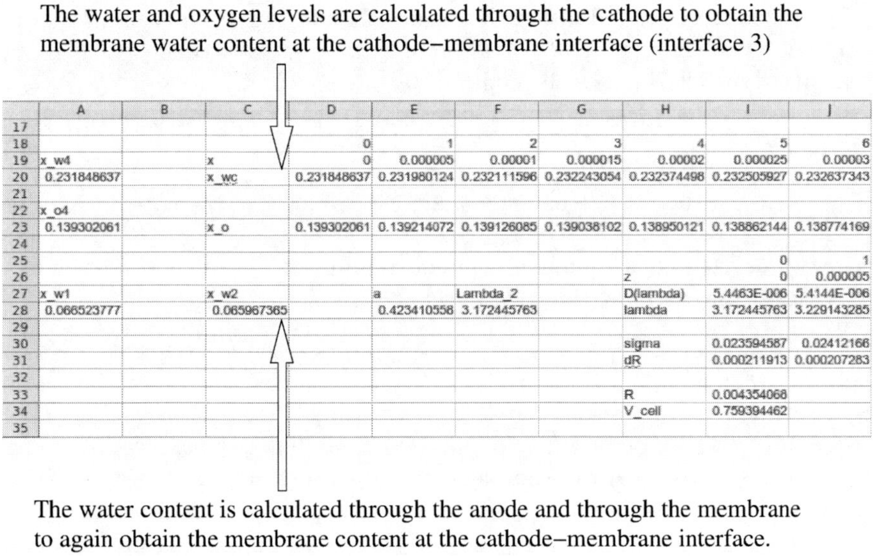

	A	B	C	D	E	F	G	H	I	J
17										
18				0	1	2	3	4	5	6
19	x_w4		x	0	0.000005	0.00001	0.000015	0.00002	0.000025	0.00003
20	0.231848637		x_wc	0.231848637	0.231980124	0.232111596	0.232243054	0.232374498	0.232505927	0.232637343
21										
22	x_o4									
23	0.139302061		x_o	0.139302061	0.139214072	0.139126085	0.139038102	0.138950121	0.138862144	0.138774169
24										
25									0	1
26								z	0	0.000005
27	x_w1		x_w2		a	Lambda_2		D(lambda)	5.4463E-006	5.4144E-006
28	0.066523777		0.065967365		0.423410558	3.172445763		lambda	3.172445763	3.229143285
29										
30								sigma	0.023594587	0.02412166
31								dR	0.000211913	0.000207283
32										
33								R	0.004354068	
34								V_cell	0.759394462	
35										

FIGURE 7.10
Spreadsheet implementation of the continuum mathematical model of a fuel cell (continued).

```
( (1-D20-D23)*(1+$H$3)/$B$12  + (0.5*D20+D23*(1+$H$3))/$B$13)
```

and corresponds with the calculation of the mole fraction of water in Equation 7.28 above, while for the mole fraction of oxygen we discretize Equation 7.29 and the code in cell E23 is of the form

```
=D23-$E$15*($B$6*$B$7*$B$10/$B$11) *

((D23*(1-$H$3)+0.5*D20)/$B$14 + (1-D20-D23)/$B$13)
```

In cell BY20, at the end of the spreadsheet to the right (not shown), the mole fraction of water at interface 3 is obtained. From this it is easy to obtain the water content, and this is then displayed in cell H5 (Figure 7.9).

In cell A28 the mole fraction of water at interface 1 is calculated, and the mole fraction of water at interface 2 is determined in cell C28. The code in cell C28 is

```
=(A28-(H3)/(1+H3))*EXP(B6*B7*B10*E7/(E5*E6)) + (H3)/(1+H3)
```

and the integration of mole fraction of water is obtained analytically. To obtain the water content at interface 3, we first calculate the water content at interface 2 and integrate this through the membrane to interface 3. The water vapor activity and water content at interface 2 are calculated in cells E28 and

F28, respectively. The diffusion coefficient of water is calculated from the water content in cells I27:AR27. For example, the diffusion coefficient of water calculated in cell J27 uses the code

```
=0.000003093814282110097*

(2.563-0.33*J28+0.0264*J28*J28-0.000671*J28*J28*J28)
```

The water content throughout the membrane is now calculated in cells I28:AR28. For example, the code in cell J28 is

```
=I28+$E$15*(2*2.5*I28/(22) - $H$3)*($B$10*$E$10/($E$11*I27))
```

which is a simple finite difference approximation of the differential equation for water content (Equation 7.38). This ultimately yields the water content at interface 3, in cell AR28 (not shown) and this is then displayed in cell H6 (Figure 7.9). By comparing these two λ's (in cells H5 and H6) we can find a value of α such that the two solutions reasonably agree, and the water content throughout the membrane.

Once we have the water content throughout the membrane, we can determine the resistance of the membrane and finally the voltage across the cell. Cells I30:AR30 calculate the conductivity of the membrane (as a function of the water content). This is then integrated in cells I31:AR31 to obtain the resistance of the membrane in cell I33. In other words, the numerical integration (essentially a summation) of the resistance of small portions of the membrane are added together to give the total resistance of the membrane. From this the voltage across the cell is obtained in cell I34. By systematically varying the current density in cell B8, and solving the above simulation to obtain the voltage across the cell, the polarization curve can be obtained. The polarization cell, obtained from the spreadsheet model, is depicted in Figure 7.11. For high currents, the concentration losses play a more important role and these are not taken into consideration here. Therefore the model is more accurate for regions of the plot with low to intermediate current. The curve, however, is just as we might expect, and this model has been successfully fitted to experimental data, validating the model to some degree.

7.4 Fluid Dynamics and the Lattice Boltzmann Model

The lattice Boltzmann model (LBM) is a powerful fluid dynamics model which incorporates the physics of mesoscopic fluid "particles" propagating and colliding on a simple lattice. This propagating and collision of "particles" mimics the interaction of atoms in a real fluid, but on a much larger scale. The upshot of this technique is that the correct fluid behavior emerges from the model

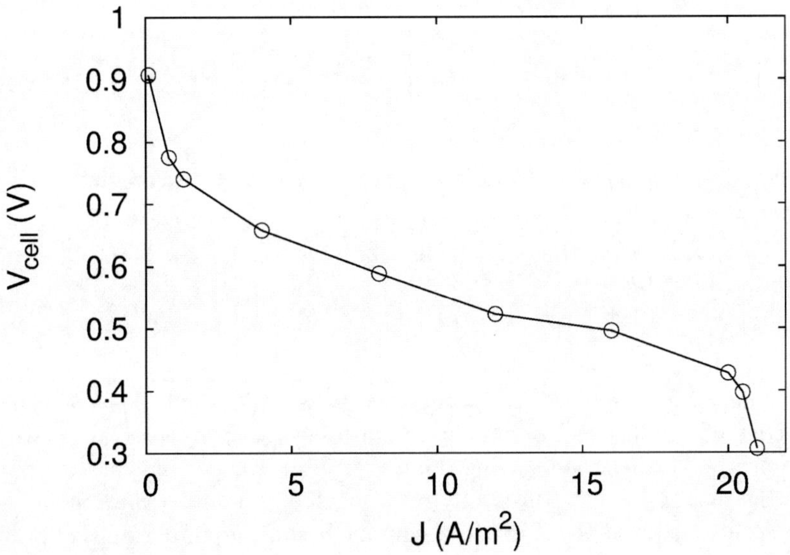

FIGURE 7.11
Current-voltage curve for a fuel cell calculated using a continuum mathematical model of a fuel cell.

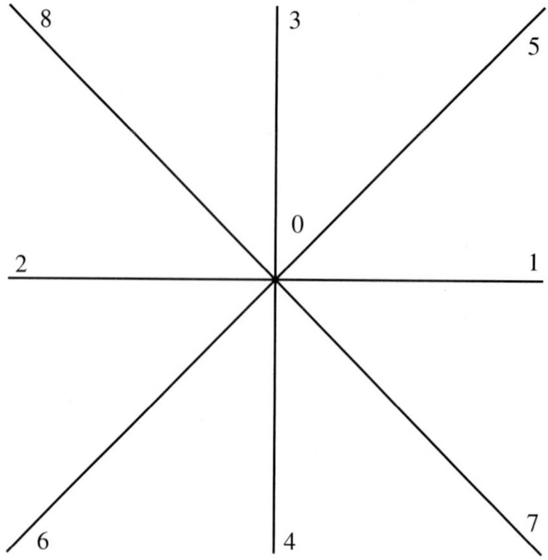

FIGURE 7.12
The nine directions for the particle distribution function in the D2Q9 lattice Boltzmann model.

without us having to necessarily worry about solving mathematically complicated fluid dynamics models. In other words, the averaged long-wavelength properties of the lattice Boltzmann system obeys the desired Navier-Stokes equations, but rather than solving the Navier-Stokes equations directly, these properties emerge quite naturally from the local interactions within the model.

As mentioned, the power of this technique lies in the simple steps of propagating fluid "particles" to neighboring lattice sites, and colliding these particles when they reach a site. Here, fluid particles are representative of mesoscopic (length scale between that of atomistic and continuum systems) portions of the fluid, and are described by a particle distribution function. The particle distribution function is a continuous variable that exists in discrete directions (either going nowhere, in the 4 $\{10\}$ directions, or in the 4 $\{11\}$ directions of a simple square lattice) such that "particles" flowing in a given direction will reach a neighboring lattice site in a single time step. Upon reaching a lattice site the particle distribution functions are collided with each other, such that the resultant particle distribution has lower hydrodynamic stresses. In particular, we can write the time evolution of the particle distribution function as

$$f_i(\vec{r} + \vec{e}_i \Delta t, t + \Delta t) = f_i^*(\vec{r}, t) = f_i(\vec{r}, t) + \Omega_i \left[\vec{f}(\vec{r}, t) \right] \tag{7.44}$$

where $\vec{f}$ is a nine-dimension vector where each component of the vector is the particle distribution function in a given direction. The component, f_i, is the

i^{th} component of this vector, whose directions are depicted in Figure 7.12. The zeroth component goes nowhere, the directions labeled from one through four go to nearest neighbors, and directions five through eight go to next-nearest neighbors. This scheme is referred to as a D2Q9 model, because it is a two-dimensional model and there are 9 components to the particle distribution function. Other models exist, in two- and three-dimensions, although we are obviously restricted here to two-dimensions if we want to solve the model in a spreadsheet. $\vec{r}$ is the position on the lattice, and $\vec{e}_i$ is a vector which points to the neighboring lattice site in the ith direction. Δt is the time step in the simulation (usually unity), and t is the current time (usually dimensionless). The first part of the above equation,

$$f_i(\vec{r} + \vec{e}_i \Delta t, t + \Delta t) = f_i^*(\vec{r}, t) \tag{7.45}$$

simply takes the post-collision particle distribution function, f_i^*, and moves it over to the next lattice site $(\vec{r} + \vec{e}_i \Delta t)$ at the next time step $(t + \Delta t)$. The second part of the evolution equation,

$$f_i^*(\vec{r}, t) = f_i(\vec{r}, t) + \Omega_i \left[\vec{f}(\vec{r}, t)\right] \tag{7.46}$$

is a more general statement that the particle distribution function is modified upon collision by a collision operator, Ω_i. This collision operator depends on all of the components of the particle distribution function at this location and at this time. However, relaxing the system such that the hydrostatic stresses go to zero at each iteration is computationally more stable (which is what we will do here). It is worth noting that this limits the viscosity that can be simulated, and if it is required to change the viscosity of the fluid then the hydrodynamic stresses must be more slowly relaxed towards zero. Furthermore, if it is required to vary the shear and bulk viscosities separately then one can relax the deviatoric (components not along the trace) and hydrostatic components (trace components) of the nonequilibrium stress tensor separately. However, as in fluid dynamics it is typical to simulate systems in dimensionless numbers and map the simulation on to real systems using characteristic quantities (speed of sound in the fluid, Reynolds number, Womersley number, etc...) this is only really an issue if the simulation has different viscosities in different locations within the same simulation (multicomponent fluids or turbulent fluid flow, for example). We can, therefore, write down the typical update of the particle distribution function in the form

$$f_i(\vec{r} + \vec{e}_i \Delta t, t + \Delta t) = f_i^*(\vec{r}, t) = f_i^{eq}(\vec{r}, t) \tag{7.47}$$

where f_i^{eq} is the distribution which, for a given density and velocity of fluid, minimizes the hydrostatic stresses. The equilibrium particle distribution function for the system considered here is given by

$$f_i^{eq}(\vec{r}, t) = w_i \rho(\vec{r}, t) \left[1 + 3\left(\vec{e}_i \cdot \vec{v}(\vec{r}, t)\right) + \frac{9}{2}\left(\vec{e}_i \cdot \vec{v}(\vec{r}, t)\right)^2 - \frac{3}{2} v^2(\vec{r}, t)\right] \tag{7.48}$$

The weights, the w's, are given by $w_0 = \frac{4}{9}$, $w_1 = w_2 = w_3 = w_4 = \frac{1}{9}$, and $w_5 = w_6 = w_7 = w_8 = \frac{1}{36}$. These values are obtained when the model is mapped onto fluid dynamic models (such as the Navier-Stokes equations), but we are not concerned about the derivation of the model here. ρ is the density of the fluid, and $\vec{v}$ is the fluids velocity. These hydrodynamic quantities (mass density ρ, and momentum density, $\rho\vec{v}$) are moments of the distribution function and can be calculated from the particle distribution function

$$\rho = \sum_i f_i \tag{7.49}$$

and

$$\rho\vec{v} = \sum_i f_i \vec{e}_i \tag{7.50}$$

In other words, the particle distribution function can be used to calculate the local density and velocity of the fluid. This in turn is used to calculate the equilibrium particle distribution function (minimizing hydrodynamic stresses). The equilibrium particle distribution function is then propagated to neighboring lattice sites. The resulting particle distribution functions are again used to calculate the local density and velocity of the fluid, and the cycle continues with the fluid dynamics being updated with each cycle of events or iteration. This is depicted in Figure 7.13.

The two-dimensional regular square lattice of the lattice Boltzmann is represented in Figure 7.13. In the first step, the particle distribution functions are represented by the circles on the lattice nodes and their directions are represented by the arrows. Only the particle distribution functions pointing towards the central node are depicted, although in the actual simulation there are particle distribution functions pointing in all directions and at all nodes. In the second step, these particle distribution functions are propagated to the next site over, which is the central node (why these particle distribution functions are the only ones depicted). Notice that the size of these particle distribution functions can be quite random at this point. The third step is to minimize the hydrostatic stresses by setting the particle distribution function equal to the equilibrium particle distribution function (which still maintains the density and fluid velocity). This would usually result in a more ordered looking set of particle distribution functions. The final, fourth, step depicted is exactly the same as that depicted in the second step. The particle distribution functions are propagated to their neighboring nodes. It is really quite amazing that the correct fluid dynamics behavior emerges from this simple system where the particle distribution functions are propagated and collided with each other.

The lattice Boltzmann model is a powerful technique which is especially useful when modeling complex boundaries, and is very popular for simulating flows through porous media. This is because the lattice Boltzmann model can include solid regions in an efficient and incredibly simple way. In particular, we need to add no-slip boundary conditions around the edge of the fluid

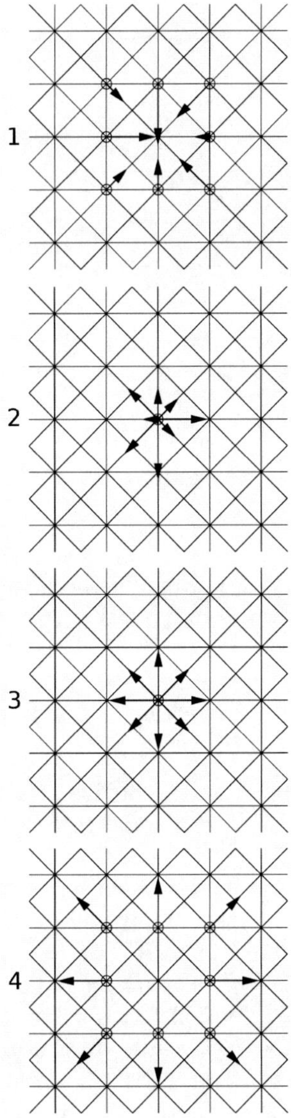

FIGURE 7.13
The propagation and collision steps in the lattice Boltzmann model are depicted. It is through the iteration of these propagation and collision steps that the fluid dynamics are captured in the model.

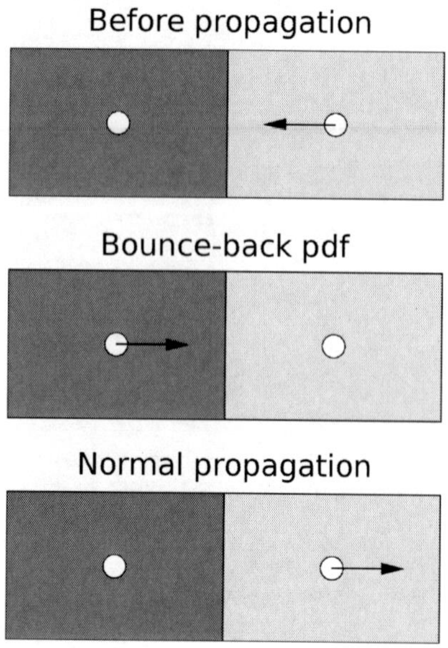

FIGURE 7.14
The bounce-back boundary conditions in the lattice Boltzmann model. The particle distribution function is propagating towards a solid boundary, and we want to bounce this back into the fluid. Therefore, we can turn it around and place it on the solid node. That way, when we propagate the particle distribution function it ends up back in the fluid domain.

domain such that the velocity of the fluid at the boundary is equal to the velocity of the boundary (here we'll take this to be zero, but moving solid interfaces can be included too). In practice, when the particle distribution function propagates from the fluid domain into a region which is solid, then the particle distribution function simply reflects, or bounces back, into the fluid domain. This is depicted in Figure 7.14.

The crucial step in the bounce-back boundary conditions is the implementation. The particle distribution functions that are propagating towards solid nodes have to bounced back, such that after propagating the particle distribution functions, the particle distribution function is on the same node but pointing in the opposite direction. The first step is to take the particle distribution function and turn it around so it points in the opposite direction, and place it on the solid node from which it is bouncing. Then we can perform the regular propagation step, moving the particle distribution functions to their neighboring nodes, and move the particle distribution functions from the solids back into the fluid domain. Mathematically, the bounce-back boundary conditions can be written in the form

$$f_k(\vec{r}, t + \Delta t) = f_i^*(\vec{r}, t) \tag{7.51}$$

FIGURE 7.15
The serpentine flow channels often seen in proton exchange membrane fuel cells.

where k represents a direction which is opposite to the direction i. In other words, if the particle distribution function is traveling in the direction i towards the wall then this particle distribution function will be bounced back by the wall to where it came from, but it will now be traveling in the opposite direction, k.

The lattice Boltzmann model allows us to simulate the hydrodynamics of Newtonian fluids in a relatively easy manner, and consider complex geometries by simply assigning some nodes as solid areas and some nodes as fluid areas. Now let's turn our attention to implementing this model within a spreadsheet to simulate the flow of gases in fuel cells.

In the fuel cells the gas enters and circulates through a serpentine arrangement of channels (at least in some designs) and this is depicted in Figure 7.15. The flow channels allow the gas to come into contact with the permeable electrodes. We can therefore apply the lattice Boltzmann model to the simulation of the fluid flow in these channels. While the real channels are three-dimensional and the flow can change as the air navigates through the serpentine channels, here we will simulate a simpler configuration. Because we are solving this model using a spreadsheet we will limit the simulation to two-dimensions and, just to see how the model works, we will only model the fluid flow as it goes around a corner. This is shown in Figure 7.16. Notice that we have to assign the density and velocity of the fluid at the boundary of the model. The density and velocity of the fluid within the simulation domain is then solved in response to these boundary conditions. Where the boundary is solid material, we can set the density and velocity to zero as these regions won't be solved in the model. For the boundary sites at the entrance and exit of the section of the channel to be simulated we have to set the density (here we set it to unity) and velocity of the fluid. The fluid velocity is assumed to obey Poiseuille flow. In other words, the velocity of flow in the channel is of

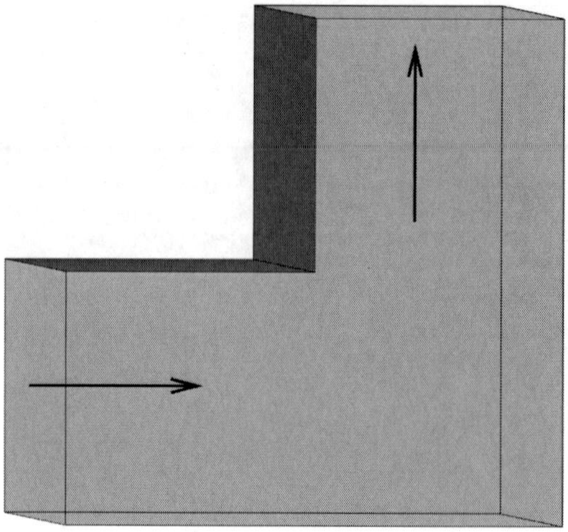

FIGURE 7.16
The simulation domain is a simple two-dimensional corner, with the fluid flowing in at the bottom and leaving at the top.

the form

$$v = v_{max}\left[\left(\frac{W}{2}\right)^2 - x^2\right] \tag{7.52}$$

where v_{max} is a constant that we can change, W is the width of the channel, and x is the distance from the center of the channel. In response to these boundary conditions we can now use a spreadsheet model to solve the fluid dynamics within the simulation domain.

The layout of the spreadsheet model is depicted in Figure 7.17. The simulation domain is replicated multiple times across the simulation (the boxes), and different aspects of the simulations are stored in these regions. For example, on the far left of the spreadsheet the top box represents an area of the spreadsheet which contains either 0's or 1's depending on whether the lattice site is fluid or solid, respectively. Below this are three boxes, each representing the same simulation domain, which store the fluid density and the velocity components. Even though the simulation size is relatively small (here we only capture 22×22 lattice sites within the simulation domain) we need multiple simulation domains in the spreadsheet, each storing different parameters within the model.

As we move across the spreadsheet, particle distribution functions are stored. The particle distribution function is a 9 component vector and so the different components are stored as we go down the spreadsheet. The first column of boxes representing the particle distribution function is the current particle distribution function in the simulation, just after the particle distribution functions have been propagated. The next column of boxes represent

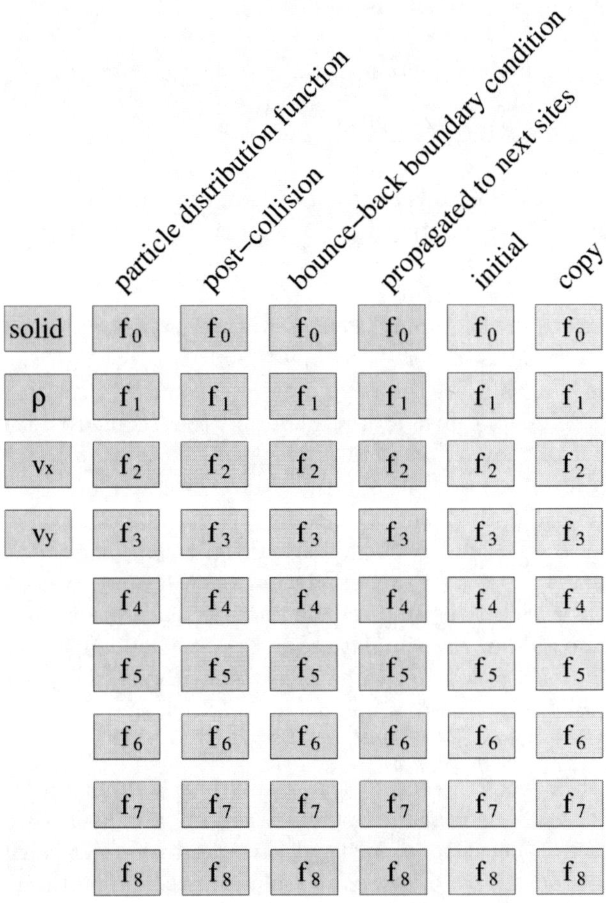

FIGURE 7.17
The layout of the spreadsheet implementation of the lattice Boltzmann model.

the post-collision particle distribution functions. Recall, that in the current model the post-collision particle distribution functions are set to the equilibrium particle distribution functions, which only depends on the current density and velocity within the fluid. The third column of boxes representing the particle distribution functions modify the particle distribution functions to account for the bounce-back boundary conditions. In particular, the particle distributions which are propagating towards a solid site are placed on that solid site, but directed back in the opposite direction towards the fluid. In the next column of boxes, the particle distribution functions are propagated to the neighboring sites in the direction of the particle distribution functions. This concludes the iteration in the model and these values have to be copied back to the first column of boxes which represents the current particle distribution functions, so that the iteration can be completed and the spreadsheet will be repopulated with calculations for the next iteration. There are two additional columns of particle distribution functions in the spreadsheet to the far right. The first is simply the initial values of the particle distribution function assuming no fluid velocity (these can be copied to the first column to reset the simulation back to the beginning). The last is a copy of the propagated particle distribution functions. Rather than copy the propagated particle distribution functions, one at a time, to the beginning of the spreadsheet to complete the iteration, the values are first stored here and then copied to the beginning. In this way all of the components of the particle distribution functions are copied across all at once. Without this step, the lower number components might be copied across and the later components would be recalculated based on newer lower components of the particle distribution functions, before being copied across. At the top of the simulation we will place two buttons which will either copy the initial values of the particle distribution functions across to the left-hand side of the simulation and reset the simulation, or copy the latest particle distribution functions across and iterate the simulation. Let's look at the actual spreadsheet implementation, with snapshots in Figures 7.17 to 7.27 taken from the spreadsheet.

Figure 7.18 depicts the top of the simulation. The maximum velocity in the center of the channels for the boundary conditions (fixed velocity and density) is assigned in cell D1. Interestingly, we need no other constants in this dimensionless model. Two buttons are included at the top of the spreadsheet. The first button runs a macro which initializes the system, and copies particle distribution functions from another section of the spreadsheet (corresponding to zero velocity and unit density) to the active part of the spreadsheet, the region where the current particle distribution function is stored (see Figure 7.22). This allows us to initialize the system and start the simulation from scratch. The second button runs a macro which copies updated particle distribution functions (subsequent to propagation and collision) back to the beginning of the iteration. In other words, the current particle distribution functions are contained within the spreadsheet, and used to obtain an updated particle distribution function. By copying this updated particle distribution

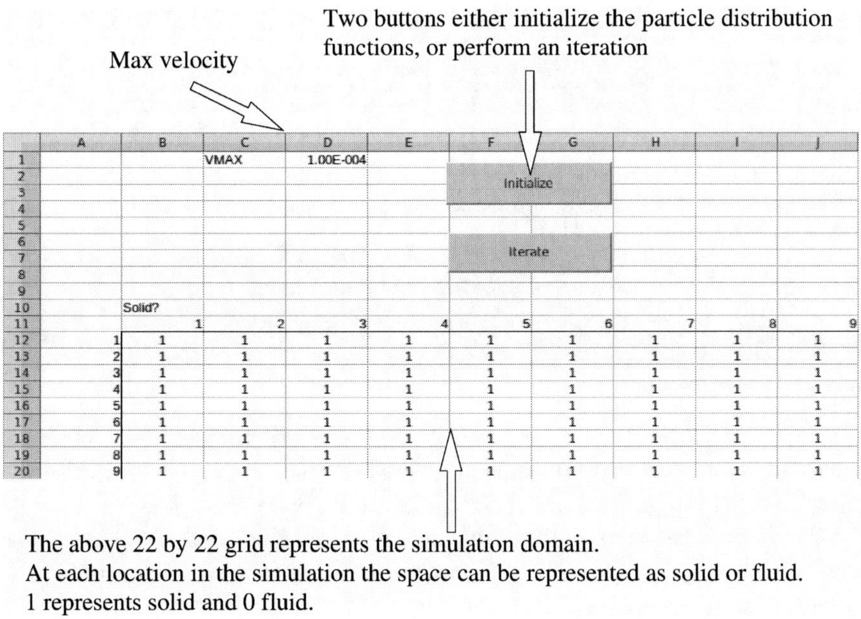

FIGURE 7.18
Spreadsheet implementation of the lattice Boltzmann model. The buttons and solid nodes.

The density around the edge of the simulation domain is set to zero if its a solid region or one if the region is fluid.

	A	B	C	D	E	F	G	H	I	J
38	RHO									
39		1	2	3	4	5	6	7	8	9
40	1	0	0	0	0	0	0	0	0	0
41	2	0	0	0	0	0	0	0	0	0
42	3	0	0	0	0	0	0	0	0	0
43	4	0	0	0	0	0	0	0	0	0
44	5	0	0	0	0	0	0	0	0	0
45	6	0	0	0	0	0	0	0	0	0
46	7	0	0	0	0	0	0	0	0	0
47	8	0	0	0	0	0	0	0	0	0
48	9	0	0	0	0	0	0	0	0	0
49	10	0	0	0	0	0	0	0	0	0
50	11	0	0	0	0	0	0	0	0	0
51	12	1	1.000873029	1.000878019	1.000834126	1.000772385	1.000706282	1.000641074	1.000579765	1.000520797
52	13	1	1.000934893	1.000889591	1.00083299	1.000768112	1.000700218	1.00063073	1.000560802	1.00048665
53	14	1	1.000982659	1.000908985	1.000838706	1.000767302	1.000695252	1.000621617	1.000546386	1.000466391
54	15	1	1.001009893	1.000923164	1.00084398	1.000767009	1.000691165	1.000614374	1.000536412	1.000454999
55	16	1	1.001020925	1.000929141	1.000845606	1.000765384	1.000687082	1.000608352	1.000529161	1.000447991
56	17	1	1.001018522	1.000926592	1.000842599	1.000761622	1.000682391	1.000602724	1.000523006	1.000442306
57	18	1	1.0010031	1.000915877	1.00083526	1.000755921	1.000677073	1.000597078	1.000516919	1.000436171
58	19	1	1.000972702	1.000898035	1.0008[illegible]5224	1.000749592	1.000671878	1.000591614	1.000510556	1.000428792
59	20	1	1.000923693	1.000876735	1.0008[illegible]486	1.000745328	1.000668408	1.0005871	1.000504192	1.000420038
60	21	1	1.000862684	1.000865569	1.000[illegible]272	1.000747486	1.000668652	1.00058475	1.000498391	1.000410383
61	22	0	0	0	0	0	0	0	0	0

In the simulation domain, the density is calculated from the particle distribution function. In particular, it is the first moment (the sum) of the particle distribution function. The code in cell C41 is

=IF(C13=1, 0, (AA13+AA39+AA65+AA91+
AA117+AA143+AA169+AA195+AA221))

FIGURE 7.19
Spreadsheet implementation of the lattice Boltzmann model. The fluid density.

function back to overwrite the current particle distribution functions, then the simulation has progressed an iteration, and the spreadsheet is automatically populated with the next updated particle distribution functions. At the bottom of Figure 7.18 is 22 × 22 size region corresponding with the simulation domain size, where cells that contain the number 1 are designated solid sites and cells that contain the number 0 are designated fluid sites. Recall that we solve the fluid dynamics in the fluid regions and have implemented fluid bounce back boundary conditions (no-slip boundary conditions) at the fluid solid interface.

The density in the simulation is depicted in Figure 7.19. Around the edge of the simulation domain the density is assigned as part of the boundary conditions, as is either 0 if the site is solid or 1 if the site is fluid. Inside the domain, the density is determined from the current particle distribution function. In particular, the code in cell C41 is

```
=IF(C13=1, 0,

(AA13+AA39+AA65+AA91+AA117+AA143+AA169+AA195+AA221))
```

In other words, if the site is solid then just put the density to zero. However,

Velocity in x–direction is set to a given value in the fluid regions of the boundary, around the edge of the simulation domain. This establishes Poiseuille flow at the boundaries.

	A	B	C	D	E	F	G	H	I	J
64	VX									
65		1	2	3	4	[illegible]	6	7	8	9
66	1	0	0	0	0	0	0	0	0	0
67	2	0	0	0	0	0	0	0	0	0
68	3	0	0	0	0	0	0	0	0	0
69	4	0	0	0	0	0	0	0	0	0
70	5	0	0	0	0	0	0	0	0	0
71	6	0	0	0	0	0	0	0	0	0
72	7	0	0	0	0	0	0	0	0	0
73	8	0	0	0	0	0	0	0	0	0
74	9	0	0	0	0	0	0	0	0	0
75	10	0	0	0	0	0	0	0	0	0
76	11	0	0	0	0	0	0	0	0	0
77	12	0.000475	0.000280487	0.000327523	0.000352163	0.000365797	0.000370323	0.000377376	0.000386017	0.000417597
78	13	0.001275	0.000940408	0.000959451	0.000978272	0.000993261	0.001001876	0.001016402	0.001038644	0.00109166
79	14	0.001875	0.001502338	0.001488616	0.001483939	0.00148668	0.001490377	0.001503776	0.001524241	0.001565293
80	15	0.002275	0.001887601	0.001861127	0.001839251	0.001828771	0.001823375	0.001828635	0.001836099	0.00185098
81	16	0.002475	0.002082959	0.002052479	0.002023175	0.002004986	0.001992106	0.00198813	0.001980662	0.001971573
82	17	0.002475	0.002084376	0.002054935	0.002026454	0.002008235	0.001993287	0.001983982	0.001966216	0.001941566
83	18	0.002275	0.001891621	0.001868156	0.001848814	0.001838743	0.001828346	0.001819646	0.001798643	0.001768323
84	19	0.001875	0.001508166	0.001498905	0.001498386	0.001503025	0.001501858	0.001498478	0.001480583	0.001453494
85	20	0.001275	0.00094646	0.000969986	0.000993609	.001012282	0.001019222	0.00102193	0.001010694	0.000993393
86	21	0.000475	0.000283881	0.000333047	0.00036041	.000377099	0.000382627	0.000387147	0.000382477	0.000378441
87	22	0	0	0	0	[illegible]	0	0	0	0

However, inside the simulation domain the velocity is obtained from the second moment of the particle distribution function (see text). The code in cell C67 is

=IF(C13=1, 0, (AA39–AA65+AA143–AA169+AA195–AA221))

FIGURE 7.20

Spreadsheet implementation of the lattice Boltzmann model. The velocity in the x-direction.

if the site is fluid then the density is given by

$$\rho = \sum_i f_i \tag{7.53}$$

the first moment of the particle distribution function. As we relax the particle distribution function directly to its equilibrium state, the post-collision particle distribution function is calculated directly from this density and the fluid velocity, which is depicted in Figure 7.20 and 7.21.

Figure 7.20 depicts the section of the spreadsheet which calculates the velocity in the x-direction, while Figure 7.21 depicts the section of the spreadsheet which calculates the velocity in the y-direction. Similar to the density, the velocity at the boundary conditions is assigned at the beginning of the simulation and is fixed for the entire duration of the simulation. In particular, if the site is solid the velocity of the fluid is set to zero but if the site is fluid then the fluid is set according to Poiseuille's flow. In other words, the velocity in the direction parallel to the channel is given by

$$u = u_{max}\left(R^2 - x^2\right) \tag{7.54}$$

where u_{max} is the maximum velocity in the center of the channel (cell D1), R is

Boundary conditions dictate the velocity in the y–direction in exactly the same way as in the x–direction.

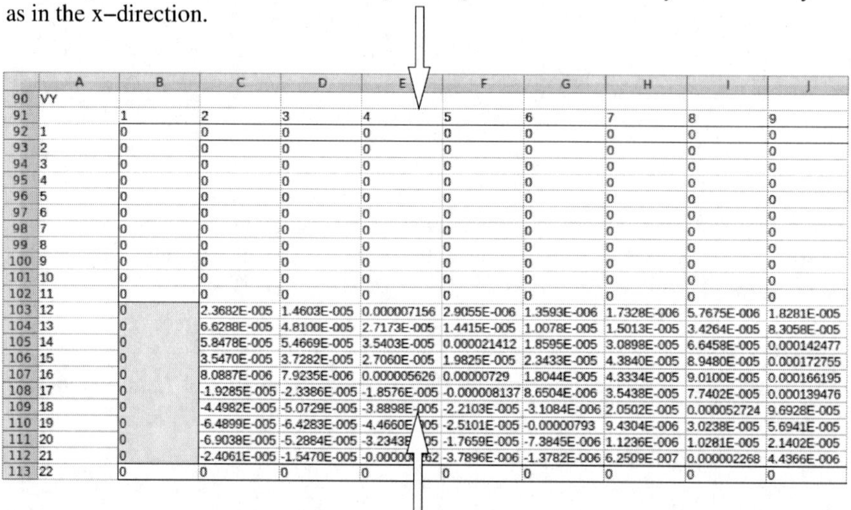

	A	B	C	D	E	F	G	H	I	J
90	VY									
91		1	2	3	4	5	6	7	8	9
92	1	0	0	0	0	0	0	0	0	0
93	2	0	0	0	0	0	0	0	0	0
94	3	0	0	0	0	0	0	0	0	0
95	4	0	0	0	0	0	0	0	0	0
96	5	0	0	0	0	0	0	0	0	0
97	6	0	0	0	0	0	0	0	0	0
98	7	0	0	0	0	0	0	0	0	0
99	8	0	0	0	0	0	0	0	0	0
100	9	0	0	0	0	0	0	0	0	0
101	10	0	0	0	0	0	0	0	0	0
102	11	0	0	0	0	0	0	0	0	0
103	12	0	2.3682E-005	1.4603E-005	0.000007156	2.9055E-006	1.3593E-006	1.7328E-006	5.7675E-006	1.8281E-005
104	13	0	6.6288E-005	4.8100E-005	2.7173E-005	1.4415E-005	1.0078E-005	1.5013E-005	3.4264E-005	8.3058E-005
105	14	0	5.8478E-005	5.4669E-005	3.5403E-005	0.000021412	1.8595E-005	3.0898E-005	6.6458E-005	0.000142477
106	15	0	3.5470E-005	3.7282E-005	2.7060E-005	1.9825E-005	2.3433E-005	4.3840E-005	8.9480E-005	0.000172755
107	16	0	8.0887E-006	7.9235E-006	0.000005626	0.00000729	1.8044E-005	4.3334E-005	9.0100E-005	0.000166195
108	17	0	-1.9285E-005	-2.3386E-005	-1.8576E-005	-0.000008137	8.6504E-006	3.5438E-005	7.7402E-005	0.000139476
109	18	0	-4.4982E-005	-5.0729E-005	-3.8898E-005	-2.2103E-005	-3.1084E-006	2.0502E-005	0.000052724	9.6928E-005
110	19	0	-6.4899E-005	-6.4283E-005	-4.4660E[illegible]05	-2.5101E-005	-0.00000793	9.4304E-006	3.0238E-005	5.6941E-005
111	20	0	-6.9038E-005	-5.2884E-005	-3.2343[illegible]05	-1.7659E-005	-7.3845E-006	1.1236E-006	1.0281E-005	2.1402E-005
112	21	0	-2.4061E-005	-1.5470E-005	-0.0000[illegible]62	-3.7896E-006	-1.3782E-006	6.2509E-007	0.000002268	4.4366E-006
113	22	0	0	0	0	0	0	0	0	0

In response to the prescribed velocities at the boundary, the fluids velocity in the simulation domain are obtained dynamically. The current velocity, calculated from the particle distribution function is obtained. The code in cell C93 is

=IF(C13=1, 0, (AA91−AA117+AA143−AA169−AA195+AA221))

FIGURE 7.21

Spreadsheet implementation of the lattice Boltzmann model. The velocity in the y-direction.

the radius of the channel, and x is the distance from the center of the channel. The velocity in the active region of the simulation domain is calculated from the second moment of the particle distribution function

$$\rho\vec{v} = \sum_i f_i \vec{e}_i \tag{7.55}$$

if the site is fluid (otherwise the velocity is set to zero). The code in cell C67 is

```
=IF(C13=1, 0, (AA39-AA65+AA143-AA169+AA195-AA221))
```

where only the components of the particle distribution function whose representative vectors e_i have components in the x-direction play a role in determining the velocity in the x-direction. In particle if e_i is oriented with a component to the right then the particle distribution function is added, while if e_i is oriented with a component to the left then the particle distribution function is subtracted. Similarly, in the y-direction, the code in cell C93 is

```
=IF(C13=1, 0, (AA91-AA117+AA143-AA169-AA195+AA221))
```

where only the components of the particle distribution function whose representative vectors e_i have components in the y-direction play a role in determining the velocity in the y-direction.

The current particle distribution function, within the spreadsheet model, is shown in Figure 7.22 (or at least the first component of the particle distribution function, f_0, with the remaining eight components in the cells below this). The particle distribution function for the boundary nodes do not change during the simulation and are calculated directly from the density and velocity. In particular, in cell Z12 the particle distribution function f_0 is calculated with the code

```
=IF(B12=1, 0, (0.44444*B40*(1-1.5*(B66*B66+B92*B92))))
```

Similarly, for the particle distribution function f_1 (shown in Figure 7.23), the code in cell Z38 is

```
=0.1111111*B40*(1 + 3*(B66) +

4.5*(B66*B66) - 1.5 *(B66*B66+B92*B92))
```

Here, we are setting the particle distribution function at the boundary condition to the equilibrium distribution based on the assigned boundary conditions (constant density and velocity). Recall that the equation for calculating the equilibrium particle distribution function, for the current D2Q9 model, is

$$f_i^{eq}(\vec{r},t) = w_i\rho(\vec{r},t)\left[1 + 3\left(\vec{e}_i \cdot \vec{v}(\vec{r},t)\right) + \frac{9}{2}\left(\vec{e}_i \cdot \vec{v}(\vec{r},t)\right)^2 - \frac{3}{2}v^2(\vec{r},t)\right] \tag{7.56}$$

where the weights, the w's, are given by $w_0 = \frac{4}{9}$, $w_1 = w_2 = w_3 = w_4 = \frac{1}{9}$,

The boundary values of the particle distribution function are obtained from the density and velocity values, which are the fixed boundary conditions. The code in cell Z12 is

=IF(B12=1, 0, (0.44444*B40*(1−1.5*(B66*B66+B92*B92))))

	Y	Z	AA	AB	AC	AD	AE	AF	AG	AH
10		F_0								
11		1	2	3	4	5	6	7	8	9
12	1	0	0	0	0	0	0	0	0	0
13	2	0	0	0	0	0	0	0	0	0
14	3	0	0	0	0	0	0	0	0	0
15	4	0	0	0	0	0	0	0	0	0
16	5	0	0	0	0	0	0	0	0	0
17	6	0	0	0	0	0	0	0	0	0
18	7	0	0	0	0	0	0	0	0	0
19	8	0	0	0	0	0	0	0	0	0
20	9	0	0	0	0	0	0	0	0	0
21	10	0	0	0	0	0	0	0	0	0
22	11	0	0	0	0	0	0	0	0	0
23	12	0.44443985	0.444832151	0.444834537	0.444814931	0.444787644	0.444758175	0.44472931	0.444701953	0.444675779
24	13	0.444438916	0.444859067	0.444839202	0.444813816	0.444785232	0.444754848	0.444724164	0.444692857	0.444659958
25	14	0.444437656	0.444879424	0.444846959	0.444815548	0.444784052	0.444751843	0.444719291	0.444685628	0.444650114
26	15	0.44443655	0.44489067	0.44485245	0.444817111	0.444783179	0.444749293	0.444715365	0.4446805	0.444644419
27	16	0.444435916	0.444895078	0.444854627	0.444817371	0.444782021	0.444747057	0.444712298	0.444676921	0.444641025
28	17	0.444435916	0.444894021	0.444853506	0.444816[illegible]35	0.444780356	0.444744978	0.444709827	0.444674239	0.44463861
29	18	0.44443655	0.444887682	0.444849244	0.44481[illegible]37	0.44477827	0.444743041	0.444707753	0.444671971	0.444636346
30	19	0.444437656	0.44487505	0.444842148	0.44480[illegible]5	0.444776214	0.444741468	0.444706049	0.444669857	0.444633768
31	20	0.444438916	0.444854151	0.444833571	0.444806[illegible]02	0.444775163	0.444740727	0.444704869	0.44466781	0.444630663
32	21	0.44443985	0.444827632	0.444829092	0.444807[illegible]92	0.44477665	0.444741499	0.444704372	0.444665883	0.444626899
33	22	0	0	0	0	0	0	0	0	0

The values of the particle distribution function in the simulation domain are later calculated in the spreadsheet, and then their values are copy and pasted back here as part of an iteration.

FIGURE 7.22
Spreadsheet implementation of the lattice Boltzmann model. The component f_0 of the current particle distribution function.

The particle distribution function for the central simulation domain is again copied from later calculations in the spreadsheet.

	Y	Z	AA	AB	AC	AD	AE	AF	AG	AH
36		F_1								
37		1	2	3	4	5	6	7	8	9
38	1	0	0	0	0	0	0	0	0	0
39	2	0	0	0	0	0	0	0	0	0
40	3	0	0	0	0	0	0	0	0	0
41	4	0	0	0	0	0	0	0	0	0
42	5	0	0	0	0	0	0	0	0	0
43	6	0	0	0	0	0	0	0	0	0
44	7	0	0	0	0	0	0	0	0	0
45	8	0	0	0	0	0	0	0	0	0
46	9	0	0	0	0	0	0	0	0	0
47	10	0	0	0	0	0	0	0	0	0
48	11	0	0	0	0	0	0	0	0	0
49	12	0.111269509	0.111269509	0.111301741	0.111317662	0.111321606	0.111318493	0.111313681	0.111307575	0.111305046
50	13	0.111536642	0.111536642	0.11152904	0.111530039	0.111530594	0.111527604	0.111523924	0.111519851	0.111520886
51	14	0.111737272	0.111737272	0.111722321	0.11170916	0.11170033	0.111692488	0.111686702	0.111681757	0.111681623
52	15	0.111871158	0.111871158	0.111854345	0.111835431	0.111819842	0.11180692	0.111797676	0.111789616	0.111784851
53	16	0.111938142	0.111938142	0.111921024	0.11190018	0.111881618	0.111865719	0.111853714	0.111842303	0.111832439
54	17	0.111938142	0.111938142	0.111921232	0.111900708	0.111882384	0.111866367	0.111853595	0.111840258	0.111826934
55	18	0.111871158	0.111871158	0.111854938	0.111836941	0.111822082	0.111808963	0.111797795	0.111784596	0.111770192
56	19	0.111[illegible]7272	0.111737272	0.111723171	0.111711338	0.111703683	0.111695893	0.111687983	0.111676514	0.111663107
57	20	0.111[illegible]642	0.111536642	0.11152983	0.111532083	0.11153392	0.111531325	0.111526247	0.11151669	0.111505352
58	21	0.1112[illegible]9509	0.111269509	0.111301754	0.111318087	0.111322553	0.11131941	0.111313709	0.111304432	0.111294972
59	22	0	0	0	0	0	0	0	0	0

The particle distribution function for the boundary is again calculated from the fixed density and velocity profile at the boundaries. The calculation here is for f_1 which is propagating to the right. The code in cell Z38 is

=0.1111111*B40*(1 + 3*(B66) + 4.5*(B66*B66) – 1.5 *(B66*B66+B92*B92))

FIGURE 7.23

Spreadsheet implementation of the lattice Boltzmann model. The component f_1 of the current particle distribution function.

and $w_5 = w_6 = w_7 = w_8 = \frac{1}{36}$. In this way the current boundary values of the particle distribution functions ensure that the velocity and density are fixed at the boundaries.

The current values of the particle distribution function in the active region of the simulation, as depicted in Figures 7.22 and 7.23, are not calculated here. The current iteration particle distribution functions are copied and pasted from the calculated values from the previous iteration (the values only are copied and pasted, so there is no code in these cells). As mentioned earlier, the current iteration particle distribution functions are used to calculate the next iteration elsewhere in the spreadsheet, and when the updated values are copied and pasted here the simulation is iterated. Let's turn our attention to how we calculate the updated particle distribution functions for the next iteration.

Figure 7.24 depicts the part of the spreadsheet calculating the post-collision particle distribution function for the component f_0. Other components are calculated in the spreadsheet below this. Recall that in the current model we are relaxing the stresses directly to local equilibrium (for simplicity and stability) and the post-collision particle distribution function is simply the equilibrium particle distribution function. The code in cell AY13 is

```
=IF(C13=1, 0, (0.444444*C41*(1 - 1.5*(C67*C67 + C93*C93))))
```

where if the site is solid then the particle distribution function is zero, else it is calculated from the equation above for the equilibrium particle distribution function. The boundary condition particle distribution functions do not change and, therefore, the values are simply copied across from the current particle distribution function. In other words, the code in cell AX12 is

```
=Z12
```

and these values remain unchanged.

Now that we have calculated the post-collision particle distribution function, we can calculate the propagation of the particle distribution functions to the next lattice site in the direction they are oriented. This process of particles propagating along in space and then colliding with each other mimics the microscopic fluid dynamics and is at the heart of why the lattice Boltzmann is such a powerful model. However, the lattice Boltzmann's main strength is the ease at which complex boundaries can be easily implemented, and before we propagate the particle distribution functions we must prepare the bounce-back boundary conditions (the no-slip boundary conditions at the solid-fluid interface).

The idea behind this step of setting the particle distribution functions for bounce-back boundary conditions is to take the particle distribution function propagating towards a solid site, and turn it around so that when we propagate the particle distribution functions it will be propagated back to where it came from, but be facing in the opposite direction. Obviously the particle distribution function components f_0 don't propagate as the zeroth direction

The post–collision particle distribution function is calculated from the density and velocity. The density and velocity are, of course, calculated from the particle distribution function. This relaxes the stresses in the fluid toward (or actually to in the current model) local equilibrium. The code in AY13 is

=IF(C13=1, 0, (0.444444*C41*(1 – 1.5*(C67*C67 + C93*C93))))

	AW	AX	AY	AZ	BA	BB	BC	BD	BE	BF
10		F_0								
11		1	2	3	4	5	6	7	8	9
12	1	0	0	0	0	0	0	0	0	0
13	2	0	0	0	0	0	0	0	0	0
14	3	0	0	0	0	0	0	0	0	0
15	4	0	0	0	0	0	0	0	0	0
16	5	0	0	0	0	0	0	0	0	0
17	6	0	0	0	0	0	0	0	0	0
18	7	0	0	0	0	0	0	0	0	0
19	8	0	0	0	0	0	0	0	0	0
20	9	0	0	0	0	0	0	0	0	0
21	10	0	0	0	0	0	0	0	0	0
22	11	0	0	0	0	0	0	0	0	0
23	12	0.44443985	0.44483196	0.444834158	0.44481464	0.444787192	0.444757811	0.444728826	0.444701574	0.444675349
24	13	0.444438916	0.444858914	0.444838758	0.444813579	0.444784725	0.444754538	0.444723635	0.444692525	0.444659489
25	14	0.444437656	0.444879229	0.444846512	0.444815288	0.444783548	0.444751518	0.444718765	0.444685285	0.444649637
26	15	0.44443655	0.444890462	0.444851982	0.444816844	0.444782661	0.444748966	0.444714823	0.444680151	0.444643916
27	16	0.444435916	0.444894849	0.44485414	0.444817093	0.444781488	0.444746722	0.44471174	0.44467656	0.444640496
28	17	0.444435916	0.444893776	0.444853	0.444815748	0.444779807	0.444744634	0.444709251	0.444673864	0.444638053
29	18	0.44443655	0.444887432	0.444848725	0.444812945	0.444777709	0.444742691	0.444707159	0.444671582	0.444635762
30	19	0.444[illegible]37656	0.444874791	0.444841624	0.444809267	0.444775644	0.444741108	0.444705441	0.444669451	0.444633163
31	20	0.44[illegible]8916	0.444853929	0.44483303	0.444806223	0.444774573	0.444740377	0.444704236	0.444667404	0.444630025
32	21	0.44[illegible]3985	0.444827361	0.444828623	0.444807145	0.444776121	0.444741081	0.444703789	0.44466541	0.444626297
33	22	0	0	0	0	0	0	0	0	0

The distributions at the boundaries remain unchanged. The code in cell AX12 is simply =Z12

FIGURE 7.24

Spreadsheet implementation of the lattice Boltzmann model. The post-collision particle distribution function.

is the rest direction (or no direction). Figure 7.25, therefore, shows the section of the spreadsheet which sets up the bounce back boundary conditions for the f_1 components. The code in cell BV38 is

```
=IF(AND(B12 = 1, C12 = 0), AY64, AX38)
```

Note that the same code is used for sites regardless of whether they are boundary or active regions, as we want the sites at the boundaries (which are solid) to store the particle distribution functions that will be propagated back into the fluid domain in the next step. However, if we are looking in the right direction, or [10] direction, then the last sites on the right of the simulation won't need to be calculated (there is obviously nothing to the right of these sites which could propagate from the left). The code in cell BV38 starts with

```
IF(AND(B12 = 1, C12 = 0)
```

which is saying if the current site is solid and the site to the right is fluid, then there will be a particle distribution function propagating from the right (in the left direction) which needs to be turned around and propagated back out to the right. If this is the case then find the particle distribution function component to the left f_2 on the lattice site to the right of the current site, and set the particle distribution function f_1 at the current site equal to this. This way when we propagate it back into the fluid it will end up on the lattice site to the right of the current lattice site but in the opposite direction (the e_1 direction from the e_2 direction). We do a similar operation, always switching the directions of the particle distribution functions propagating into the solid sites, for all components of the particle distribution function. In other words, at the end of this step, we have particle distribution functions at all other solid sites adjacent to fluid sites, ready to propagate the required particle distribution functions back into the fluid.

The final step in the iteration is to propagate the particle distribution functions in the directions towards the neighboring lattice sites. (In other words, propagate the f's in the directions of the $\vec{e}$ vectors to the next lattice site.) Figure 7.27 depicts the sections of the spreadsheet which calculate the propagation of the particle distribution function in the e_1 direction. Note that only the particle distribution functions in the active region are required here, as this is what we will be copying and pasting back to the current particle distribution functions as depicted in Figure 7.22. The propagation step is really quite simple, as all we have to do is move the particle distribution functions over. So in the $\vec{e}_1$ direction the particle distribution functions are copied to the right. The code in cell CU39 is

```
=BV39
```

which is the particle distribution function in the same direction, f_1, but from the lattice site to its immediate left (from the bounce-back boundary condition step). Similarly, particle distribution functions are propagated in the

The bounce–back boundary conditions are implemented at the interface between the fluid and the solid. Before we propagate the particle distribution functions we modify the distributions at the interface. In particular, if the distribution is going to be propagated into a solid, then place the distribution on the solid node facing back towards the fluid. The code in cell BV38 is

=IF(AND(B12 = 1, C12 = 0), AY64, AX38)

This sets the distribution up, ready to be propagated at the next step back into the fluid.

	BU	BV	BW	BX	BY	BZ	CA	CB	CC	CD
36		F_1								
37		1	2	3	4	5	6	7	8	9
38	1	0	0	0	0	0	0	0	0	0
39	2	0	0	0	0	0	0	0	0	0
40	3	0	0	0	0	0	0	0	0	0
41	4	0	0	0	0	0	0	0	0	0
42	5	0	0	0	0	0	0	0	0	0
43	6	0	0	0	0	0	0	0	0	0
44	7	0	0	0	0	0	0	0	0	0
45	8	0	0	0	0	0	0	0	0	0
46	9	0	0	0	0	0	0	0	0	0
47	10	0	0	0	0	0	0	0	0	0
48	11	0	0	0	0	0	0	0	0	0
49	12	0.111269509	0.111301606	0.111317863	0.111321207	0.111318891	0.111313049	0.11130815	0.111304215	0.111308196
50	13	0.111536642	0.111528933	0.111530251	0.111530235	0.111528016	0.111523329	0.111520439	0.111520079	0.111529532
51	14	0.111737272	0.111722208	0.111709393	0.111699985	0.111692933	0.111686128	0.111682392	0.111680841	0.111685642
52	15	0.111871158	0.111854234	0.111835677	0.111819504	0.111807396	0.111797116	0.111790298	0.111784085	0.111779966
53	16	0.111938142	0.111920911	0.111900438	0.111881283	0.111866223	0.111853157	0.111843025	0.111831672	0.111819553
54	17	0.111938142	0.111921118	0.111900976	0.111882045	0.111866891	0.111853028	0.111841007	0.111826147	0.111808873
55	18	0.111871158	0.111854821	0.111837216	0.111821732	0.111809496	0.111797206	0.111785357	0.111769371	0.111750202
56	19	0.111737272	0.111723046	0.111711614	0.111703314	0.111696425	0.111687361	0.111677271	0.111662239	0.111644053
57	20	0.111536642	0.111529708	0.11153234	0.111533523	0.111531834	0.111525581	0.111517424	0.111504429	0.11148927
58	21	0.111269509	0.111301589	0.111318323	0.111322088	0.111319895	0.111312971	0.111305147	0.111293981	0.111282844
59	22	0	0	0	0	0	0	0	0	0

FIGURE 7.25

Spreadsheet implementation of the lattice Boltzmann model. The implementation of bounce-back boundary conditions.

other directions to complete the iteration. We now have to copy the values of the propagated particle distribution functions back to the current particle distribution function (see Figure 7.22). However, if we simply copy each component of the particle distribution function in turn, then some components will be automatically updating while the first components to be copied across are copied across. For this reason, we copy the values for all of the particle distribution functions to a separate location in the spreadsheet (all the way over to the right) and then copy these values one at a time into the current particle distribution functions. The act of copying and pasting all of these values within the spreadsheet is stored in a macro that can be run at the push of a button, and each time the button is pushed the simulation iterates. The lattice Boltzmann model is a dynamics model and so we can see how, in response to the boundary conditions, the flow field within the simulation domain evolves. Figure 7.26 shows the fluid flow within the lattice Boltzmann simulation described above. The fluid velocity is depicted as arrows, where the size of the arrows indicate magnitude and the direction of the arrows the direction of fluid flow. The fluid velocity is assigned at the boundaries and then the lattice Boltzmann model is used to calculate the fluid velocity inside the fluid domain of the active region of the simulations (the region where the fluid is updated).

Typically, the lattice Boltzmann model might be solved in three-dimensions and incorporate a larger simulation domain. The small two-dimensional model implemented here is for instructional purposes, to demonstrate how simple and powerful this technique is. Furthermore, the lattice-Boltzmann can be used to capture multi-component fluids, turbulence, or flow through porous media and is a wonderful model for capturing the fluid dynamics associated with fuel cells and many other alternative energy systems.

7.5 Exercises

1. The photovoltaic and electrochemical equivalent circuit model couples a solar cell with an electrochemical cell. However, there are a number of parameters which can depend on the the exact system. Vary the parameters (within realistic ranges) and maximize the efficiency of these devices.

2. Implement the simple electrochemical model of a fuel cell. Determine how sensitive the current-voltage curve is to ohmic losses and limit current.

3. The current voltage curve can be obtained using the continuum electrochemical model of a fuel cell. Obtain such a current-voltage curve and fit this to the simple electrochemical model of a fuel cell. For example, to get a good

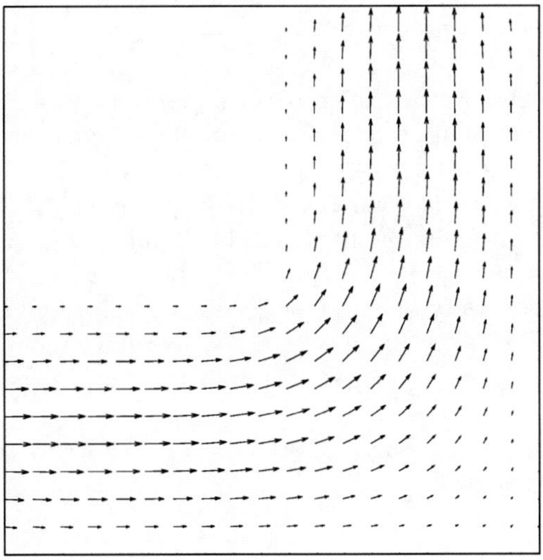

FIGURE 7.26
Fluid flow around a corner in a serpentine fuel cell channel, calculated using the lattice Boltzmann model.

The particle distribution function is now propagated to the next lattice site over. In particular, the distribution f_1 is propagated to the right, and the new distribution in cell CU39 is simply

=BV39

which is the same distribution f_1 but from the lattice site to its immediate left.

	CS	CT	CU	CV	CW	CX	CY	CZ	DA	DB
36		F_1								
37		1	2	3	4	5	6	7	8	9
38	1									
39	2		0	0	0	0	0	0	0	0
40	3		0	0	0	0	0	0	0	0
41	4		0	0	0	0	0	0	0	0
42	5		0	0	0	0	0	0	0	0
43	6		0	0	0	0	0	0	0	0
44	7		0	0	0	0	0	0	0	0
45	8		0	0	0	0	0	0	0	0
46	9		0	0	0	0	0	0	0	0
47	10		0	0	0	0	0	0	0	0
48	11		0	0	0	0	0	0	0	0
49	12		0.111269509	0.111301606	0.111317863	0.111321207	0.111318891	0.111313049	0.11130815	0.111304215
50	13		0.111536642	0.111528933	0.111530251	0.111530235	0.111528016	0.111523329	0.111520439	0.111520079
51	14		0.111737272	0.111722208	0.111709393	0.111699985	0.111692933	0.111686128	0.111682392	0.111680841
52	15		0.111871158	0.111854234	0.111835677	0.111819504	0.111807396	0.111797116	0.111790298	0.111784085
53	16		0.111938142	0.111920911	0.111900438	0.111881283	0.111866223	0.111853157	0.111843025	0.111831672
54	17		0.111938142	0.111921118	0.111900976	0.111882045	0.111866891	0.111853028	0.111841007	0.111826147
55	18		0.111871158	0.111854821	0.111837216	0.111821732	0.111809496	0.111797206	0.111785357	0.111769371
56	19		0.111737272	0.111723046	0.111711614	0.111703314	0.111696425	0.111687361	0.111677271	0.111662239
57	20		0.111536642	0.111529706	0.11153234	0.111533523	0.111531834	0.111525581	0.111517424	0.111504429
58	21		0.111269509	0.111301589	0.111318323	0.111322088	0.111319895	0.111312971	0.111305147	0.111293981
59	22									

FIGURE 7.27
Spreadsheet implementation of the lattice Boltzmann model. Propagation of the particle distribution function.

fit you could change the phenomonological constants (ξ_1, ξ_2, ξ_3, and ξ_4), the ohmic losses (R_S), or the limit current (I_L).

4. The nature of fluid flow through porous membranes might also be modeled using a lattice Boltzmann model. For example, when fluid travels towards a solid wall in the lattice Boltzmann model, it is entirely bounced back to where it came from by the solid wall. However, in porous media some of the fluid would continue on its journey. In particular, fluid flow through porous media can be represented using Darcy's law which states that the flow rate of a fluid depends on the porousity of the structure it is flowing through. In the lattice Boltzmann model this can be incorporated by allowing some of the particle distribution function to propagate through the porous media (equal to the porosity) while some of the fluid bounces back (equal to one minus the porosity). Simulate fluid flow though a porous wall. How easily can the fluid move through this porous wall?

5. Could the lattice Boltzmann model be used to look at the aerodynamics of air flow in wind turbine studies, either as a way of looking at the air flow over land, such that the best locations could be identified for wind turbines or through the actual simulation of air flow over the wind turbine blades? This would presumably require three-dimensional modeling which would be very impractical using a spreadsheet.

Additional Reading

Larminie, J., Dicks, A., & McDonald, M. S. (2003). Fuel cell systems explained (Vol. 2). New York: Wiley.

Reece, S. Y., Hamel, J. A., Sung, K., Jarvi, T. D., Esswein, A. J., Pijpers, J. J., & Nocera, D. G. (2011). Wireless solar water splitting using silicon-based semiconductors and earth-abundant catalysts. Science, 334(6056), 645-648.

Succi, S. (2001). The Lattice-Boltzmann Equation. Oxford University Press, Oxford.

Index